A series of student texts in

CONTEMPORARY BIOLOGY

₈ₛ.   As my uncle

General Editors:
Professor E. J. W. Barrington, F.R.S.
Professor Arthur J. Willis

Principles of Animal Physiology

Second Edition

Dennis W. Wood

B.Sc., Ph.D.

Senior Lecturer in Zoology, University of Durham

American Elsevier Publishing Company, Inc.
New York

155101

© Dennis W. Wood, 1974

American Elsevier Publishing Company, Inc.
52 Vanderbilt Avenue, New York, N.Y. 10017

First published in Great Britain by
Edward Arnold (Publishers) Ltd.

Cloth edition ISBN: 0–444–19534–3
Paper edition ISBN: 0–444–19533–5

Library of Congress Catalog Card Number: 74–4090

Printed in Great Britain by
William Clowes & Sons, Limited
London, Beccles and Colchester

Preface to the Second Edition

The welcome accorded to the first edition suggests that it filled a definite gap in the provision of textbooks in this field, and in revising it I have tried to resist the temptation to add too much material in such a way that its scope and content would be altered. At the same time, new work and the experience of using the book in the teaching of undergraduates have made it necessary to carry out some revision of most chapters; and more extensive revision of those parts that have been the subject of considerable advances in recent years, such as cell structure and function and the structure and function of muscle. Apart from the revision of existing material, some sections have been deleted and others added. The latter include material on enzyme kinetics, renal organs, air-breathing aquatic animals, hormones and cyclic AMP. There are several new figures, and some other figures have been revised.

With some reluctance all units have been changed to the Système International. The formulation of the SI system appears to have been dominated by physical scientists, and I regret the failure to retain the Ångstrom unit (or a named equivalent) and the calorie. I have taken advantage of the permission granted during the transitional phase of change to SI units to include in brackets the calorific equivalent of joules where this seemed appropriate.

I am grateful to all those who took the trouble to write and point out errors in the first edition, or to suggest improvements; and to my colleagues Drs Anstee, Bowler and Hyde, who read certain old or proposed new sections and gave me the benefit of their specialist knowledge and advice.

D.W.W.

Durham
1974

Table of Contents

I

Introduction

Physiology is the analysis of function in living organisms. Because these are composed of chemical substances, the analysis involves the application of the laws of chemistry and physics, and there is no need to invoke a special 'vitalistic principle' to explain how living organisms work, although this does not mean that other ways of looking at life are either wrong or invalid.

Physiological processes are exceedingly complex, being the sum of many interrelated reactions. For this reason, animals have often been able to change some of the details of their physiology through the play of selection on genetic mutation, although there remains a striking uniformity of principle and often of detail in fundamental physiological phenomena. This has enabled some authors to make a distinction between *general* physiology—the study of fundamental processes in terms of chemistry and physics; and *comparative* physiology—the attempt to discern functional similarities and differences between gross systems of whole organisms. The distinction may sometimes be convenient, but it is artificial because it is impossible to understand the functioning of the whole organism without a thorough knowledge of fundamental processes, and this book makes no attempt to separate the two approaches.

Thus, the initial approach of the physiologist must be to analyse physiological processes in terms of the individual reactions which comprise them, and then to put them together to make an intelligible picture of the whole process. This approach has yielded valuable information, but it is necessary to remember that the integration of individual reactions into a complete process may involve their modification, and the development of properties that are peculiar to the overall process but which cannot necessarily be deduced from a study of the individual reactions. A chemical parallel is the way in which molecules possess properties of their own which are not exhibited by the atoms they contain when these exist independently.

One way of trying to understand a complex system is to formulate a *model* that will exhibit the same properties. Thus we may describe an excitable membrane in terms of an equivalent electrical circuit or draw analogies between the contractile mechanism of a muscle and a damped spring. Such models may be misleading if the analogies are pressed too far but, properly used, they can help to clarify thought, and to suggest further experiments which will aid in the testing of the concepts on which they are based.

Physiology, like all natural sciences, lives and feeds on experiment. We begin the study of a system by accumulating descriptive information about it, and when there is a sufficient quantity of that we erect a hypothesis to explain it. This hypothesis must then be tested by experiment, and if the results of our experiments conform to our hypothesis, it will become a theory which gains a wide measure of acceptance, and which will stand unless and until some piece of information is obtained which does not seem to be in accordance with it. Then it must either be modified or a new theory formulated. In biological research we rarely know all the facts about a situation, and hypotheses are erected on the basis of the weight of the evidence, which may shift as more data is accumulated. The topics dealt with in this book are often in this exciting and stimulating state, and new information about many of them is published daily. Nevertheless, it is sometimes surprising to find how many of the basic ideas remain unchallenged over the years.

Physiological processes often possess an intrinsic interest of their own for anyone who enjoys the feel of precision and order, and may even be held to exhibit a kind of elegance and beauty for this reason. But the primary aim of physiologists must be to understand the relationship between the living organism and its environment. This book therefore begins with a consideration of the energy relationships between organisms and their environment, because it is their use of energy that distinguishes animate organisms from inanimate matter. This leads us naturally to the nutritional requirements of animals, which obtain their energy by ingesting solid food, from which we turn to the organization and functioning of the cells since it is their job to deal with the raw materials obtained through feeding. This involves a consideration of the cellular oxidation of metabolites, and thence of more general topics relating to metabolism such as respiration and excretion. With this basis, the relationship between animals and various environmental factors such as temperature, pH and the supply of water can be examined.

One feature that emerges from the consideration of such topics is that animals tend to be adapted to a particular kind of environment, and their behaviour is therefore directed towards remaining in this environment as far as possible. To do this they must be able to detect various environmental characteristics and to react to changes in them with appropriate responses. The final chapters are therefore concerned with the mechanisms of nervous integration and the various effector systems used by animals in responding to their environment. The last chapter seeks to look back in general terms over the ground covered and to find whether there are some features of the physiological processes through which animals maintain more or less constant living conditions which are common to them all.

2

Energy and Food

Animals need food for growth and for the energy to maintain their bodies in the living state, and hence to behave in those ways that we consider to be characteristic of living organisms. Since the processes responsible for growth also require a supply of energy to drive them, it is evident that the fundamental requirement of the food intake of any animal is that it should contain foods which may be used as fuels, for the liberation of energy. Elementary books of biology often attempt to define 'life' in descriptive terms, but the real difference between living organisms and non-living matter lies in the way they utilize energy. Although both obey the same physico-chemical laws, they differ in the way these apply to them.

ENERGETICS OF MATTER

2.1 Energy

Energy is often defined as the capacity to do work, although this definition applies fully only to systems not encountered in practice. *Work* is defined as the product of a given force acting through a given distance, which implies that movement takes place when work is done. Two factors are involved in the production of this movement: an intensity factor (the force), and a capacity factor (the distance). The intensity factors in, for example, stretching, the passage of an electrical current, and expansion, are tension, voltage and pressure respectively. The corresponding capacity factors are length, charge and volume. These definitions apply equally well to other forms of work, but not to heat, although this is also a form of energy. Heat possesses no capacity factor to match its intensity factor, temperature. Nor does the production of heat obviously result in the performance of work, although it can certainly be converted into mechanical energy, as in a steam engine, and thence into other forms of energy. The conversion, however, as we shall see below, is never 100% efficient. These preliminary considerations are sufficient to show that heat occupies a special position in energetics.

The energetics of systems is governed by the laws of *thermodynamics*. These are three in number, but the third is really an extension of the second, and being concerned with events at absolute zero temperature, will not concern us here. It should be emphasized that no special ability is required for an understanding of the basic concepts of thermodynamics, which exhibit the stark simplicity of most fundamental scientific laws.

2.2 First Law of Thermodynamics

This states that *the energy of the universe remains constant*. Thus, although one form of energy may be converted to another, none is lost in the process. This does not mean that a given system may not lose energy to its surroundings, for the law is concerned simply with the total energy of the universe. Nor does it imply that all forms of energy are equally useful, a subject covered by the Second Law. It is convenient to distinguish between *potential* energy, which is the inherent capacity of a system to do work, and the *kinetic* energy that is actually expended in the performance of work. The difference is well illustrated by a car at rest and in motion. When the car is at rest, and provided that it has petrol in its tank, it possesses potential energy although it is not doing work. When the car is moving, the potential energy is being converted into kinetic energy, and at the end of the journey there is less potential energy in the system than there was at the beginning, a loss largely represented by a decrease in the quantity of petrol in the tank. It should be noted that the potential energy of the system has been converted into heat as well as into kinetic energy. It is universally true that under normal conditions the conversion of potential energy into kinetic energy always involves the production of heat, which escapes into the surrounding medium and hence represents a waste of some of the potential energy.

2.3 Second Law of Thermodynamics

This law, which may be stated in a variety of forms, is concerned with the *probability* that a particular energy change will take place. It states that matter tends towards disorderliness rather than orderliness, because orderliness in matter is always accompanied by a high degree of potential energy; and there is a perpetual tendency for the potential energy of systems to become converted into kinetic energy and heat. The kinetic energy will be used to perform work, but in the course of this work, it will become effectively lost to the system. If it is used to convert a compound into some other compound with some potential energy of its own, the conversion will have been accompanied by the loss of some of the original potential energy as heat, and so the potential energy of the new compound will always be less than that of the original compound. And the new compound, in turn, will follow the same trend as the one that produced it. The heat produced could, in theory, be used to provide more work, by means of a heat engine, of which the steam engine is one example. But this conversion can never be 100% efficient, because a heat engine functions by taking heat from a source and rejecting it to what is known as a 'sink', i.e. to an area of less heat. The very mechanism of conversion therefore depends on the loss of some of the heat. The trend is clear: unless heat is supplied to a system from an external source, its potential energy will be reduced through its conversion into other forms of energy which will be lost.

2.4 Entropy

The transformation of order into disorder involves the breakdown of an ordered pattern into randomness. Translated into terms of the breakdown of molecules, it implies the tendency of large molecules to break down into smaller ones, until ultimately the original potential energy of the system has been converted into the random movement of small molecules. Energy in a random disorganized state in which it simply functions as the kinetic energy of the random movement of molecules is known as *entropy*; and a common way of stating the Second Law of Thermodynamics is that *all processes tend towards maximum entropy*. Any reaction that will increase entropy will therefore tend to be thermodynamically *spontaneous*. This does not mean that it will happen automatically and immediately, but that if the conditions are right it is this reaction that will proceed and not the opposite one. Thus, the Second Law tells us the *direction* in which a reaction will normally proceed.

Chemists normally deal with what is known as a *closed* system—one in which the amounts of the initial reactants is specific and fixed. In such a system, the result of an energy change in accordance with the Second Law is given by

$$\Delta E = Q + W \qquad (2.1)$$

where ΔE is the change in the energy of the system, Q represents the heat given off, and W the work done. The heat Q will exchange with its surroundings, since a closed system has fixed reactants but is not insulated from its environment. Equation *2.1* is sometimes written with a minus sign in front of W, an alternative which emphasizes that a change in energy may be positive or negative (gained or lost), depending on the conditions in which the reaction takes place. The clear conclusion from equation *2.1* is that any flow of heat between a system and its surroundings must be accompanied by changes in the energy content of the system and in the work done by it on its surroundings.

2.5 Enthalpy

A change in heat between a system and its surroundings is called enthalpy, and given the symbol ΔH. In terms of enthalpy, equation *2.1* can therefore be rewritten

$$\Delta E = \Delta H + W \tag{2.2}$$

The performance of work in a chemical system implies that changes in pressure and volume must be occurring. If the volume and pressure can be kept constant, however, then

$$\Delta E = \Delta H \tag{2.3}$$

In fact, volume and pressure changes are negligible in living cells, and it is this equation which applies to them. It can also be arranged, by suitable techniques, to apply to the combustion of substances in the process of *direct calorimetry*, in which the potential energy of a substance is converted entirely into heat, which is measured. Thus, the energetics of living cells and the energy potential of individual compounds can both be considered in terms of heat changes. For this reason, it is customary to refer to changes in the energy content of a biological system as changes in heat. Formerly, the calorie was the unit of measurement (the amount of heat needed to raise 1 g of water by 1°C, between 14.5°C and 15.5°C), but this has now been superseded by the *Joule* (1 cal $\equiv$ 4.19 J). Biological reactions usually involve large heat values, and these are dealt with in kiloJoules (kJ).

2.6 Free energy

It will now be appreciated that every physical and chemical event is dictated by its energy relationships, and that the Second Law gives guidance about the probability that a given change will take place. It is often difficult to work out the full energy relationships of such a change, and for this reason a further concept has been introduced, that of *free energy*, denoted by F. Free energy is the energy that can be obtained as useful work during a change in a system. In other words, it is the energy that can be 'freed' or liberated during the course of such a change. The useful energy obtained during a reaction will be equal to the total energy change less the entropy, i.e.

$$\Delta F = \Delta E - T \Delta S \tag{2.4}$$

where ΔF is the change in the free energy of the system, or the extent of the conversion of useful energy, and $T\Delta S$ represents the change in entropy (the energy degraded to heat). S is the symbol for entropy and T refers to the temperature of the reaction.

In practice, a modified concept of free energy is often used, known as *standard free energy* and denoted variously in textbooks by the symbol $F°$, $°G$ and F'. In the generalized reaction

$$A \rightleftharpoons B \qquad (2.5)$$

in which the reactant A is converted into the product B, the free energy change may be expressed by the equation

$$\Delta F = \Delta F° + RT \log_e \frac{B}{A} \qquad (2.6)$$

where R is the gas constant (8.3 J mole^{-1} degree^{-1}), T the absolute temperature and A, B, the molar concentrations of the reactant A and the product B. Let us now consider the special case when the system is at equilibrium. In this situation there is no conversion of A to B, and from this two consequences follow. The first is that the reactant A and the product B can be represented by an equilibrium constant, i.e.

$$\frac{B}{A} = K_{eq} \qquad (2.7)$$

Secondly, since no conversion is taking place, there is no further change in free energy in the system, and $\Delta F = 0$. Thus, for the system at equilibrium, we may rewrite equation *2.6* as

$$0 = \Delta F° + RT \log_e K_{eq}$$

which when re-arranged, becomes

$$\Delta F° = -RT \log_e K_{eq}.$$

Since R is a constant, T is constant for a given temperature, and Napierian logarithms can be converted to base 10 logarithms by multiplying by 2.303, the equation can be further simplified:

$$\Delta F° = -(8.3 \times 2.303) \, T \log_{10} K_{eq}$$
$$= -19.1 \, T \log_{10} K_{eq} \qquad (2.8)$$

It will be evident from equations *2.7* and *2.8* that if the concentrations of reactants and products in a system at equilibrium are known, K_{eq} and hence $\Delta F°$ can be calculated.

In the ideal case in which 1 mole reactant is converted to 1 mole of product, the K_{eq} would be unity, and since the logarithm of one is zero, in this situation $\Delta F° = 0$. When K_{eq} is less than unity, since the logarithm of a fraction is negative, ΔF° must be positive; and the smaller the value of K_{eq} the greater will be the positive value of ΔF°. Conversely, if the K_{eq} is greater than unity, ΔF° will be negative, increasing in negative value as K_{eq} increases. Tables showing the numerical relationship between ΔF° and K_{eq} can be found in many text-books of biochemistry.

Reference to equation 2.7 shows than when ΔF° is positive, more A is present than B, in other words the reaction is proceeding in the *backward* direction; when ΔF° is negative, the reaction is proceeding in the *forward* direction. Reactions in which ΔF° is positive (K_{eq} less than unity) are known as *endergonic*, because they will proceed spontaneously in the reverse direction, starting with unimolar concentrations of reactants and products. To make them proceed in the forward direction, energy must be supplied to the system from an external source. When ΔF° is negative (K_{eq} greater than unity) the reaction will proceed spontaneously in the forward direction, with liberation of energy, and is then termed *exergonic*.

A practical example will help to illustrate these concepts. One of the stages in the breakdown of glucose in the cell is the rearrangement of the phosphorylated glucose-1-phosphate configuration to the glucose-6-phosphate configuration, by the action of an enzyme *phosphoglucomutase*. For a substrate concentration of 0.02 M at 25°C, and provided excess enzyme is present, it is found that the final equilibrium mixture contains 0.001M glucose-1-phosphate and 0.019M glucose-6-phosphate. Exactly the same molar concentrations result if the substrate is glucose-6-phosphate and the reaction is allowed to proceed the other way round. Hence,

$$K_{eq} = \frac{0.019}{0.001} = 19$$

and since this is greater than unity, we shall expect ΔF° to be negative. Using equation 2.8,

$$\Delta F^\circ = -19.1 \times 298 \times \log_{10} 19$$
$$= -7305 \text{ J} \ (-1745 \text{ calories})$$

This is therefore a reaction that tends to proceed in the forward direction.

From the point of view of thermodynamics, then, reactions will tend to take place spontaneously in the direction indicated by the sign of ΔF°. It has already been pointed out that such reactions are not necessarily spontaneous in the literal sense. Some are, but it will be found that the value of ΔF° in these cases is relatively high. Many reactions which have a negative ΔF° nevertheless require some help to get them going. In the majority of cellular reactions, a catalyst is required for the change to occur, and in all living systems the necessary catalysts take the form of *enzymes*. The nature of enzymes, and the way they facilitate reactions, is discussed in §3.13. For the present, we need note only two facts about them. The first is that although they alter the energetics of the reaction, they do not constitute an external supply of energy. The second is that they do not alter the value of $\Delta F°$, nor the equilibrium point of the reaction. Thus, from both points of view, they are thermodynamically irrelevant, although biologically they are essential.

2.7 Application to biological systems

It should now be apparent that thermodynamics is concerned with ends and not with means, as the thermodynamic irrelevance of enzymes demonstrates. Because of this, thermodynamic considerations may be applied at any level of a system, from individual steps in a process to the process as a whole, to whole organisms, and even to populations. Living organisms constitute physico-chemical systems no less than non-living matter, and in accordance with the Second Law of Thermodynamics they are degrading energy continuously.

In investigating cellular mechanisms, biochemical systems tend to be considered in isolation from one another, whereas in the intact cell they often form part of a continuum, in which a substance produced in one system may be involved in another, and its concentration within the cell will depend on both. Furthermore, classical thermodynamics deals with *closed* systems which do not exchange matter with their surroundings. In contrast, living organisms, and the cells of which they are composed, appear to act in opposition to the Second Law. They consist of highly organized matter, and their activities are directed to the maintenance of this organization. In this sense, they create and maintain order out of disorder. They are able to do this, not by flouting the Second Law, but by engaging in the exchange of matter with their surroundings. Animals take in matter with a high potential energy content, and they degrade it in accordance with the laws of thermodynamics, utilizing the free energy to maintain their orderly structure. The matter taken in as fairly complex molecules is therefore broken down, and finally excreted in the form of simple molecules with a low potential energy. It is for this reason that living organisms are often described as *open* thermodynamic systems, in contrast to the closed systems employed by chemists. For this reason, Classical thermodynamics is inadequate to deal with living systems, and an extension of thermodynamic theory known as *non-equilibrium* or *irreversible* thermodynamics has been developed, in recognition of the fact that living cells and organisms represent, not closed equilibrium systems but open *steady-state* systems that exchange matter with their surroundings.

The complex substances taken in by animals are derived from plants, through the kinds of food chains with which the reader will already be familiar. Green plants are the primary synthesizers, and the external source of energy which they must have for this purpose is obtained by trapping a small proportion of the sun's radiant energy. This energy is used by green plants to build up carbohydrates from carbon dioxide and water, and thence to form proteins from carbohydrates and simple sources of nitrogen like nitrates. The total energy turnover of living organisms in what is often termed the *biosphere* is shown in Fig. 2.1. It should be clear from this figure that living organisms contribute to the overall degradation of the energy of the universe in the same way as non-living matter, as predicted from the laws of thermodynamics.

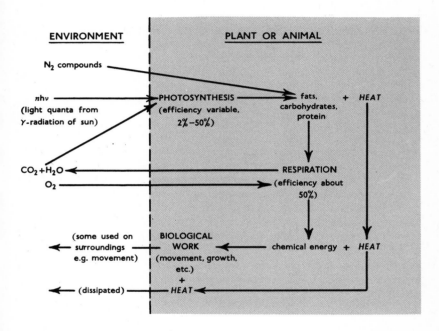

Fig. 2.1 Energy flow in the biosphere. Some additional exchange of matter takes place with the environment through defaecation following ingestion. Both plants and animals die, and their bodies are devoured by bacteria for their own energy requirements. Neither point is included in the diagram.

The synthesis by cells of macromolecules from simpler precursor molecules is known as *anabolism*. Since it is a process which represents a decrease in entropy, it requires a supply of free energy. The opposite process, in which relatively large nutrient molecules are broken down into smaller simple molecules, is known as *catabolism*. It represents an increase in entropy, and therefore involves the release of free energy. This energy is conserved in the form of ATP (§2.11), which is then available to provide the energy required for anabolism. The two processes of anabolism and catabolism are together termed *metabolism*, which is therefore the sum of all the reactions taking place in a living organism.

2.8 The energetics of atoms

The atom is the basic particle involved in chemical change. Atoms of the same element have the same chemical properties, and these differ

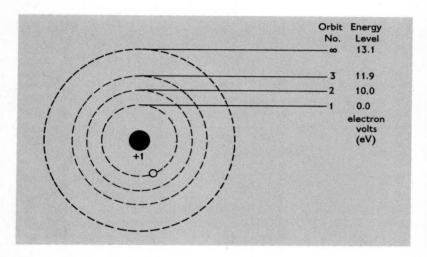

Fig. 2.2 Pictorial representation of the hydrogen atom. The single electron is normally in the innermost shell, but may absorb energy and be transposed to a shell farther from the nucleus. If the energy absorbed exceeds that of the outermost shell, the electron is removed from the atom, which becomes a hydrogen ion.

from those of the atoms of other elements. Chemical compounds are formed by the unions of atoms of different elements in simple numerical proportions that are always the same for any given element. The atom is accordingly the smallest particle of an element that can take part in a chemical change. The properties of atoms are therefore important to an understanding of the biochemistry of living organisms.

An atom may be pictured as a central nucleus of relatively massive dimensions and bearing a positive charge, surrounded by a number of negative particles or electrons whose number is such that the charge on the nucleus is neutralized (Fig. 2.2). The nucleus is a complex of positive *protons* and negative *electrons*, the latter being combined with some of the protons to form neutral particles or *neutrons*. The mass—which for our purposes can be regarded as equivalent to the weight—of both protons and neutrons is very close to unity. For example, the sodium atom has 11 protons and 12 neutrons in its nucleus, and a corresponding atomic weight of 23. The uncombined protons in the nucleus give it its positive charge, and this in turn determines the number of electrons that surround the nucleus in what is often termed the *electron cloud*.

The electrons in the electron cloud may be visualized as being in orbit around the nucleus (Fig. 2.2). An atom may possess a number of concentric orbits or *shells*, which are numbered 1,2,3 . . . . etc. outwards from the nucleus. According to the quantum theory, each shell has a definite energy level, or rather energy range, and any electron in a given shell possesses an energy level appropriate to that shell. The precise energy level of an electron is determined by the *sub-shell* that it occupies within the orbit of its shell. Only two electrons can occupy each sub-shell, and there is a maximum of four sub-shells in each shell, so that the latter cannot contain more than eight electrons, although it may contain less. It is the number and position of these electrons that determines the chemical properties of an atom. A detailed discussion of these is beyond the scope of this book, but one deserves mention. It is possible for the atoms of many elements in appropriate circumstances to gain or lose electrons from their electron clouds, and hence to form a charged particle or *ion*, e.g.

$$Na \rightleftharpoons Na^+ + e \text{ (an electron)}$$
$$Cl + e \rightleftharpoons Cl^-$$

Some ions are formed not from single atoms but from aggregations of atoms that bear an overall charge that is due to the electronic content of one or more of its atoms, but the principle is the same.

The electrons in the electron cloud are thus at different energy levels, and occupy an orbit and sub-shell appropriate to that energy level. According to the quantum theory, the differences in energy level between successive shells are not continuously variable but stepwise, because energy must be thought of as occurring in discrete packets or *quanta*. The energy level of the atomic shells increases the farther away they are from the nucleus, and it is also easier to remove an electron from an outer shell than from an inner one. It should be understood that this pictorial representation of an atom, although it serves to illuminate many atomic properties, is not entirely satisfactory and a proper understanding is only achieved through the mathematical formulations of wave-mechanics, which treats electrons as waves rather than as particles. Nevertheless, for biological purposes, the pictorial representation can be a useful one in understanding the energetics of biochemical reactions.

The concept of atomic shells at different energy levels helps us to picture how an external source of energy can raise an electron in the electron cloud from a lower energy level to a higher one, a process which involves the transfer of the electron from a shell proximal to the nucleus to a more distal shell. According to the Second Law of Thermodynamics, the tendency then will be for this electron to lose its acquired energy and return to its unexcited or *ground* state. If the energy liberated during this process can be suitably harnessed, it can be used to perform work, and this is what happens in photosynthesis. A quantum of light energy, known as a *photon*, strikes an atom within the plant cell and raises one of its electrons to a higher level, and hence to a more distal empty shell. The excitation energy of this electron is then used to build up carbohydrate from carbon dioxide and water, and the electron returns to its ground state, chlorophyll serving as a catalyst for the conversion of electrical energy into chemical (bond) energy. We shall encounter another use of an excited electron when considering the oxidation of carbohydrate within the cell (§3.21). Vision involves a similar process, the electronic excitation energy being used to generate a response in the photoreceptor cell. The mechanism whereby electronic excitation energy is harnessed is unknown.

2.9 Radioactive isotopes

As we have seen, the chemical properties of an element are governed by the arrangement and number of electrons in the electron cloud, the latter being related to the positive charge on the nucleus produced by its uncombined protons. Many elements can exist in more than one form, or *isotope*, because it is possible for the number of neutrons in their nuclei to vary. Since these do not affect the charge on the nucleus, there is no difference in chemical properties between isotopes, but merely a difference in atomic weight. For example, chlorine gas contains two isotopes, one of which contains eighteen neutrons, the other twenty neutrons. Their corresponding atomic weights are 35 and 37, and because they always occur in constant proportions with Cl^{35} predominating, the atomic weight of chlorine is 35.5.

Protons and neutrons are bound together in an atomic nucleus with considerable energy. The stability of a nucleus depends on the balance between these two particles, the greatest stability being achieved when they are equal in number or nearly so, as in the case of sodium mentioned earlier. If the ratio between them departs from unity by a considerable factor, the nucleus becomes unstable. An excess of neutrons leads to the breakdown of some of them, thereby restoring a better balance between protons and neutrons, but resulting in the emission from the nucleus of electrons (β-particles). An excess of protons results in the emission of positively-charged particles, although these are more than protons, being helium nuclei (α-particles). Both α- and β-emission involve considerable energy changes in the nucleus, and these are accompanied by the emission of electromagnetic or γ-ray radiation, which is characterized by a very short wavelength and considerable penetrating ability, being very similar to X-rays. Thus, naturally-occurring isotopes may be stable, as in the case of chlorine, or unstable, and in the latter case exhibit radioactivity. It is possible to excite the nuclei of stable elements by radioactive bombardment, utilizing either a *nuclear pile* containing naturally radioactive substances, or in a machine known as a *cyclotron*. By this means, extra neutrons are introduced into the nuclei of the hitherto stable irradiated element, and the binding energy needed results in the excitation of the nucleus to a higher energy level. This tends to return to its ground state, and in the process it emits γ-radiation, and sometimes even protons, and sometimes α-radiation.

Thus, artificial isotopes can be formed which are not found in nature, which are radioactive. Some of the most important biological elements, carbon, nitrogen and phosphorus among them, will form such radioactive isotopes, and their use has revolutionized many aspects of experimental biochemistry. All types of radiation produce ionization of atoms in their path, and this ionization effect is both proportional to their concentration and capable of being measured in minute quantity. Since radioactive isotopes are identical in chemical properties with their corresponding non-radioactive natural isotopes they can be mixed with the latter in a known concentration and the mixture used in various biochemical situations. This procedure is known as *labelling* an element, and it can be used to follow the passage of the normal element into and out of cells and tissues under various experimental conditions, their concentrations being measured at any required stage because of the radioactivity they possess. In order to be useful in experimental situations, a radioactive isotope should have a sufficiently long life for its detection to be possible over a reasonable period of time. In practice, the *half-life* of the radioactivity produces an acceptable limit of usefulness, and fortunately the half-lives of the elements mentioned above, and of other biologically important elements, is sufficiently great for them to be used.

2.10 The energetics of molecules

The energetics of molecules are more complex than those of atoms. Molecules are by definition composed of more than one atom, being the smallest particle of a specific substance that can exist in the free state. The aggregation of a number of atoms into a single system results in the development by that system of properties not possessed by the individual atoms from which it is formed. Since biological organisms are highly organized, and utilize many complex molecules, this fact has considerable biological significance.

Molecules can possess at least three different kinds of energy in addition to the electronic energy of their atoms. *Resonance* is a property of molecules which exhibit a structural symmetry in relation to the single and double bonds that bind their atoms together, and which permits them to exist in more than one structural form, although their chemical formula remains the same. For resonance to occur, this structural symmetry must be accompanied by a rough equivalence in the energetics of the alternative structures. Where resonance occurs, it is found that the energy content of the molecule is greater than the sum of the bond energies that link the individual atoms together. The commonest example of resonance, which is described in all textbooks of chemistry, is benzene C_6H_6, but a number of biologically important compounds exhibit resonance, or this is a property of their breakdown products. In the next section we shall refer to the substance ATP, a triphosphate that can lose its terminal phosphate to give rise to its diphosphate (ADP). ADP possesses more resonance forms than ATP, and since resonance energy tends to make a molecule more stable, the result is that ATP is all the time inclined to give up its terminal phosphate group, an important property since this substance acts as the immediate energy source of living cells, and it is the splitting off of the terminal phosphate that liberates the energy utilized by the cell. The interested student is advised to consult a textbook of biochemistry for other examples of resonance in biological compounds.

Other forms of energy may be associated with molecules, such as the *rotational* energy, which results from the rotation of the molecule about its axis; and *vibrational* energy, due to the vibration of the constituent atoms about their mean position in the molecule. Their relationship to biological phenomena is beyond the scope of this book. What it is important to grasp is that the energy content of a molecule can change either because of the supply of energy from an external source, which may affect individual atoms; or through rearrangement of the molecule without alteration of the chemical formula. A corollary of the latter is that an apparently similar type of chemical structure may mask quite different energetic properties. At first sight, the substance ATP referred to above seems unnecessarily complicated, for there appears to be no reason why an inorganic triphosphate should not serve the same function. But the coupling of the complex organic substance adenine with phosphate groups and the configuration permitted by their combination with a pentose sugar confers additional properties on the molecule. The student will find that the adenine-pentose-phosphate complex occurs again and again in cellular systems, because the unique phosphate linkage it affords give it specific properties in relation to the transfer and liberation of energy.

2.11 Energy-rich compounds

As we have seen, the energy contained within a molecule may be greater than the sum of the energies that were required to form each individual chemical bond between the atoms in the molecule. In addition, the structure of a particular molecule may be such that it is more easily broken down by a certain type of reaction than by other types. If this type of reaction leads to a greater number of resonance forms, the reaction will be especially facilitated. The substance *adenosine triphosphate* (ATP) may be thought

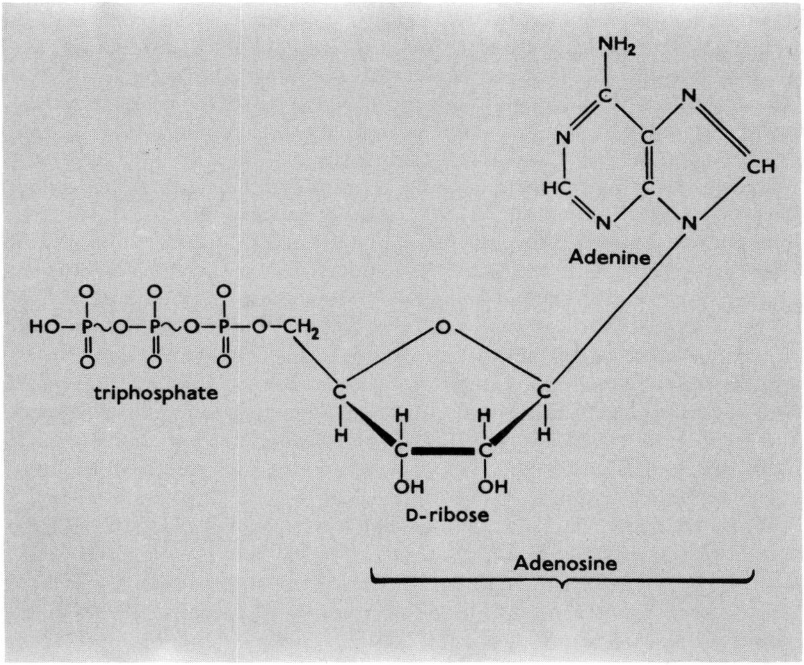

Fig. 2.3   Chemical structure of adenosine triphosphate (ATP).

of as a molecule of adenine linked through a pentose (5-carbon) sugar to three molecules of orthophosphoric acid, H_3PO_4 (Fig. 2.3). It was mentioned above that the terminal phosphate group could be removed from the molecule of ATP to produce adenosine diphosphate (ADP) and inorganic phosphate. This reaction is a hydrolysis, proceeding by the addition of water to the molecule, and because of the resonance considerations already given, it proceeds with relative ease. The removal of the terminal phosphate group from ATP results in the liberation of energy equivalent to about 30 558 J (7300 cal) per phosphate group per mole. It is also possible to hydrolyse the terminal phosphate group of the resulting ADP, although with less ease, with the release of the same amount of energy; but the remaining phosphate group offers considerable resistance to its removal in this way, and when this group is removed, the yield is only about 12 558 J (3000 cal) per mole. For this reason, ATP and ADP are described as *energy-rich* or *high-energy* compounds, and this fact is often denoted symbolically by writing the links between the two terminal phosphate groups as A—P$\sim$P$\sim$P. Although the bond $\sim$ is often referred to as a *high-energy bond*, it should be noted that this is merely a convenience. The extra energy available results from the properties of the molecule as a whole, and not from the characteristics of a single chemical bond.

The reader will find many different values given in books for the standard free energy change that results from the breakdown of the energy-rich bonds of ATP. This is due to the kinds of difficulties already briefly referred to as being involved in the determination of standard free energy changes in actual biological systems rather than ideal laboratory systems. The use of different temperatures and pH values, and the difficulty of determining the precise equilibrium point have led to a number of different figures. The one given above represents the standard free energy change at pH 7.0 and 37°C with excess Mg^{2+} present, and based on unimolar concentrations of reactants and products. In living cells, conditions can be very far from these 'ideal' ones, and result in a value for $\Delta F°$ as large as $-52\ 365$ J ($-12\ 500$ cal). Although we must compare different substances from the ideal standpoint, and it will be seen below that this approach yields an important fact about cellular energetics, this qualification must always be borne in mind.

There are many other phosphates which may be described as 'energy-rich' in the same way as ATP. If all such phosphates are arranged in order according to the size of the standard free energy change of their hydrolysis, it is found that they form a continuous series in which some have a much greater standard free energy than ATP and some much less, ranging from phosphoenolpyruvate with a $\Delta F°$ of -62 kJ (-14.8 kcal) to glycerol-1-phosphate at -9.2 kJ (-2.2 kcal). ATP fits into the *middle* of this series, and it is because of this that it is so important in cellular energetics. It is able to act as an intermediate carrier of phosphate groups from substances of high phosphate bond energy to those of lower phosphate bond energy. Transfer of phosphoryl groups in this way is normally mediated in the cell by ATP alone, and there is no direct transfer between other phosphate compounds. For example, in the intact cell the substance 1,3-diphosphoglycerate, with a $\Delta F°$ of -49.4 kJ (-11.8 kcal) is transformed into 3-phosphoglycerate by the loss of one of its phosphate groups, and this substance has a $\Delta F°$ of -18.8 kJ (-4.5 kcal). This change is accomplished by the transfer of the phosphate group to ADP to form ATP. It is convenient, though again not strictly accurate, to call ATP and any phosphates with a greater $-\Delta F°$ 'high-energy' compounds, and any phosphates with a smaller $-\Delta F°$ 'low-energy' compounds, on the basis of the central position enjoyed by ATP.

Although ADP is the product of many cellular reactions which use ATP as energy source, some reactions are encountered in which both terminal phosphate groups are removed together as a single inorganic *pyrophosphate* molecule that is itself a high-energy compound. The two ways in which ATP may be broken down are thus:

(i) ATP —— $ADP + P_i$ (orthophosphate)
(ii) ATP —— $AMP + P P_i$ (pyrophosphate)

Because of these two pathways, two enzyme systems must also be present in the cell, by which ATP may be reconstituted. ADP is re-formed into ATP through the energy-yielding reactions of glycolysis (§3.16) and oxidative phosphorylation (§3.19), using orthophosphate (P_i). AMP is first transformed into ADP by transfer of a high-energy phosphate group from ATP, under the influence of the enzyme *adenylate kinase*, thus leaving two molecules of ADP to be dealt with in the usual way. The $P P_i$ is hydrolysed through the action of the enzyme *inorganic pyrophosphatase* to two molecules of P_i, which seems a waste of phosphate bond energy, but in fact within the cell this reaction can be a means of ensuring that certain synthetic reactions are irreversible. A special form of AMP, known as *cyclic 3', 5'* AMP has excited interest in recent years. It is referred to in §13.18.

Some high-energy phosphates act specifically as reservoirs of phosphate bond energy for the quick reconversion of ADP to ATP. Notable among such *phosphagens* are the substances *creatine phosphate* (*phosphorylcreatine*) and *arginine phosphate*, which will be encountered again in connection with the energetics of muscular contraction (§11.3). Since the 'high-energy' bond depends on molecular structure, such bonds are not, of course, confined to phosphates. Certain sulphur-containing compounds act as energy-rich substances and two of these, *lipoic acid* and *co-enzyme A* are important in the cellular oxidation of carbohydrates (§3.17).

THE CONSTITUTION OF LIVING MATTER

2.12 Protoplasm

Living cells consist of *protoplasm*, a heterogeneous material whose matrix or ground substance is a solution of *protein* in water, which gives to protoplasm its viscous nature. Water is fundamental to life: probably all vital processes take place in solution, and water is the basis of the transporting media of animals, either in the form of blood pumped around a circulatory system, or as tissue fluid through which gases and solutes can diffuse, or both. It is therefore not surprising to find that water is the largest single constituent by weight of living matter. The amount present varies according to species and environmental conditions, but is usually within the range 65–90% by weight, the normal level for most animals being 75–85% by weight. Protein accounts for a further 10–20%, *lipids* (fats and fatty substances) for less than 5%, and *carbohydrates* and *mineral salts* for only about 1%.

2.13 Proteins

Apart from forming, with water, the ground substance of protoplasm, protein is an important constituent of the various cellular membranes in conjunction with lipids, and many biologically active compounds are proteins, including enzymes (§3.12). Proteins may also be used as fuels to provide energy, but because they cannot be stored as lipids and carbohydrates can, their use for this purpose is limited.

Proteins can be broken down into their constituent units in the laboratory by heating with strong acids or alkalis. These units all have the same basic chemical nature, and are known as *amino acids*. This name is derived from a basic —NH_2 or *amino* group, and a terminal acidic —COOH or *carboxyl* group. With one exception, all amino acids obtained from living tissues have their amino group attached to the carbon atom that is nearest to the carboxyl group. For this reason, they are known as α-amino acids and can be represented by the general formula $R \cdot CH(NH_2) \cdot COOH$, in which R may be one of a variety of organic chains or rings. There are about twenty amino acids which are used by living organisms to form proteins, although they are not all necessarily found in every protein.

The breakdown of a protein into its amino acids involves a hydrolysis, water being added to the molecule. The synthesis of proteins requires the opposite process of *condensation*, because water is removed when two amino acids are joined together:

$$
\begin{array}{c}
H_2N-CH-CO\cdot(OH) \;+\; (H)-N-CH-COOH \longrightarrow \\
\quad\;\; | \qquad\qquad\qquad\qquad\; | \quad\; | \\
\quad\;\; R \qquad\qquad\qquad\qquad\; H \;\; R \\
\\
\qquad\qquad H_2N-CH-C-N-CH-COOH \;+\; H_2O \\
\qquad\qquad\qquad\;\; | \quad\; \| \;\; | \quad\; | \\
\qquad\qquad\qquad\;\; R \quad\; O \;\; H \;\; R
\end{array}
$$

The —OC·NH— link so formed is known as a *peptide link*. The conjunction of two amino acids in this way produces a *dipeptide*. The addition of a third amino acid results in the formation of a *tripeptide*, and so on until a *polypeptide* chain is formed. When a polypeptide contains some fifty or so amino acids, it begins to exhibit the characteristic properties of proteins. The amino acid *glycine* $CH_3CH(NH_2)COOH$ has the smallest molecule of any amino acid, but even if fifty glycine molecules were joined together by peptide links, the molecular weight of the resulting compound would be over 3500. In practice, other amino acids with molecular weights much larger than that of glycine would be involved, and a protein with fifty amino acids would have a molecular weight of at least 6000. This would be a very small protein, however, and many naturally-occurring ones have molecular weights running into hundreds of thousands. When it is realized that an English dictionary is based on only twenty-six letters, the theoretical diversity of proteins based on some twenty possible components can be appreciated, especially when the enormous size they can attain is taken into account.

Many proteins assume a helical form, and can be likened to a spiral piece of wire or, if there is more than one helix to a protein, to the strands of wire in an electric flex, wound round one another. The analogy should not be taken too far. The individual helices coil and fold back on themselves, and adjacent pieces produced by this configuration cross-bond to one another by means of hydrogen bonds, or links between amino acids with a particular affinity with one another, or through the presence of hydrophobic (water-repelling) groups that mutually repel the solvent water and are thereby drawn closer to one another. Because its configuration depends on its chemical composition, each protein possesses a three-dimensional configuration that is unique to itself.

The external shape of a protein molecule, as distinct from its detailed configuration, is either spherical, when it is known as a *globular* protein, or like a long rod, when it is a *fibrous* protein. Fibrous proteins tend to be stable and rather insoluble, and so are found, for example, in muscle and connective tissue. Biologically active proteins, such as enzymes and antigens, seem usually to be globular.

Proteins tend to polymerize readily, the basic molecule linking up with a number of similar molecules to form a much larger one. This tendency makes the determination of their molecular weight difficult, because it is impossible to be sure that the protein analysed *in vitro* is polymerized to the same extent *in vivo*. Indeed, the study of proteins presents many formidable problems, for their complexity makes the elaboration of their structure hard and laborious.

2.14 Analysis of proteins

Proteins can often be separated from one another by the process of *electrophoresis*. In solution, the —NH_2 and COOH groups ionize to NH_3^+ and COO^- ions respectively, and some of the R side chains may also bear charged groups. Except at one precise pH, the *isoelectric point*, the charges due to these groups do not cancel one another out, and so a protein be-

haves as if it was a mosaic of positive and negative charges, in which one
or the other predominates to give it an overall positive or negative charge.
The latter can be varied by altering the pH of the solvent medium, but
proteins with different overall charges will maintain their relative differences in such media. The media used in electrophoresis mostly cause proteins to be negatively charged, and when an electric current is passed across
a solution of this sort, the protein moves towards the positive pole, or
anode. In a mixture of proteins, the movement of each individual protein
is governed by the strength of its charge and the size of its molecule. The
differences in the speed of movement of individual proteins under these
conditions can be used to separate them. The protein mixture is applied

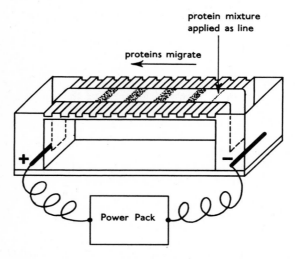

Fig. 2.4 A simple apparatus for the electrophoresis of proteins. For explanation
see text.

to one end of a strip of filter paper or other suitable material, whose ends
are made to dip into compartments containing solvent medium at either
end of the apparatus (Fig. 2.4). The compartments are separated from
one another, and each contains one of the two electrodes. The filter paper
is saturated with solvent, allowing a current to pass from one electrode
to the other. As the current flows, the proteins move at different rates along
the paper, thereby becoming separated from one another. The current
is stopped after a suitable period, and the areas of paper containing the
different proteins can be cut apart. Nowadays, some electrophoretic
apparatus allows continuous separation of the mixture, which is applied

and the separated proteins run off the whole time the apparatus is functioning.

Proteins separated by this means can be subjected to purification techniques, and they are then ready for further analysis. The amino acid sequence of the molecule can be determined by its controlled disintegration into a number of polypeptide chains. This can be accomplished either by differential chemical attack, the success or otherwise of which will itself provide clues to the structure of the molecule, or by using specific enzymes known to attack only certain amino acids. The individual polypeptide chains are then further broken down by similar means, and they are also hydrolysed to provide a mixture of their constituent amino acids.

The analysis of a mixture of amino acids may be achieved by using the process of *chromatography*. A number of different chromatographic methods is available, but the student is most likely to encounter *paper* chromatography. This method utilizes a large sheet of filter paper, which is suspended from a trough of solvent placed at the top of a tank. The solvent runs slowly down through the paper, carrying with it any substance that may have been placed on it and which will dissolve in it, such as amino acids and sugars. The mixture of amino acids is placed at the top of the paper in the form of a small spot, and spots of individual known amino acids can be placed alongside it for purposes of comparison. For a given solvent, amino acids are carried down the paper at rates specific to them, thereby converting the single spot of the original mixture into a series of spots strung out down the paper (Fig. 2.5). The single amino acids placed on the paper for comparison produce one spot at the appropriate point on the paper, and this can be used to identify the corresponding acid in the mixture, which takes up the same relative position on the paper. The spots can be coloured by a suitable chemical reaction, the usual reagent employed being *ninhydrin*, which imparts a blue/purple colour to amino acids. Alternatively, the spots may be viewed under ultra-violet light, when they fluoresce. A variety of individual 'spot tests' is also available for amino acids. The solvents used in paper chromatography are never equally suitable for all amino acids, for they all fail to separate some of them. It is therefore necessary to use several different solvents, and a suitable way of doing so is to carry out *two-dimensional* chromatography. In this technique, the spot is first allowed to run in one direction with a given solvent; and the paper is then turned through a right angle and run with a different solvent. Any spots which still contain more than one amino acid may then be separated by the action of the second solvent.

Such methods of analysis are quick to describe, but may take a long time to produce results in practice. It took ten years to determine the content and sequence of the small *insulin* molecule, with a molecular weight of about 6000. Nor do these methods tell us anything about the shape or configuration of the molecule. For these, other methods such as X-ray diffraction must be employed, which are beyond the scope of this book.

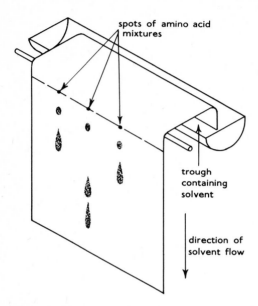

Fig. 2.5 Diagram to illustrate the technique of paper chromatography. The paper is suspended from the trough over a rod, and the entire assembly is placed in a glass tank. The amino acids are carried down by the solvent, as shown.

2.15 Properties of proteins

Because protein molecules are so large they form suspensions, or *colloidal* solutions, in water. Colloids are often defined in relation to their large particle size, but this is not entirely satisfactory, and it is better to define them by their other properties. Those that are of biological importance are: (i) they diffuse slowly, for example egg albumen (white) diffuses through water at only 1/20 000 of the speed of sodium chloride; (ii) their large size prevents them from diffusing across natural membranes; (iii) they form an *interface* or boundary between themselves and their solvent medium, the properties of which will be dealt with below; and (iv) proteins can exist in either a *sol* form, which is hydrated and takes the form of a viscous liquid; or a *gel* form that contains less water than the sol, and is more solid. Cooking gelatine is a sol when warmed, and a gel when cold. Gelation probably results from the formation of numerous strong cross-linkages between the protein molecules with the elimination of water; solation from the imbibition of water and the breaking-down of some of the cross-linkages.

The molecules in a homogeneous liquid are free to move in all directions, except at a boundary or interface with air or another liquid. In the latter situation the molecules in the homogeneous liquid are more attracted to one another than to those of the other liquid, and so they form a membrane at the interface between the two media. Proteins are large particles with considerable mutual attraction, and in an aqueous medium they tend to form a homogeneous packet of molecules that produces a strong interface with the medium. The mutual attraction of any molecules in a homogeneous liquid is such that the membrane they form at an interface with another liquid possesses a definite binding energy, which we call *surface tension*. The surface tension at a protein/water interface is very large. Any reaction that tends to reduce this surface tension will, according to the Second Law of Thermodynamics, be facilitated. One way in which it may be reduced is by the attachment of substances to the molecules that form the interface membrane, because this decreases their mutual attraction. The energy change involved in this process of *adsorption* can be very large, and it is probably a factor in enzyme action (§3.13).

Proteins are fairly stable in solution, because they attract water molecules to themselves, and therefore bind themselves to their solvent. However, they are susceptible to a number of outside influences. Heat, changes in pH, a wide variety of chemicals, and certain other circumstances tend to produce the *denaturation* of proteins, a condition in which the ordered arrangement of the molecule becomes less orderly. This process, which is presumed to involve the breakage of some of the cross-linkages in the molecule, produces changes in some of the properties of proteins, including the inactivation of enzymes, and a tendency to be less soluble. Adjustment of the pH of the solvent to the isoelectric point of the protein causes it to become *salted out*, a kind of precipitation which can be reversed if the pH is again altered fairly radically. If a protein is heated, denaturation is succeeded by coagulation of the protein (as in the boiling of an egg) and this is irreversible.

2.16 Lipids

The term *lipid* is applied not only to fats and waxes, which are clearly fatty in nature, but also to other substances which are soluble in fat solvents, and used in this way it includes a rather heterogeneous group of substances. Thus sterols (§13.8), and carotenoids, which include the precursors of visual purple (§9.22), are classed as lipids.

Neutral fats are a combination of the trihydric alcohol *glycerol* with three molecules of *fatty acid*:

$$CH_2(OH) \quad (H)O \cdot CO \cdot R_1 \qquad CH_2O \cdot CO \cdot R_1$$
$$CH \cdot (OH) + (H)O \cdot CO \cdot R_2 \longrightarrow CH \cdot O \cdot CO \cdot R_2 + 3H_2O$$
$$CH_2(OH) \quad (H)O \cdot CO \cdot R_3 \qquad CH_2O \cdot CO \cdot R_3$$

glycerol   3 fatty acid   triglyceride
                molecules

The R in the above formulae represents the side-chain of the fatty acid, the commonest naturally-occurring fatty acids being *palmitic, oleic* and *stearic* acids. Because glycerol is a trihydric alcohol, bearing three side groups each of which terminates in an —OH radical, the resulting fat is known as a *triglyceride*. It will be noted that the formation of fats is, like that of proteins, a condensation process in which water is eliminated. Conversely, their breakdown, whether by chemicals or enzymes, involves their hydrolysis. Since triglycerides are neutral in pH and inert, they are useful as a storage medium for carbon compounds, and this appears to be their main function. As we shall see, the fatty acids and glycerol they contain can be broken down for the liberation of energy to be used by the cell (§3.18).

Phospholipids are like triglycerides in which one fatty acid molecule has been replaced by a phosphate group. In most cases, the phosphate group is combined in turn with an organic base, e.g. *lecithin*:

$$CH_2O \cdot CO \cdot R_1$$
$$CH \cdot O \cdot CO \cdot R_2$$
$$CH_2O \cdot P \cdot O \cdot (CH_2)_2N(CH_3)_3$$
$$\diagup\diagdown \qquad\qquad |$$
$$O \quad OH \qquad OH$$

The importance of phospholipids is that they possess both hydrophilic (water-attracting) and hydrophobic (water-repelling) groups, and hence can act as binding agents between water-soluble and fat-soluble substances. This appears to be one of their roles in cell membranes, all of which seem to contain both a protein and lipids (§3.2).

Waxes are similar in structure to triglycerides, but instead of glycerol they contain higher aliphatic alcohols, and the fatty acids combined with these are larger and have longer side-chains than those found in neutral fats. They are fairly tough and tend to have reasonably high melting points, and so they function as protective and impermeable coverings, as in the cuticular waxes of a wide variety of animals.

2.17 Carbohydrates

These are substances with an empirical formula $(CH_2O)n$, where n may be 3, 4, 5 (pentoses) or 6 (hexoses); or they are compounds formed by the aggregation of these substances. Pentoses and hexoses are often depicted as open-chain compounds, but in their biologically active state some of the carbon atoms are normally arranged in a continuous ring. In the case of the most commonly occurring hexose, *glucose*, five of the carbon atoms form the ring, and OH groups are attached to three of these (Fig. 2.6).

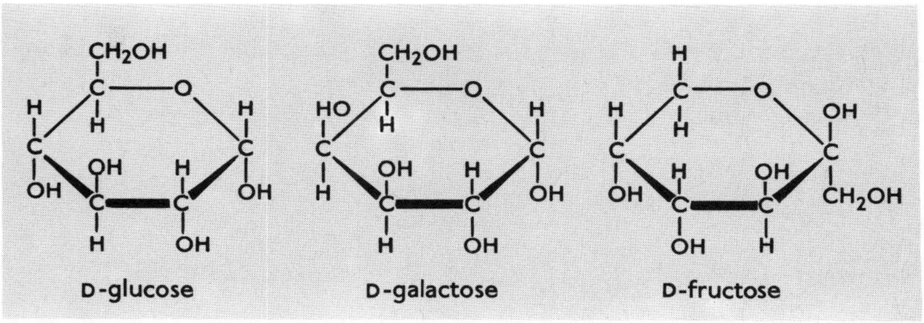

Fig. 2.6 Structure of glucose, galactose and fructose, to illustrate the characteristics of the ring structure of hexose sugars. Glucose and galactose are isomers.

It is possible for the position of an —OH group to differ in its orientation, and this means that a hexose like glucose may have a counterpart or *isomer*, which is identical in chemical structure but different in orientation. Glucose and galactose are both hexoses, but differ in this way (Fig. 2.6).

A carbohydrate with a single ring of carbon atoms, i.e. a hexose or a pentose, is known as a *monosaccharide*. Two monosaccharides can be joined together to form a disaccharide. An —OH group of one monosaccharide joins with one of the carbon atoms of the other to form a *glycoside bond*, with the elimination of water, the process once again being a condensation e.g:

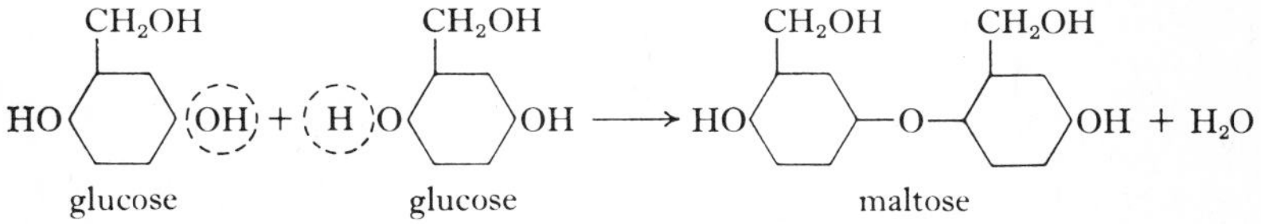

The process can be repeated almost indefinitely, to produce *trisaccharides* and ultimately large *polysaccharides*. The latter have the general formula $(C_6H_{10}O_5)n$, and the value of n may be 200 or more. Because of their large size, polysaccharides form colloidal solutions and even solids, and will not pass across natural membranes. As they are chemically inert and do not ionize, they are ideal substances for use as energy reserves, and are stored for this purpose as *starch* by plants and *glycogen* by animals. They are also used for protective or skeletal structures like the matrix of connective tissue, the cement between cells (both in conjunction with protein), and the cellulose cell walls of plants and the cellulose test of tunicates. In combination with proteins, polysaccharides form *mucopolysaccharides*, which are found in mucus and act as lubricants and in feeding processes.

Polysaccharides take the form of strings of hexose units joined together by condensation, and the strings may be either simple or branched (Fig. 2.7). Starch contains both the unbranched *amylose* and the branched

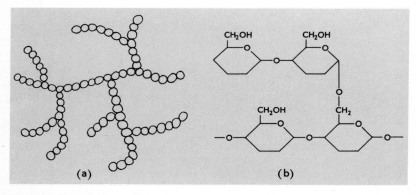

(a)   (b)

Fig. 2.7 Branched configuration of glycogen. The circles in (a) symbolize hexose units. The method of branching is shown in (b).

amylopectin, but glycogen is entirely composed of the latter. In the vertebrate liver, the formation of glycogen requires the presence of a *primer*, a short length of at least three or four glucose units, and further units are added to this by the enzyme *glycogen synthetase*. The individual glucose molecules must first be in an activated form, produced by their phosphorylation by an energy-rich phosphorylated nucleotide (§3.6), *uridylic diphosphate* (UDP). The glucose becomes attached to the UDP, and it is as a UDP-glucose complex that it is acted on by the glycogen synthetase to be joined to the growing primer, with the uridylic acid and inorganic phosphate being split off. Glycogen and other polysaccharides may be broken down enzymically and chemically by hydrolysis, but the pathway for their breakdown in living cells is not a simple reversal of their formation. When the stored glycogen is to be utilized, it is phosphorylated by the enzyme *phosphorylase a*, using inorganic phosphate, to form the energy-rich glucose-1-phosphate (§3.14).

Thus, the three major constituents of living matter (if we except water) are all capable of being built up from simple units by a process of condensation, or broken down into those units by hydrolysis. They may be ingested by animals in either form, but as the multiple units are generally too large to pass across natural membranes, they must be broken down into their constituent units before being absorbed into the animal. This is the task of the hydrolytic enzymes secreted by animals for this purpose, and already familiar to the student in connection with mammalian digestion.

THE NUTRITIONAL REQUIREMENTS OF ANIMALS

2.18 Quantities

The largest amounts of food required by animals are the energy-yielding compounds in the form of carbohydrates, lipids, and to a lesser extent proteins. These compounds are needed in quantities which are measured in terms of g/kg body weight daily. Some amino acids and fats are needed for special purposes in much smaller quantities, measured in mg/kg body weight daily. In addition, specific growth factors like vitamins are required, but in minute quantities of the order of μg/kg daily. Whereas the need for the major foods and for special fats and amino acids is virtually constant throughout the animal kingdom, the requirements for growth factors are much more diverse, and reflect the diversity of the evolution of cellular mechanisms. We shall therefore consider the requirements of animals for only the first two groups of substances, and larger texts should be consulted for the detailed requirements for those falling into the third group.

2.19 Carbon and nitrogen requirements

As we have seen, all animals ultimately obtain their carbon- and nitrogen-containing foods from green plants. Any organisms that, like green plants, can form carbohydrate from carbon dioxide and water, are termed *phototrophic*. Obviously, this term is normally restricted to green plants, but in the borderland area between plants and animals there are organisms claimed by zoologists that exhibit phototropism. These are all green flagellates, like *Euglena*. Animals generally are *heterotrophic*, being able only to utilize the compounds that plants have already made.

Bacteria and some plants can *fix* atmospheric nitrogen, but this ability is uncommon. Photosynthetic plants generally utilize nitrate or ammonia as their nitrogen source, and combine these with carbohydrate to form amino acids and proteins. Organisms that can use inorganic nitrogen in this way are known as *autotrophs*. Those which can use a single amino acid as their sole source of nitrogen or an amino acid plus ammonia, are *mesotrophs*; and *metatrophs* are those which need more than one amino acid. It should be realized that the application of these terms to a particular organism reflects laboratory experiments to determine the simplest form of nitrogen utilization of which it is capable, and does not imply that it lives solely upon such a source in natural conditions. Autotrophs can act like mesotrophs, and mesotrophs can act like metatrophs. It is probably the case that animals which can use simple compounds in the laboratory also use more complex ones in their normal habitats.

In general, phototrophic organisms are also autotrophic. These are generally green plants. However, there are some colourless protozoans which, although they are not phototrophic, can utilize fairly simple carbon compounds, such as fatty acids or their salts, and which are also autotrophic. Many colourless flagellates are able to grow with acetate as their carbon source, like *Chilomonas* and *Polytoma*. *Tetrahymena geleii* seems to be intermediate between this condition and that of most animals in that it can utilize some acetate, but also needs some hexose sugar. A few colourless flagellates like *Chilomonas* and *Astasia* can use ammonia as their nitrogen source, but most of them are mesotrophic like *Euglena deses* or metatrophic like most other species of *Euglena*. The synthetic ability of some flagellates is evidently very high, and the autotrophic forms can actually synthesize all the amino acids found in higher animals from their inorganic nitrogen source. Mesotrophic animals presumably need only the —NH$_2$ group.

All other animals appear to be metatrophic, having lost this synthetic power to varying degrees. They need a mixture of amino acids in their diet, although the particular ones needed may vary between species. *Essential* amino acids are those which cannot be synthesized, and therefore must be supplied in the diet. The remaining *non-essential* amino acids can be synthesized, either by using ammonia, or the —NH$_2$ group from another amino acid which is joined with carbohydrate, or from specific precursor amino acids. The fact that some amino acids are used specifically for conversion into others means that the balance of amino acids in the diet must be correct. If a diet is deficient in a precursor its derived amino acid cannot be made, and the precursor itself then becomes an essential amino acid. Thus, the amino acid *tyrosine* is not essential for the chick, but becomes so if the animal's diet is deficient in its precursor *phenylalanine*.

On average, about ten or twelve amino acids are essential for most metatrophs, and about seven of these are common to nearly all, a further three being common to most. This basic similarity in amino acid requirements suggests that the pattern of the ability to synthesize some, and the loss of ability to synthesize other amino acids, was established fairly early in evolution. However, the quantities of each amino acid required varies considerably between species, a factor which often makes the culture of animals on artificial media difficult.

The lipid requirements of many animals seem to vary, at least so far as the need for neutral fats is concerned. This is not because the animals which do not need them in their diet do not use them in their bodies, but because they are readily made from carbohydrate and even protein. Fat is widely used for storage purposes even when it is not an essential dietary factor. However, there seems to be a special need by some animals for one or two unsaturated fatty acids, i.e. fatty acids with double bonds instead of single bonds between some of the carbon atoms. These appear to be essential for proper growth and maturation in those animals that need them. The most widely-needed is probably *linoleic acid*,

$$CH_3(CH_2)_4CH{=}CH \cdot CH_2 \cdot CH{=}CH(CH_2)_7COOH.$$

Larvae of the moth *Ephestia* need this fatty acid if the imago is to emerge successfully from the pupa and proper development of the wings is to take place; and this is the case in certain other insects. Rats will not grow properly without it, and although man can synthesize a certain amount, this must be supplemented in his diet. It is probable that the need for linoleic acid and similar fatty acids will prove to be widespread.

3

Cellular Organization and Function

Cellular material is continuously being replaced, even after an animal has reached maturity and its growth has ceased. The extent of the replacement varies between different cells, since tissue that is highly vascularized and active has a more rapid turnover than lightly vascularized tissue. For example, the liver proteins of mammals have a very short life, in some cases of only a few hours, whereas the proteins of tendons remain unchanged for long periods. Other cellular components are also exchanged with the environment. It has been shown that injected phosphate is rapidly incorporated into the calcium phosphate of such an apparently static tissue as bone. And there are changes that are more subtle and little understood, such as the decrease in the salt and water content of cells as they age.

Observations like these emphasize that there is a dynamic balance within cells, and it should be the primary aim of the study of cells to elucidate the mechanisms that are responsible for it. We require to know two types of facts: the structure and distribution of the cellular components, from which inferences can often be made about their function; and the chemical interactions within cells.

3.1 The study of cellular structure

The structure of cells is nearly always investigated by some form of microscopy. In microscopy the cells are examined by passing some kind of illumination through them, and the picture presented by their interference with the passage of this illumination is resolved and magnified by the microscope. Resolution depends on the wavelength of the illumination employed, and the wavelength of white light is such that objects closer together than about 0.25 μm cannot be resolved as separate objects by the light microscope, which limits its useful magnification to about 1500×.

In order to achieve better resolution and greater magnification, illumination of shorter wavelength must be employed. Electrons are similar to light in that they behave as waves as well as particles, and their wavelength is such that when they are employed for microscopic illumination a resolution of 1–2 nm is possible with biological material. This gives the *electron microscope* in which they are utilized a useful magnification over 250 times greater than that of the light microscope. For this reason, the use of the electron microscope has revolutionized the study of cells.

In the electron microscope (Fig. 3.1) a certain type of metal is induced

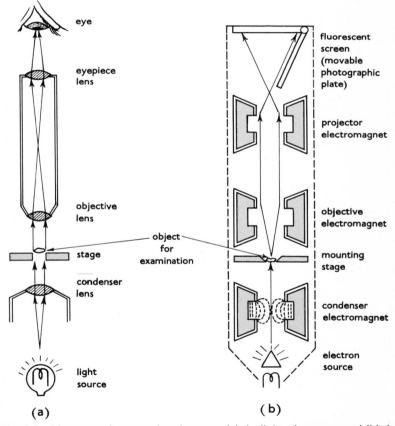

(a)   (b)

Fig. 3.1   Diagrammatic comparison between (a) the light microscope and (b) the electron microscope. The lines of force are shown for only one magnet of the latter. The paths indicated for the illumination are not precise; in the electron microscope the electrons actually take a very complex path.

to form a cloud of electrons at its surface by being warmed. A set of positively charged plates, suitably orientated, draws the electrons away in the form of a beam and focuses them on the object to be examined, like the condenser of a light microscope. The electrons that pass through the specimen are drawn out by a further set of plates, and the extent to which this is done determines the magnification of the object. Finally, the electrons impinge on a fluorescent screen to form a visual picture, or are allowed to strike a photographic plate for a permanent record. Because electrons are scattered by gas molecules it is necessary to evacuate the apparatus before use, with the specimen in position.

In electron microscopy, either potassium permanganate $KMnO_4$ or osmium tetroxide (osmic acid) OsO_4 is normally used as both fixative and stain. These substances combine with some of the cellular components, and since osmium and manganese are heavy atoms they absorb electrons and increase the contrast between these components and the rest of the cell. The parts that absorb electrons appear dark on the screen and are described as *electron-dense*.

The reader will be familiar with the generalized structure of an animal cell (Fig. 3.2). There is a limiting layer, the *plasma membrane*, that encloses the *cytoplasm*, a heterogeneous material which consists of a proteinaceous ground substance containing a variety of inclusions. Set apart from the cytoplasm by its own membranous envelope is the *nucleus*. We shall examine the nature and function of each of these in turn.

3.2 The plasma membrane

Cellular organization seems to be linked with the presence of a number of membranous components, in which the membrane appears to be an important part of the functional unit. Most of these membranes—the plasma membrane, the endoplasmic reticulum, the nuclear envelope and the mitochondrial membrane, and those surrounding smaller inclusions— look similar under the electron microscope, and are generally believed to possess a basically similar structure. There is good evidence that the nuclear envelope and rough endoplasmic reticulum are continuous, but the other membranous components do not appear to be joined together in one structural system.

Under the electron microscope the plasma membrane appears as two electron-dense lines about 2.5 nm thick, separated by a gap of similar thickness (Plate 1). Despite some uncertainty over the details, there is general agreement about the physico-chemical nature of the plasma membrane. Its basis appears to be a double layer of lipid material, each layer consisting of a sheet of phospholipid one molecule thick. It will be recalled that phospholipids are similar to neutral fats in being compounds of fatty acids and glycerol, but are essentially diglycerides, one fatty acid having been replaced by a phosphate group (§2.16). The simplest phospholipid extracted from plasma membranes is *phosphatidic acid*, in which the glycerol residue has a nitrogenous group attached in addition to the phosphate group (Fig. 3.3). The other extracted phospholipids have more complex groups attached at this point. Examples of these are *phosphatidyl inositol* (inositol is a substance related to sugars), *phosphatidyl serine* (serine is an amino acid), and *phosphatidyl choline* (*lecithin*). The most abundant of these phospholipids in the plasma membrane is lecithin.

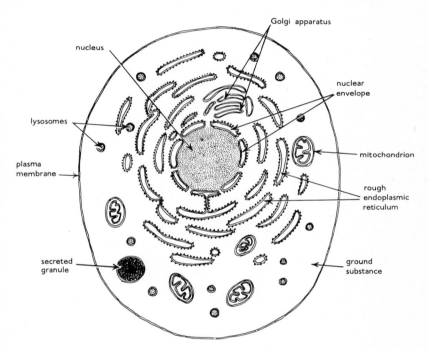

Fig. 3.2 Generalized diagram of the animal cell in section, based on the findings of electron microscopists. Not to scale.

As can be seen from Fig. 3.3, these phospholipids consists essentially of two long hydrocarbon chains, and a terminal or *polar* group that contains the glycerol/phosphate residue. The hydrocarbon chains are insoluble in water, or *hydrophobic*, whereas the polar group is soluble in water, or *hydrophilic*. Because the extra- and intracellular fluids are aqueous, the polar groups come to lie in them, with the result that two layers of phospholipid are formed with their polar groups pointing outwards and their

36

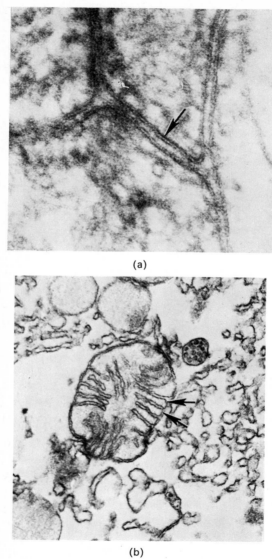

(a)

(b)

Plate 1 (a)–(d) Electronmicrographs of sections of cells to show the cellular components. (a) Plasma membranes of pedal support cells of *Limnaea*. Each adjacent membrane gives the appearance of two dense lines (arrowed). × 158 000. (b) A mitochondrion. The arrows point to an area in which the inner unit membrane can be seen to project into the organelle as cristae. × 13 000. (Courtesy of A. Peat, University of Durham.) *See opposite for (c) and (d).*

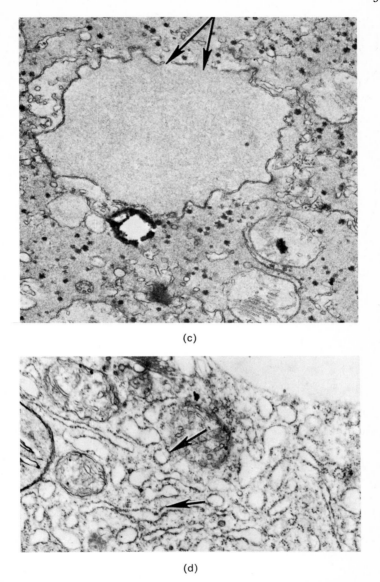

(c)

(d)

Plate 1 continued   (c) Pores in the nuclear envelope. Two pores are arrowed, but others are visible. × 8000. (d) The endoplasmic reticulum of a rat liver cell, showing the cisternae bounded by granular ribosomes (arrowed). × 19 000. (Courtesy of A. Peat, University of Durham.)

long chains pointing inwards towards each other (Fig. 3.4). At the pH of living cells, lecithin forms a balanced zwitterion with a neutral charge (§7.3) and the molecules produce a closely-packed layer because there is no electrostatic repulsion between them. The other phospholipids possess a net negative charge, and if placed adjacent to one another would produce bad packing through electrostatic repulsion. They are therefore thought to be interspersed among the lecithin and related choline-containing phospholipids.

The phospholipids are bound into the plasma membrane in combination with proteins or peptides. Some of the protein is structural, and

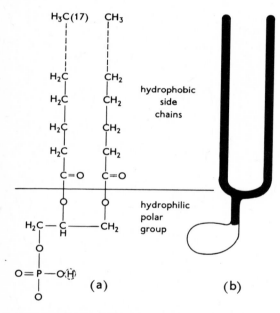

Fig. 3.3   (a) The chemical structure of phosphatidic acid and (b) a symbolical representation of it used in Fig. 3.4

confers both stability and elasticity upon the membrane; and some is present in the form of enzymes. In addition, each molecule of phospholipid is associated with a molecule of cholesterol, which is thought to aid in the packing together of the lipid molecules. There is also a quite high concentration of carbohydrate associated with the membrane. The precise way in which these substances are related to one another within the membrane is not certain, but Fig. 3.4 gives one of the more likely patterns.

Because some cellular membranes differ in thickness and detailed composition, and carry out different functions, other models than that in Fig. 3.4 have been proposed. Some retain the lipid bilayer, but exhibit different positions and configurations for the protein component, including one in which there are protein-lined channels connecting the two sides of the membrane. Another adopts a 'micellar' structure, with the lipids arranged in globular subunits, but its properties would be such that it is unlikely to be common, if it exists at all. There is considerable agreement that the protein does not form a precise layer, but is interspersed among the other components.

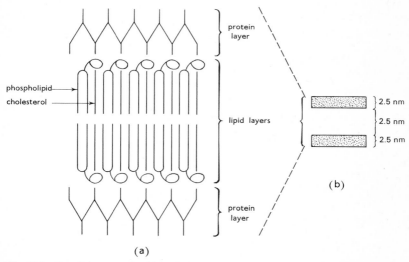

Fig. 3.4   Possible organization of the plasma membrane. The electron-dense lines in (b) are generally supposed to represent the polar heads of the lipid components. (Modified from Finean, 1961, *Int. Rev. Cytol.*, **12**, 303)

3.3 The permeability of cells

The ease with which material passes across the plasma membrane is a measure of its *permeability* to that material. Elementary courses in biology sometimes leave the impression that membranes are semi-permeable in the chemical sense, but this is obviously untrue. The truly semi-permeable membrane passes only water, whereas natural membranes also permit solutes to pass. The passage of solutes is selective, and results in the formation of characteristic concentrations of solute particles within the cell. Consequently, if a cell is placed in a fluid that contains a different total particle concentration from itself it will shrink or swell, depending on whether the external concentration is higher or lower. This osmotic effect is often used to determine the permeability of the plasma membrane to water, by measuring the change in volume of the cell. Using this criterion, it has been calculated that it takes about six weeks for 1 cm^3 of water to pass through 1 cm^2 of the outer surface of the freshwater protozoan *Zoothamnium* in its natural habitat. Since this animal lives in a medium much less concentrated than itself, it is in a permanent condition of osmotic stress in which water is tending to move in continuously. The passage of water into *Zoothamnium* is thus quite slow, but is thought to be typical of plasma membranes in general. Despite this fact, it can be shown on theoretical grounds that such a rate of entry is still greater than would be expected from the passive entry of water, and so the concept of *solvent drag* has been advanced in explanation. The idea is that as the water molecules pass through the membrane they exert a dragging force on those behind, and the water moves in faster, possibly taking solute particles with it.

Almost all natural membranes are freely permeable to water, and to neutral substances that are fat-soluble. Neutral substances that are water-soluble also pass through such membranes, but rather less readily. This is not surprising, since a substance that will dissolve in or combine with one or more of the materials of which the membrane is constructed is likely to pass rather more easily than one which will not. Fat-soluble substances pass more quickly because the core of the membrane is lipid, the aqueous phase being represented by the polar heads of the lipid, and the protein. In general, too, very large molecules of any kind will not be able to pass through natural membranes, but this statement needs qualifying in view of the phenomenon known as *pinocytosis*. Cells are capable of taking in drops of the surrounding medium by enclosing them in a piece of the plasma membrane which becomes separated as a small vacuole and moves into the cytoplasm. The process may be thought of as a kind of phago-cytosis of the surrounding fluid, but since this may contain solutes these will also be ingested, and obviously quite large molecules can be taken into the cell in this way. It is not known how far the process is used for this purpose, nor is its significance in the economy of the cell very clear.

Given the ability of the particle to cross the plasma membrane, it will do so under the influence of the concentration gradient between the cell and the surrounding fluid. The process is one of simple diffusion from the more concentrated solution to the less concentrated one.

When particles are charged, a further complication is introduced. The entry of electrolytes into cells is generally found to be slower than that of non-electrolytes of comparable size. Weak electrolytes, which dissociate into ions to only a limited extent in solution, enter more rapidly than strong electrolytes, and the higher the valency of an ion, the slower is its penetration. These facts indicate that the passage of a charged particle through the membrane is slowed in proportion to its charge. However, this statement requires modification because ions tend to attract a sphere of water around themselves, and they do so to differing extents. The size of the ion then depends on the amount of water surrounding it, and this affects its rate of penetration. The point is important, because the behaviour of membranes to ions is such that they are often pictured as sieve-like structures, although this analogy should not be pressed too far. Unlike a sieve, the membrane is a dynamic structure and not a static one. We cannot be sure that all the holes or pores are the same size, nor do we know whether they remain the same size in all circumstances. Some workers believe that alterations in permeability to a given substance are due to changes in the size of the pores.

The passage of ions across the membrane is also affected by the presence of a charge on the membrane. All living cells exhibit such a charge, the origin and maintenance of which is considered in §9.2. Since the outside of the cell is always positive and the inside negative, and since like charges repel one another, it should follow that negative ions (anions) will pass more easily into the cell than positive ions (cations), and this is generally found to be true. However, the membrane is probably best thought of as a mosaic of positive and negative charges in which one is predominant overall. This is a reasonable concept, because the outside of the membrane is believed to be proteinaceous. As a result, the entry of cations is not completely blocked, but merely slower than that of anions. Some workers believe that the pores in the membrane are lined with protein, and these, too, will possess either a net positive charge or a net negative charge, according to the nature of the lining protein. Thus, some pores will pass cations more easily, others anions, a characteristic that has an important bearing on the generation of postsynaptic potentials (§9.9).

3.4 Active transport

In some cases, the movement of water and solutes cannot be explained solely as a passive phenomenon, and in these cases the concept of *active transport* is usually invoked. Active transport is normally assumed whenever solutes are transported against their concentration gradient, or water is transported against the prevailing osmotic pressure. It may also be involved in *facilitated diffusion*, a term applied to the passage of a solute in the direction predicted from its concentration gradient, but at a faster rate than can be accounted for by passive diffusion.

A common method of studying active transport is by the use of metabolic inhibitors, of which the commonest ones used are probably potassium cyanide KCN and 2,4-dinitrophenol (DNP). Both interfere with the formation and utilization of ATP, and it is generally assumed that if their use results in the abnormal exit or entry of substances, then these sub-stances must be actively transported in normal circumstances. This is probably too simple a view, because it is more than likely that the main-tenance of the normal structure and pore size of a membrane requires metabolic support, even though no work may be involved in the actual movement of substances, and permeability will thus depend on the avail-ability of ATP.

Active transport of solutes undoubtedly occurs, as in the uptake of salts by freshwater animals (§6.5), but it has been questioned whether active transport of water does occur. Many workers maintain that solutes are always secreted first, water following osmotically. Pinocytosis on a minute scale has been advanced as one explanation for the active transport of water, and it receives some support from electron-microscope studies that seem to show tiny fluid-filled vacuoles within transporting cells. The most generally accepted mechanism for the active transport of solutes assumes the presence in the membrane of lipid-soluble *carriers*. These are sub-stances that can form loose combinations with a solute, and being fat-soluble they are able to move across the membrane with their solute attached, and then to give it up on the other side. An example is the trans-port of sodium out of excitable cells (§9.2).

3.5 The nucleus

The nucleus is separated from the cytoplasm by the *nuclear envelope*, a pair of membranes enclosing a space of about 14 nm and interrupted at intervals by circular pores some 40–70 nm in diameter. Estimates of the thickness of the two membranes vary from 4–9 nm each, the lower figure probably being more typical. The pores are large enough to permit the passage of quite big molecules.

The hereditary material is carried in the chromosomes, which consist of *nucleoprotein*, a combination of protein and the substance *deoxyribo-nucleic acid* (DNA). For a long time, it was uncertain whether the protein or the DNA, or both, was the carrier of the hereditary characters, but there is now no doubt that this is the role of DNA. The functions of the protein are obscure, but it cannot be entirely ruled out that it might play a very minor part in the carriage of hereditary information.

3.6 Replication of DNA

The DNA molecule consists of two cross-linked chains or strands wound round one another in helical form (Fig. 3.5a). Each of the two strands is made up of a large number of units known as *nucleotides* strung out in a longitudinal sequence like the beads on a necklace. There are something like 10 000 000 000 nucleotides in a single nucleus. A nucleo-tide is composed of a molecule of *phosphoric acid* attached to the pentose sugar *deoxyribose*, which in turn is joined to a nitrogenous base that is either a *purine* or a *pyrimidine* (Fig. 3.5b). A DNA nucleotide contains either one of the two purines *adenine* and *guanine* or one of the two pyrimidines *cytosine* and *thymine*. The phosphate sugar rings are arranged around the outside of each strand of the DNA double helix, and the bases are turned inwards to the axis of the molecule and form cross-linkages between the two strands (Fig. 3.5a).

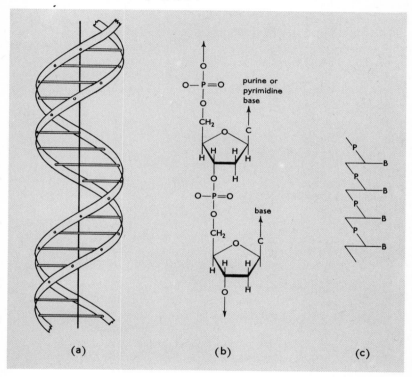

Fig. 3.5 (a) Double-stranded configuration of DNA. (After Watson and Crick 1953, *Nature*, **171**, 737); (b) the chemical structure of DNA; (c) symbolical representation of DNA.

If DNA is the carrier of the hereditary characters, it must be able to reproduce itself precisely and without aberration (except in the case of a mutation). The nature of the sugar and phosphate in the DNA molecule does not vary, and therefore the genetic information must reside in different combinations of the four bases. The cross-linkages between the bases are restricted, because the adenine (A) on one strand of the double helix can only link across with thymine (T) on the other strand. Similarly, cytosine (C) can only link with guanine (G). Thus, if the sequence of bases in a given section of a DNA strand is, say, CGAGGAGTT, the sequence in the corresponding section of the other strand can only be GCTCCT-CAA:

$$-C-G-A-G-G-A-G-T-T-$$
$$\quad | \quad | \quad | \quad | \quad | \quad | \quad | \quad | \quad |$$
$$-G-C-T-C-C-T-C-A-A-$$

This obligatory cross-linking ensures that the DNA can multiply itself and yet preserve its base sequence, a process known as *replication*. In replication the DNA molecule is split into its two strands, and the bases on each strand pick up their corresponding bases (as part of nucleotides) from the nuclear fluid. The obligatory cross-linking ensures that the bases picked up are the same as those lost through splitting:

nucleotides from the nuclear fluid

C—G—G—C—A
T—G
C
T—C
G—A—G—T—T—
C—T—C—A—A—
G—A
C
G
A—C
G—T
C—C
G

This diagram only summarizes what happens. The process, which is incredibly rapid, actually involves the breaking of one of the strands into short lengths, so that its replication is discontinuous, in contrast to that of the other strand, which is replicated in one continuous length. In both cases, the enzyme *DNA polymerase* joins the attached nucleotides together, but the short lengths of the broken strand are joined by the enzyme *DNA ligase*, which otherwise functions as a 'correcter' of faulty DNA. This method of replication is made necessary by the difference in chemical polarity between the two strands, for details of which a text book of biochemistry should be consulted.

3.7 Coding

The function of the hereditary substance is the direction and control of the formation and activity of the other components of the cell. Cellular processes are initiated and controlled by enzymes, which are proteins, and therefore the base sequence of DNA must somehow be linked with the structure and synthesis of proteins. Proteins are formed by the sequential joining of amino acids, and the problem therefore resolves itself into how the DNA base sequence *codes* for the amino acid sequence of the protein. Proteins are made from a maximum of about twenty amino acids, and if each of the four bases coded for one amino acid the system could handle only four amino acids. The code must involve a combination of bases for each amino acid, and the combination for a given acid must be different from that for any other acid. It is apparent that a code of two bases per acid would provide only $4^2 = 16$ different combinations, which is still too few. A code of three bases (*triplets*) would provide $4^3 = 64$ combinations, which is more than sufficient. There is now ample evidence that the triplet hypothesis is correct.

It is known that most amino acids are coded for by more than one triplet, a condition known as *degeneracy*, but this still leaves a number of 'spare' triplets. At least some of these act in punctuating the code. It is important that the triplets should be correctly read in a sequence of bases, especially since not all possible triplets code for an amino acid. For example, suppose that one sequence of bases was GACATTCGCATA. It might be correct to read the triplets from G, to give GAC ATT CGC ATA, but equally it might be correct to begin reading at A, to give (G) ACA TTC GCA (TA), the bases in brackets being linked to others on either side of the triplets distinguished. Clearly, it is essential to have some kind of *marker* that will indicate where the triplets for a given sequence begin and end. The spare triplets probably function as such markers. Thus, a gene appears to be represented by what is termed a *cistron*, a unit of DNA delineated by a marker at either end, which codes for a peptide chain or protein.

3.8 Messenger RNA

Most metabolic and synthetic processes of the cell take place in the cytoplasm. If the code for proteins resides in the nucleus, it must somehow be transferred to the cytoplasm for translation into action. The nuclear code is borne into the cytoplasm by a form of *ribonucleic acid* (RNA), which because of its function is known as *messenger RNA* (mRNA).

RNA is very similar in structure to DNA, but its sugar is ribose instead of deoxyribose; and instead of thymine, it contains as one of its bases the closely related *uracil*. It never appears to form more than a single strand, and its mode of production by DNA is far from clear. However, there is no doubt that mRNA possesses an equivalent base sequence to the section of DNA that produces it.

3.9 Cytoplasm

Cytoplasm is not a homogeneous substance. Under the light microscope a variety of granules and other bodies can be seen which, when viewed with the electron microscope often prove to be complex in structure. The study of the functions of cytoplasm requires the separation of these inclusions from one another, which has been achieved particularly by the process of *differential centrifugation*. The cells are broken up by a suitable technique such as homogenization, and this material is then placed in a centrifuge, which spins round at high speed. The particles in the material are thrown outwards by the centrifugal force of the machine to an extent which differs largely according to their size. By removing the sediment that forms at different speeds of rotation the material can be separated into fractions of particles, each with a definite size range. The particles in these fractions can be examined under the electron microscope and compared with those in the intact cell, and can also be subjected to biochemical investigation. The centrifuge can be used further to fragment the larger particles by subjecting them to higher speeds of rotation, and this is a useful way of extracting any enzyme systems they may contain.

3.10 The endoplasmic reticulum

In spite of its name, the endoplasmic reticulum is not confined to the endoplasm of most cells, but is distributed throughout the cytoplasm. In section, under the electron microscope, it looks like a number of disconnected tubules or canals (Fig. 3.2 and Plate 1). The membranes that bound these canals are double, and are continuous with the nuclear envelope. The system is usually divided into two parts, the *rough* endoplasmic reticulum (ergastoplasm) in which the membrane on the cytoplasmic side has granular particles about 15 nm in diameter attached to it; and the *smooth* endoplasmic reticulum, which lacks these granules. The granules are known as *ribosomes* and contain ribosomal RNA (rRNA). An especially concentrated and organized region of the smooth reticulum is distinguished as the *Golgi apparatus*.

One of the functions of the endoplasmic reticulum is the formation of secretory material, and it is particularly well developed in cells that specialize in secretion. For example, in the exocrine cells of the pancreas, which secrete the precursors of protein-digesting enzymes, secretory granules first appear in the canals or *cisternae* of the rough reticulum. They then disappear, to reappear in the vacuoles of the Golgi apparatus, from which they enter the cytoplasm as *zymogen granules*. The zymogen granules are presumably nipped off from the Golgi apparatus, since the secreted protein is enclosed by a membrane identical with that of the apparatus. The granules pass to the plasma membrane, from which they are discharged by fusion of the membrane of the granule with the plasma membrane. The process emphasizes the relationship between the two parts of the endoplasmic reticulum, even though many of its details are not understood. There is also no doubt that in certain circumstances rough reticulum can be transformed into smooth. The mode of discharge of the granules from the cell indicates the essentially similar nature of the reticular membranes and the plasma membrane.

The rough endoplasmic reticulum is concerned with the production of proteins to be secreted, and possibly also in the synthesis of proteins used within the cell.

3.11 Protein synthesis

Three kinds of RNA are needed for the synthesis of protein. The mRNA possesses a fairly large molecule, as would be expected from the fact that it codes for a peptide chain. Its molecular weight is 300 000 or more, and it contains upwards of 500 nucleotides. Ribosomal RNA is much larger, with a molecular weight between one and a half and two million, and is combined in the ribosome with protein to form a nucleoprotein. The third type of RNA is *soluble* or *transfer* RNA (sRNA or tRNA), and its molecule is much smaller than that of the other types of RNA, being about 25 000 to 30 000 and representing 75–80 nucleotides. As we have seen, the mRNA carries a code from the nucleus for a sequence of amino acids which, when joined together, form a peptide chain. It is the task of the tRNA to pick up these amino acids from the cytoplasm and attach (or transfer) them to the mRNA. Ribosomal RNA joins up the sequence of amino acids so produced to form a peptide chain. For this action to be accomplished there is a different molecule of tRNA for each amino acid that occurs in the cell, and this fits only into its specific place along the length of the mRNA molecule. A single amino acid may be coded for by more than one triplet, so there is more than one tRNA specific for each amino acid, i.e. there is one specific tRNA for each coding triplet or *codon*.   It may be that such complementary tRNA molecules differ in intracellular location.

The molecule of tRNA appears to exist in what is known as a 'clover-leaf' configuration (Fig. 3.6). In the straight 'arms' of the tRNA molecule at least some of the bases pair across as in the two strands of DNA, but at one point on one of the loops there are three unpaired bases and it is believed that these unpaired bases form the triplet that matches the corresponding codon of mRNA. One free end of the tRNA strand ends in the base sequence CCA, and the other in the single base guanine, and these do not pair across with one another.

In protein synthesis, the amino acids are first picked up by their specific tRNA molecules, the energy for this being derived from ATP with the release of pyrophosphate. The energy-rich amino acid becomes attached to the end of its tRNA molecule that bears the CCA sequence, AMP and the phosphorylating enzyme being liberated at the same time. The bases in tRNA include a number not represented in other nuclei acids, although

they are chemically related to them. These bases are normal when the tRNA is produced in the nucleus, but are changed subsequently by enzyme action, and the specificity of a tRNA molecule for its amino acid may well reside in a sequence of these 'strange' bases.

Each tRNA 'charged' with its amino acid must now be aligned on the mRNA strand, by virtue of its coding triplet (*anti-codon* or *nodoc*) 'recognizing' its corresponding codon on the mRNA. But first, the mRNA must

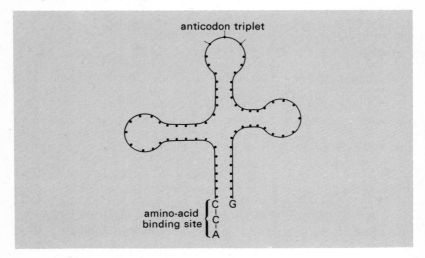

Fig. 3.6 Diagrammatic representation of the secondary structure of a tRNA molecule. The dots represent bases which are often linked across as in other nucleic acid molecules.

be bound to a ribosome. Ribosomes are composed of two subunits, and must be dissociated into these in order to pick up the mRNA. This is the task of the smaller of the two subunits, and once it has been accomplished the two subunits are re-united to form a ribosome/mRNA complex. The 'charged' tRNA molecules then attach to the MRNA in accordance with its coding sequence, as the mRNA strand is moved along the ribosome. This movement requires energy which is obtained by the breakdown of GTP (guanine triphosphate) to GDP, the terminal guanine of the tRNA having been phosphorylated earlier at the expense of ATP. As the mRNA strand is moved on, the now discharged tRNA molecule is released, its amino acid having been joined to those of earlier tRNA molecules (Fig. 3.7.).

Very careful separation of ribosomes from other cellular constituents leads to the recovery of what appear to be clumps of individual ribosomes.

Under the electron microscope it can be seen that the ribosomes in these clumps, which are sometimes termed polysomes, are joined by thin threads. It is believed that these threads represent a single strand of mRNA on which the ribosomes are aggregated, and for this and other reasons it is thought that more than one ribosome utilizes a strand of mRNA at the same time.

All three kinds of RNA are produced in the nucleus, mRNA and tRNA from the nuclear DNA and ribosomes from the nucleolus. It is therefore not surprising to find that protein synthesis already appears to be occurring as these nucleic acids are passed into the cytoplasm. This means that the endoplasmic reticulum is not essential for the synthesis of protein, and in

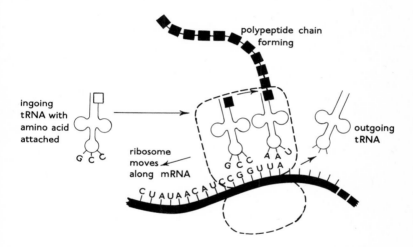

Fig. 3.7   Diagram to illustrate the formation of a peptide chain. The ribosome is indicated by a broken outline. (After Korner, 1966, *Discovery* (now incorporated with *Science Journal*), **27** (5), 30)

fact ribosomes are also found loose in the cytoplasm. It is believed that the reticulum is primarily a system for the production of protein for export, and that the cytoplasmic ribosomes are the main source of proteins for intracellular use. The suggestion that protein might be conveyed directly to the outside of the cell by means of a connection between the reticulum and the plasma membrane is not well founded. Secretory cells generally function like the pancreatic cells referred to earlier, in which zymogen granules pass through the cytoplasm to the plasma membrane. Nor are the cisternae of the reticulum normally continuous with the exterior.

3.12 Mitochondria

These are spherical, ovoid or rod-shaped bodies, variable in size but of the order of 3–6 μm in length and 0.5 μm in diameter. They move constantly about the living cell, and over a period of time they appear to change in shape and fragment, or even disappear. Their origin is unknown, but since they contain DNA it is conceivable that they are at least partly self-producing.

Mitochondria are bounded by two membranes, and these each consist of two electron-dense lines like the plasma membrane, but differ from one another in detailed composition and function. The membranes are separated by a gap of about 10 nm, and the inner membrane projects into the central space of the organelle to form partitions or *cristae* (Fig. 3.8). The cristae

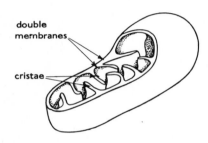

Fig. 3.8 A mitochondrion, partly cut away to show the internal projections or cristae. (From *Cell Biology*, 4th ed., W. B. Saunders, Philadelphia), De Robertis, *et al.*, 1965

do not form complete partitions, and therefore allow the granular substance within the mitochondrion to be continuous. The number and shape of the cristae varies between tissues and even within the same tissue during the life history of an animal. In general, the more active a cell the more numerous are the mitochondria and their cristae within it, which suggests that they have a metabolic role.

Mitochondria contain enzymes that break down fatty acids and participate in the breakdown of carbohydrates, including carbohydrate fractions derived from protein metabolism. Since the initial stages of such metabolism occur in the cytoplasm outside the mitochondrion, the latter must be regarded as part of a wider cytoplasmic system. The function of this system is to release the energy stored in metabolites derived from the animal's food, and to change it into a form in which it may be readily utilized by cells.

3.13 Enzymes

The word *enzyme* simply means 'in yeast', and reflects the fact that our early knowledge of enzymes came from the study of the fermentation of yeast.

It was pointed out in §2.6 that biological reactions invariably require a source of energy, even those that are thermodynamically spontaneous or exergonic. In the laboratory the energy required is most often supplied in the form of heat, and sometimes pressure. Living cells cannot tolerate the level of heat necessary, and there is little pressure inside them, and so some other kind of activating system is necessary. This takes the form of an organic catalyst, the *enzyme*. All known enzymes are proteins, and are therefore affected by such external factors as pH, temperature, and the ions of certain heavy metals.

Enzymes seem to possess what are known as *active sites* on their surfaces, which are endowed with a configuration that is specific to the *substrate*, the substance to be changed. In some way that is not understood the substrate becomes adsorbed on to the enzyme at the active site. The stress of adsorption, which appears to be reflected in changes observed in the configuration of the active site, and may be equivalent to considerable applied pressures in the laboratory, somehow increases the energy of the adsorbed substrate, which is said to have been *activated*. Because the configuration of the active site is so specific for the substrate, or for a particular type of substrate, the products of the reaction are not attracted to it, and they rapidly separate from the enzyme and free it for further action.

Because the enzyme enters into the reaction itself, by combining temporarily with the substrate, the relative concentrations of enzyme and substrate affect the velocity of the reaction. If the enzyme is in excess at the start of the reaction, and the substrate concentration is varied, the initial reaction velocity (v) increases with increase in substrate concentration until a limiting value (V_{max}) is reached, which is dependent on the enzyme concentration (Fig. 3.9a). If there is an excess of substrate, and the enzyme concentration is varied, the amount transformed is less the lower the enzyme concentration (Fig. 3.9b). These facts help to emphasize that enzyme-catalysed reactions are no simple chemical conversions, and they are in theory reversible. This means that the rate at which they proceed depends on the ratio between the concentration of the substrate and that of the products derived from it. If the products are removed, the reaction will proceed in the direction of the formation of further reaction products. If the products are allowed to build up in concentration, there will be a tendency for the reaction to be pushed in the opposite direction, and it will slow down, stop, or even reverse. These general principles require qualification, however, because they are subject to the energetics of the system. If the thermodynamics of the reaction are such that it tends to go in a particular direction, the energy required may be too great for any significant reversal of direction to take place.

The process of enzymic conversion can be visualized as one in which enzyme E and substrate S form a complex ES, which then breaks down to release the products P of the reaction, and free the enzyme for further activity:

$$\text{(i)} \quad [E] + [S] \underset{k_2}{\overset{k_1}{\rightleftharpoons}} [ES]$$

$$\text{(ii)} \quad [ES] \underset{k_4}{\overset{k_3}{\rightleftharpoons}} [E] + [P]$$

k_1, k_2, k_3 and k_4 are velocity constants for the reactions as shown. The

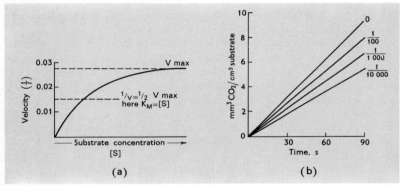

Fig. 3.9 Graphs to show the effect on enzyme activity of (a) an excess of enzyme and (b) an excess of substrate (fractions are enzyme dilutions). The enzyme is carbonic anhydrase, which causes CO_2 to be liberated from bicarbonate

brackets represent molar concentrations, not percentage concentrations. Enzymes are proteins, and therefore have large molecular weights; so 1% enzyme concentration would be much smaller in terms of active concentration than 1% substrate.

If we assume the reaction system to be in a steady state in which the rate of formation of ES is equal to its breakdown, i.e. the concentration of [ES] remains constant, it can be shown that

$$\frac{[S]\,([E]-[ES])}{[ES]} = \frac{k_2 + k_3}{k_1} = K_m \qquad (3.1)$$

The 'combined constant' K_m is known as the *Michaelis* or *Michaelis–Menten* constant. The reaction involving k_4 is normally so slight that it is ignored. By a suitable rearrangement and substitution, equation *3.1* can be expressed as

$$v = \frac{V_{max}\,[S]}{K_m + [S]} \qquad (3.2)$$

and this is known as the *Michaelis–Menten* equation. If we take the special case when $v = \frac{1}{2}V_{max}$ (see Fig. 3.9a), equation 3.2 becomes

$$\frac{V_{max}}{2} = \frac{V_{max}\,[S]}{K_m + [S]}$$

and dividing throughout by V_{max},

$$\frac{1}{2} = \frac{[S]}{K_m + [S]} \quad \text{and therefore,} \quad K_m + [S] = 2[S]$$

$$\text{and } K_m = [S]$$

In other words, the substrate concentration at which the velocity is half-maximal is equal in value to K_m. This means that if substrate concentration is plotted against reaction velocity, K_m can simply be read off the graph.

It will be seen that the Michaelis constant K_m is independent of enzyme concentration, although it may vary with pH and temperature. With these qualifications, its value is characteristic for a given enzyme and substrate. In cases in which an enzyme can act on more than one substrate, the enzyme has a different K_m value for each substrate.

Most enzyme systems are more complex in practice than the above treatment suggests, and involve two or three enzyme-substrate complexes, e.g.

$$E + S \rightleftharpoons ES \rightleftharpoons E_2 \rightleftharpoons EP \rightleftharpoons E + P$$

in which E_2 represents a transition state complex and EP an enzyme-product complex. Furthermore, there is often more than one substrate molecule and more than one reaction product. In all these complex systems, analysis on the above lines becomes much more complicated, but the starting-point for such analysis remains the Michaelis–Menten equation. In experimental work, the equation is often transformed in ways that allow the experimental results to be plotted in a straight line, and the interested reader should look up the derivation and use of the *Lineweaver–Burk* and *Eadie-Hofstee* plots in a text book of biochemistry.

The three-dimensional configuration of the active site limits the kind of substrate molecule with which the enzyme will react. For this reason, enzymes are said to be *specific* for a particular substrate. For example, the enzyme that splits ammonia from (deaminates) the amino acid D-alanine has no effect on its isomer L-alanine, although the only difference between these two chemically identical substances is their configuration. In practice, the extent of enzyme specificity may be variable, and may relate to a group of similar substances rather than a single substance or an isomer. The hydrolytic digestive enzymes of the vertebrate gut provide examples of both types. The enzyme *maltase* will attack only maltose and no other disaccharide. On the other hand, proteases and lipases are relatively non-specific and will attack a wide variety of proteins and fats respectively.

Enzymes concerned in cellular metabolism are usually highly specific for a given substrate, as would be expected in a system that is both highly complex and well organized. These enzymes always consist of a *protein carrier*, and a *prosthetic group* that is an organic radical. If the two components are separated, neither can act as an enzyme. Most enzymes probably do not dissociate into these two components in the cell, but the enzymes known as *dehydrogenases* are an exception. Their prosthetic group is normally dissociated from the protein carrier, and combines with it only when acting on a substrate. Such independent prosthetic groups, which are known as *co-enzymes*, can remove hydrogen from a substrate in conjunction with their protein carrier, and then dissociate from the protein, still carrying the hydrogen. The advantage of this system is that a single co-enzyme can act as the prosthetic group for a number of different protein carriers, each specific for a given substrate, enabling a variety of metabolic pathways to end at the same point. There are two co-enzymes in particular that function in this way in the dehydrogenation of a wide range of substrates. These are *co-enzyme I* or nicotineamide-adenine dinucleotide, NAD; and *co-enzyme II* or nicotineamide-adenine dinucleotide phosphate, NADP.* For the sake of brevity, we usually speak of the hydrogen being removed by the co-enzyme, but it should be understood that this is always done in conjunction with a specific protein carrier.

As their full name indicates, NAD and NADP are nucleotides. Another very important co-enzyme which, like NAD and NADP will be encountered in connection with the cellular oxidation of carbohydrates, is co-enzyme A, which has already been referred to in relation to high-energy compounds. Although it contains a pentose sugar and a phosphate group, co-enzyme A is not a nucleotide but contains a non-cyclic molecule derived from pantothenic acid (vitamin B complex), which is united to the sugar by two further phosphate groups. The pantothenic acid derivative bears a terminal sulphydryl —SH group, and it is at the sulphur atom that the biochemically important reactions take place. For this reason the co-enzyme is often written CoA—SH for convenience.

3.14 Temperature and pH sensitivity of enzymes

Like any other reaction, those due to enzymes are sensitive to changes in temperature. In addition, the fact that they are proteins means that they become increasingly denatured as the temperature rises, and this

---

* These are the internationally agreed names and abbreviations. In older literature, NAD may be referred to as diphospho-pyridine nucleotide (DPN), and NADP as triphosphopyridine nucleotide (TPN).

has certain consequences. Initially, an increase in temperature results in a corresponding increase in the reaction rate, as in other chemical reactions. Above a certain temperature, the rate of reaction begins first to flatten out and then to fall (Fig. 3.10a), because the enzyme becomes increasingly denatured and hence loses its activity. At first, the loss of enzyme is more than balanced by the increase in reaction rate, but as the concentration of the enzyme decreases this balance alters. The optimum reaction rate occurs at the temperature at which the maximum kinetic rate is just exceeding the effect of denaturation, i.e. at the top of the curve in Fig. 3.10. Most enzymes are denatured irreversibly between 50°C and 80°C, but some are known which undergo reversible denaturation until temperatures in excess of 100°C are reached.

It is believed that the active sites of enzymes must have a specific electric charge distribution if they are to function properly. Since enzymes

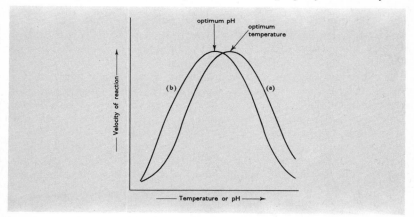

Fig. 3.10 The effect of (a) temperature and (b) pH on enzyme activity.

are proteins, the acidity or alkalinity of the medium will affect the ionic character of their amino and carboxylic groups and of any charged side groups attached to them, and thereby also alter the catalytic activity of the active sites. If the substrate is electrically neutral, the activity of an enzyme may be relatively unaffected by pH, e.g. *invertase*, which breaks down the neutral molecule sucrose to glucose, exhibits a constant level of activity over the range pH 3.0–7.5.

However, more commonly, enzymes are affected in such a way that there is an *optimum pH* for each enzyme, and the commonest kind of curve with such an optimum is shown in Fig. 3.10b. The effect of the optimum pH is increased by the fact that as the pH of the medium departs from the optimum the enzyme becomes increasingly denatured and hence falls off in activity. It should be noted that many enzymes do not function *in vivo* at the optimum pH determined *in vitro* and it may be that cell pH plays a part in limiting the rates of action of some enzymes.

3.15 Multi-enzyme systems

Intracellular enzymes usually work together in sequences in which the product of one enzyme becomes the substrate of the next enzyme. Such chains of enzymes may function in one of three different ways. The individual enzymes may be in solution as separate entities; they may be physically associated into enzyme complexes; or they may be associated with some other large supramolecular structure such as a membrane or a ribosome. The advantage of the two latter types of system is that the substrate sequences can be precisely organized. In the case of enzyme complexes, the substrate molecules are passed round within the complex, so that the distance over which they must be transferred is kept very small. When enzymes are bound to a membrane, as for example the enzymes of the electron transport chain (§3.19), which are bound to the inner membrane of the mitochondrion, this must facilitate the precision and ease of transporting the substrates through the chain.

Each enzyme member of a multi-enzyme system must have its own characteristic K_m for its substrate and co-factor. Usually, one member of the sequence limits the rate at which the system can function as a whole, either because its enzyme concentration or its substrate concentration is rate-limiting. In many cases, the end-product of a multi-enzyme system can inhibit the first enzyme in the chain, so that the rate of the entire sequence then depends on the steady-state concentration of the end-product. The first enzyme is then known as a *regulatory enzyme* and the inhibitory end-product as the *effector* or *modulator*. The system is an economical one, since once the first reaction in the sequence has started, the others must follow; and its inhibition means that energy will not be wasted nor metabolites used unnecessarily in the ensuing steps. It is significant that regulatory enzymes appear to be large molecules consisting of more than one polypeptide chain; and that they show an atypical relationship between reaction velocity and substrate concentration, so that they do not exhibit a simple Michaelis–Menten relationship, being generally much more sensitive to small changes in substrate concentration than would be expected on this basis.

Sometimes, enzymes appear to exist as multiple molecular forms within an animal, and even within the same cell. These are known as *isoenzymes* or *isozymes*. For example, *lactic dehydrogenase* in the rat is known to exist in at least five different major forms. Each form catalyses the same reaction, but has a different K_m value for the substrate. The different forms exist in different proportions in different types of tissue, and the proportions are under genetic control. Isozymes thus represent yet another kind of cellular control system.

3.16 Oxidation of carbohydrates

Carbohydrate is stored by animals in the form of glycogen (§2.16), in the liver of vertebrates and in the muscles of animals generally. It is transported around the body in the form of the hexose sugar glucose. In the utilization of both glycogen and glucose, the initial step is a phosphorylation by an enzyme. Under the influence of *phosphorylase*, the glycogen molecule is split into its constituent monosaccharide units in the form of *glucose-1-phosphate*, an energy-rich substance whose formation involves the utilization of inorganic phosphate. The product is then converted by an enzyme termed an *isomerase* to *glucose 6-phosphate*. Glucose is also phosphorylated, using ATP, this time by the enzyme *hexokinase* (and in certain circumstances *glukokinase*) and also becomes glucose-6-phosphate.

The glucose 6-phosphate from both sources, after conversion by another isomerase to *fructose 6-phosphate*, undergoes a second phosphorylation, involving ATP. The enzyme responsible, *phosphofructokinase*, is a regulatory enzyme for this series of reactions. At this point, the molecule, which contains six carbon atoms and two high-energy phosphate groups, splits into two parts, each part containing three carbon atoms and one phosphate, and hence loosely known as *triose phosphate*. The two triose phosphates differ in configuration (Fig. 3.11) and only one, *glyceraldehyde 3-phosphate*, can proceed to the next step, but the other is converted into it by yet another isomerase. Each triose phosphate molecule now gains a second phosphate group, derived from inorganic phosphate. A dehydrogenase next removes two hydrogen atoms from the molecule, which then undergoes a succession of molecular rearrangements. In the course of these a molecule of water is lost, and the phosphate groups are used to phosphorylate ADP to ATP, the eventual product being *ketopyruvic acid*. This completes the first phase of carbohydrate oxidation, which is known as *glycolysis*, and is summarized in Fig. 3.11. It will be observed that no oxygen has been needed for this process, oxidation having been achieved by the removal of hydrogen. Glycolysis is therefore said to be *anaerobic*.

As a result of glycolysis there is a net gain of two molecules of ATP for each molecule of glucose, since two molecules of ATP were required for the phosphorylation of the glucose, and two molecules of ATP were produced by each molecule of triose phosphate. The other products of glycolysis, two pairs of hydrogen atoms and two molecules of ketopyruvic acid, now follow separate paths in which each is oxidized aerobically to produce more ATP. Ketopyruvic acid undergoes *oxidative decarboxylation*, in which oxygen derived from water is used to break down this substance to carbon dioxide. In the course of this decarboxylation further

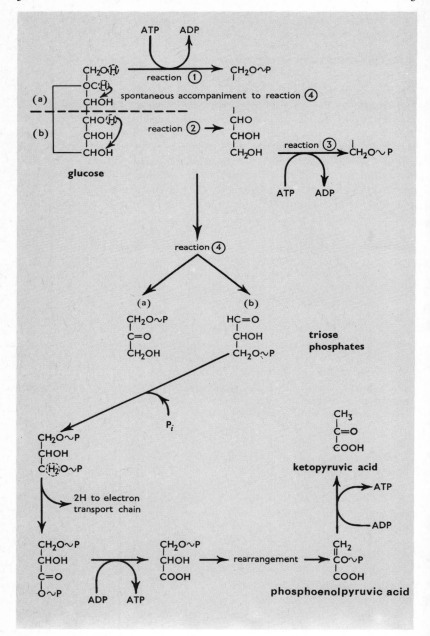

Fig. 3.11 Summary of glycolysis (anaerobic splitting of glucose).

hydrogen atoms are produced, which join those derived from glycolysis in undergoing oxidation by molecular oxygen derived from respiration, a process known as *oxidative phosphorylation*. Oxidative decarboxylation and oxidative phosphorylation both occur within the mitochondrion.

3.17 Oxidative decarboxylation

Ketopyruvic acid $CH_3CO \cdot COOH$ may be regarded as a combination of *acetyl* radical $CH_3 \cdot C{=}O$, carbon dioxide and a hydrogen atom.

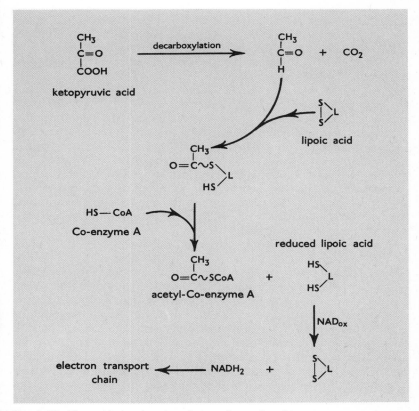

Fig. 3.12 The oxidative decarboxylation of pyruvic acid.

Oxidative decarboxylation involves not only the removal of the carbon dioxide and the hydrogen atom, but also the joining up of the acetyl radical to co-enzyme A. The carbon dioxide is first removed by a decarboxylating enzyme, and the acetyl radical with the hydrogen atom attached is taken up by a carrier substance, *lipoic acid*. The terminal sulphur group of lipoic acid can be reduced by hydrogen, with the formation of an energy-rich compound (§2.11). The energy-rich acetyl-lipoic acid-H complex now reacts with co-enzyme A to form acetyl-CoA, the lipoic acid being split off with its hydrogen atom still attached. Finally, the reduced lipoic acid yields up its hydrogen atoms to the co-enzyme NAD and the lipoic acid is restored to its original state (Fig. 3.12).

3.18 Tricarboxylic acid (Krebs) cycle

The acetyl-CoA resulting from oxidative decarboxylation now enters into a cyclic series of reactions, and because some of the key substances in the series are tricarboxylic acids, the cycle is known as the *tricarboxylic acid* (TCA) *cycle*, or sometimes the *Krebs cycle* in honour of the biochemist who, more than any other, elucidated it. It is convenient to begin consideration of the cycle with *oxaloacetic acid*, which contains 4 carbon atoms (the entire cycle is given in Fig. 3.13). The acetyl-CoA gives up its acetyl radical, a 2-carbon residue, to the oxaloacetic acid, with the result that a 6-carbon acid is formed. This passes through several internal rearrangements and is then dehydrogenated and decarboxylated, carbon dioxide being evolved, and the 5-carbon α-ketoglutaric acid results. The latter is once more decarboxylated and dehydrogenated, so that a four-carbon acid, *succinic acid*, is once again produced. Finally, the succinic acid undergoes 2 further dehydrogenations but no more decarboxylations, with the result that it retains its 4 carbon atoms but is converted once more into oxaloacetic acid.

It will be seen that in one complete turn of the cycle there are 4 dehydrogenations. Each of these results in the liberation of a pair of hydrogen atoms, and if we add the pair of hydrogen atoms removed from ketopyruvic acid in the formation of acetyl-CoA, the net yield from oxidative decarboxylation is 5 pairs of hydrogen atoms. A further 2 pairs of hydrogen atoms having been liberated during glycolysis, the products of the entire process up to this stage, if we neglect the carbon dioxide, are 12 pairs of hydrogen atoms and two molecules of ATP. Since the latter represents a useful energy store of only 61.2 kJ (14 600 cal), whereas the oxidation of glucose should theoretically yield about 2872 kJ mol^{-1} (685 000 cal), it is evident that little of the energy potential of the carbohydrate has so far been realized. But by setting aside 12 pairs of hydrogen atoms, the necessary steps have been taken for the liberation of a great deal more usable energy in the final phase of oxidation. The entire process up to this point has been anaerobic, oxidation having proceeded by the removal of hydrogen atoms. Now, the hydrogen atoms undergo the final aerobic oxidative phase.

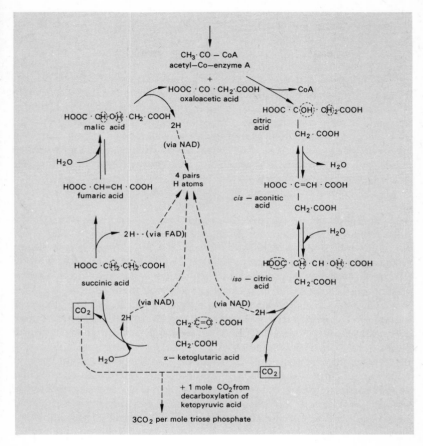

Fig. 3.13   The tricarboxylic acid (TCA) cycle.

3.19 Oxidative phosphorylation

Still within the mitochondrion, the 12 pairs of hydrogen atoms are united with 12 atoms of molecular oxygen, derived from the respiratory processes of the animal, with the formation of water. The oxidation of the hydrogen atoms involves a series of oxidation-reduction components which are often referred to collectively as the *electron-transport chain*. The reason is that the major role in the system is taken by substances called *cytochromes*. Cytochromes have a molecular structure similar to that of chlorophyll and haemoglobin. Like the latter they contain iron, and in the case of cytochromes this can exist either as Fe^{2+} (reduced) or Fe^{3+} (oxidized):

$$Fe^{2+} \rightleftharpoons Fe^{3+} + e \text{ (an electron)}$$

By the successive reduction and oxidation of a cytochrome through its ability first to take up and then to lose an electron, the substance can act as a carrier of electrons. It should be noted that cytochromes carry electrons and not hydrogen atoms. They therefore differ from NAD which carries electrons in conjunction with H atoms, in the form $NADH + H^+$. The electrons taken up by cytochromes are derived from the hydrogen atoms liberated in the anaerobic phase of oxidative metabolism described above, and the separation of an electron from a hydrogen atom involves the formation of a hydrogen ion:

$$H \rightleftharpoons H^+ + e$$

The hydrogen ions go into solution but, as we shall see, they are required again at the end of the chain.

The hydrogen atoms from the TCA cycle enter the electron transport chain attached to NAD, except for one pair. The exception is the pair derived from the oxidation by dehydrogenation of succinic acid to fumaric acid, which are handed on to a *flavoprotein*. Flavoproteins are enzymes that, like NAD and NADP, can take up hydrogen atoms. All the hydrogen atoms derived from carbohydrate dehydrogenation reach the same point eventually, since those attached to NAD also hand on their hydrogen atoms to flavoproteins (denoted by FP in Fig. 3.14). This step is necessary because the flavoproteins are enzymes that split the hydrogen atoms into hydrogen ions and electrons. They pass on the electrons to the cytochromes of which at least five are known—b, c_1, c_2, a, a_3. Some of these cytochromes require the presence of other elements or substances for their proper action. Since the details of the electron transport chain are by no means all settled, it is quite possible that other components and factors are involved. It should be noted that the pair of H atoms derived from glycolysis do not enter the mitochondrion attached to extramitochondrial NAD, because the latter, in common with a number of other substances, including ADP and ATP, is unable to cross the inner mitochondrial membrane. This membrane contains a number of carrier systems that facilitate the passage of such substances. In the case of the H atoms from glycolysis, there is a 'shuttle system' that removes them (or, strictly, their excited electrons) from their NAD and hands them over to a flavoprotein.

The final cytochrome in the chain is acted upon by the enzyme *cytochrome oxidase* in such a way that the electron carried is used to excite an atom of oxygen, causing it to pick up hydrogen ions from the medium to form water. In a sense, therefore, the hydrogen ions and electrons that were separated from one another before entering the chain of cytochromes are re-united in the formation of water at the end of the chain.

It can be calculated that the change in free energy for the oxidation of, for example, reduced NAD by molecular oxygen is approximately $-52\,000$ kcal. This energy is represented by electrons in the hydrogen atoms that have been raised to a higher energy level than the ground state (§2.8) when the latter were separated off in the earlier dehydrogenations. Much of the energy in the original glycogen or glucose has thus been used to

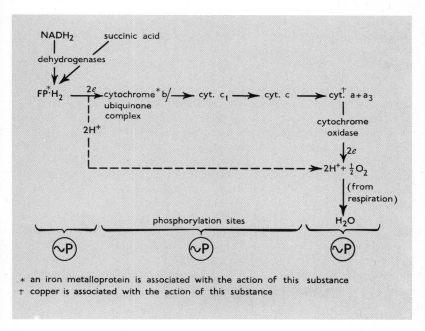

* an iron metalloprotein is associated with the action of this substance
† copper is associated with the action of this substance

Fig. 3.14 Oxidative phosphorylation. (Based on Griffiths, 1965, in Campbell and Greville (eds.). *Essays in Biochemistry*, Vol. 1, Academic Press, New York and London)

excite electrons, which give up their energy during their passage through the electron transport chain for the production of ATP, and finish up at the end of the chain at the ground energy level. There are in theory five places in the chain at which the phosphorylation of ADP to ATP could occur. The value of ΔF° is greater at three of these than at the remaining two, and it is probable that the maximum normal output of the electron transport chain is three molecules of ATP for each pair of hydrogen atoms transported, but there is at least one case known in which four molecules are produced and others may yet be discovered. Some of the H atoms, including those from glycolysis and from the succinic dehydrogenase of the TCA cycle, pass into the electron transport chain after the first phosphorylation site (Fig. 3.14) and in these cases only two molecules of ATP are generated for each pair of hydrogen atoms.

The way in which phosphorylation takes place is only incompletely understood. It is clear that a number of steps are involved, and that at some stage a *coupling factor*, a highly active protein specific for each site of phosphorylation in the system, is necessary. The substrate of reduced cytochrome is coupled enzymatically to its factor by a high-energy bond whose energy is derived from the excited electron, and this complex then picks up inorganic phosphate from the medium to form factor$\sim$P, with oxidation of the substrate. The factor$\sim$P then hands on its high-energy phosphate group to ADP, to form ATP, with reconstitution of the factor.

We can now see how much useful energy is obtained from the oxidation of a glucose molecule. In glycolysis, two molecules of ATP were generated, and two pairs of hydrogen atoms were removed that eventually yielded four molecules of ATP. During oxidative decarboxylation, two pairs of H atoms were passed to NAD, and these produced six molecules of ATP. Three of the pairs of H atoms from the TCA cycle yield 3ATP each, and one pair yields 2ATP (i.e. a total of 22 ATP for a glucose molecule). To these must be added the 2 ATP generated during the conversion of α-ketoglutaric acid to succinic acid in the TCA cycle. This gives us a total of 36 moles of ATP for each mole of glucose. Since each mole of ATP represents an energy store of 30.56 kJ (7.3 kcal), the minimum yield for the whole process will be 36×30.56 kJ = 1100 kJ (262.8 kcal). The theoretical free energy change for the oxidation of a mole of glucose is about 2872 kJ, so the conversion efficiency of the cell is about 38%. Under certain circumstances, the efficiency may be higher.

Oxidative phosphorylation requires an adequate supply of oxygen. Under anaerobic conditions the process cannot take place, and the pyruvic acid produced in glycolysis is converted into lactic acid without further release of energy. Lactic acid retains as many hydrogen atoms per carbon atom as glucose, so its carbon atoms are still at the same oxidation level. The ultimate products of oxidative metabolism are CO_2 and water, much simpler compounds. So oxidative metabolism is much more efficient than anaerobic metabolism, and is the preferred mechanism of the great majority of living organisms.

The liberation of energy from carbohydrate may by now appear to the reader to be unnecessarily slow and complex. In fact, such a system has a number of advantages. It results in the production of a large number of relatively small units of stored energy in the form of ATP, which in turn permits a more controlled use of stored energy with less tendency for it to be wasted. It is a physico-chemical fact that spontaneous reactions in the normal sense of the adjective result in the production of considerable heat energy, which is lost. By spreading the overall reaction over a large number of steps, so that only a small change in free energy is permitted at any stage, the degradation of the available energy into heat is lessened, and this accounts for the quite striking efficiency achieved. Furthermore, a process that has many steps of varied type is more susceptible to fine control. Such control may be exercised through variation in enzyme activity, and possibly facilitated by the fact that enzyme activity is somehow bound up with the organization of the mitochondrial membrane. Finally, the diversity of the system enables other metabolic substances to be incorporated into it at appropriate points, as we shall presently see.

Oxidation via the TCA cycle is the path followed by most of the glucose metabolized, but some follows another route, the *pentose phosphate* pathway. The purpose of this pathway is to produce pentose sugars, and especially D-ribose for the synthesis of nucleotides; and to cause an increase in cytoplasmic reduced NADP in certain cells in which this substance is required for synthetic processes. Thus, in cells of the liver, adrenal cortex and fat depots, as much as 30% of the glucose may be metabolized through this pathway during active synthesis, whereas in other cells the percentage following this route is so low as to be negligible. The details may be found in a textbook of biochemistry.

3.20 Metabolism of fats

Higher animals and plants can store large amounts of neutral fat as a fuel reserve. Neutral fat has a high calorific value—37.7 kJ/g (9 kcal/g) against 16.7 kJ/g (4 kcal/g) for glycogen or starch. Fats are stored largely as triglycerides, so must be also mobilized as glycerol and fatty acids. Glycerol is very similar in structure to triose phosphate without its phosphate group (see Fig. 3.11 and §2.15) and since the cell is able to phosphorylate glycerol, the resulting triose phosphate then follows the normal path for carbohydrate metabolism.

Although other pathways exist, the great majority of fatty acids undergo β-*oxidation* in which the long chain of the acid is split at every second (β) carbon atom. The terminal carboxyl group of the acid is first linked with co-enzyme A, for which energy is supplied by the cleavage of ATP to AMP. At this stage the fatty-acid-CoA complex is carried across the mitochondrial membranes linked with the substance *carnitine*. It now undergoes dehydrogenation, the H atoms being passed on first to FAD and then to a second flavoprotein (see Fig. 3.14):

$$FAD + R \cdot CH_2 \cdot CH_2 \cdot CH_2 \cdot CO—CoA \rightleftharpoons$$
$$R \cdot CH_2 \cdot CH{=}CH \cdot CO—CoA + (FADH + H^+)$$

Water is added enzymatically to form a molecule that is then further dehydrogenated, the H atoms this time being removed by NAD which now enters the first stage of the electron transport chain:

$$R \cdot CH_2 \cdot CH = CH \cdot CO - CoA + H_2O \longrightarrow R \cdot CH_2CH \cdot OH \cdot CH_2 \cdot CO -$$
$$CoA \cdot NAD + R \cdot CH_2 \cdot CH \cdot OH \cdot CH_2 \cdot CO - CoA \rightleftharpoons$$
$$R \cdot CH_2 \cdot C = O \cdot CH_2 \cdot CO - CoA + (NADH + H^+)$$

Under the influence of *thiolase*, the two terminal carbon atoms are detached from the rest of the chain as acetyl-Co-enzyme A, $CH_3 \cdot CO - CoA$, the hydrogen atom required to form the methyl group CH_3 being derived from water; and at the same time the new terminal carbon of the shortened chain of the acid is joined to more co-enzyme A and the whole process is repeated until the entire chain has been broken up and oxidized. The acetyl-Co-enzyme A enters the TCA cycle (Fig. 3.15).

3.21 Metabolism of proteins

In the metabolism of proteins, the constituent amino acids first lose their nitrogenous group by deamination. The products of deamination are ammonia and a carbohydrate fraction. The fate of the ammonia is described in Chapter 5. The carbohydrate fraction is a keto acid in the case of ornithine, histidine, glutamic acid and several others, which produce α-ketoglutaric acid on deamination; and aspartic acid forms oxaloacetic acid. These products can enter directly into the TCA cycle at the appropriate point. Other amino acids generally emerge from deamination as ketopyruvic acid, which is then subjected to oxidative decarboxylation. Glycine, which has the smallest molecule of any amino acid, is transformed into an acetyl residue, and this joins with CoA to form acetyl-CoA. In these latter cases deamination takes place by dehydrogenation (§5.1), and the hydrogen atoms liberated are handed on to NAD for oxidative phosphorylation.

Since the metabolism of fats and proteins is always linked into the pathway for carbohydrate metabolism (the relationships are summarized in Fig. 3.15) it might be thought that simple reversibility of the reactions would provide a basis for the interconversion of substances within the cell, and also for their synthesis from simple components. Essentially, this is true; but in some cases, as in the synthesis of fatty acids by the cell, the reverse process is much more complex because it requires a supply of energy to drive it, with the result that the picture is complicated by a number of intermediate steps. The details of these reactions are beyond the scope of this book, and a textbook of biochemistry should be consulted for them.

3.22 Lysosomes

Lysosomes are small spherical organelles, about 0.25–0.8 μm in diameter, and bounded by a double membrane similar to those of other organelles. They contain acid hydrolytic enzymes analogous to the digestive enzymes of the gut, although there is some uncertainty about their precise identity. They certainly include a number of proteases, and probably some lipases and amylases. Because of their appearance and contents, many workers

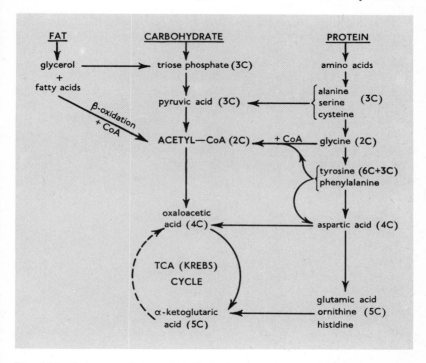

Fig. 3.15 Summary of the metabolic pathways and interrelationships of fats, carbohydrates and proteins.

believe that lysosomes are formed by the rough endoplasmic reticulum and Golgi apparatus in a similar manner to zymogen granules.

Ideas about the functions of lysosomes are still rather tentative. They undoubtedly discharge their contents into the vacuoles produced by the plasma membrane during the phagocytosis of foreign bodies or the ingestion of food particles, and they do so by fusion of their membranes with those of the vacuole. It is therefore presumed that they play a part in intracellular digestion. The spent vacuoles that result from this process are discharged from the cell if it borders a medium suitable for their being transported away. In addition, it is known that viable lysosomes can be discharged from cells, which suggests that they might also be responsible for at least some extra-cellular digestion.

Lysosomes undergo lysis in damaged or dying cells, and thereby aid in their digestion and removal, and they function in a similar manner in the removal of cells during the metamorphosis of anuran tadpoles and possibly other animals such as insects. Sometimes *autophagy* of cells is observed, in which an area of the cytoplasm is isolated by the formation of a limiting membrane within which various cytoplasmic inclusions can be seen in various stages of degradation. These isolated areas contain hydrolytic enzymes, and most workers think these are derived from lysosomes and even that the area is first formed by the fusion of lysosomal membranes. It is not certain whether there is a definite selection of material for autophagy, or whether it is a regularly occurring random process. Actions similar to lysis or autophagy might account for the removal of unwanted parts of a cell, as in embyrological differentiation, or of groups of cells as in uterine cycles or moulting. It is also possible that lysosomes are utilized by spermatozoa in their invasion of the ovum.

4

Respiration and Metabolism

Oxidative metabolism is the most efficient mechanism available to animals for the release of energy from the foodstuffs they ingest. All animals use this mechanism if the supply of oxygen is sufficient and, indeed, most of them are probably incapable of prolonged anaerobic respiration, the exceptions being certain gut parasites. An adequate supply of oxygen is therefore essential for the vast majority of animals.

The term *respiration* includes both oxidative metabolism within the cell, and the means by which oxygen reaches the cell. Since all oxidative metabolism involves the production of carbon dioxide, the transport and elimination of this gas is normally considered at the same time as the transport and utilization of oxygen, although it is really a form of excretion. Cellular respiratory processes have been considered in Chapter 3, and in this chapter we shall confine ourselves to the exchange and transport of the respiratory gases. We shall discuss at the same time some general aspects of the metabolism of the whole animal, since these are often measured in terms of the consumption of oxygen.

4.1 Availability of oxygen

The amount of oxygen available to an animal differs in different environmental media. At a total atmospheric pressure of 760 mmHg (1 atm), the fraction of the air's gases due to oxygen, known as its *partial pressure*, is 159.2 mmHg; or, in terms of volume, about 200 cm^3/l (approximately 21% of the total volume). The partial pressure of oxygen in water may be very little less than in air; but it is very sensitive to reduction by dissolved solutes and rise in temperature (see below). The oxygen content of natural waters varies, for the reasons given below, between 0.04 and 10 cm^3/l, or about 1% or less of the total volume. For these reasons the amount of water that must pass across the respiratory surface of an aquatic animal is often greater than the volume of air that must pass across the respiratory surface of a land animal, in order to extract the same amount of oxygen from the medium. In addition, water is a viscous medium compared with air, and therefore cannot be moved over the respiratory surface at the same rate as air without the expenditure of energy.

The solubility of oxygen in water is reduced by an increase in temperature, which is why we boil liquids to remove gases. At $37°C$ the quantity of oxygen dissolved in distilled water falls to about 5 cm³/l, or about 0.5% of the total volume; but since the rate of diffusion (see below) increases with a rise in temperature, the one virtually compensates for the other at biologically viable temperatures.

Much more important is the effect of solutes on the quantity of dissolved oxygen, for the presence of salts results in a decrease in the dissolved oxygen concentration. For example, the solutes in sea-water reduce its saturation level by oxygen by nearly one-third in comparison with distilled water. The effect is largely of environmental importance, since the salinity of body fluids is such that none of the common gases is reduced in solubility by more than 10% through it, and in the case of oxygen the figure is only about 2.5%

4.2 Respiratory surfaces and gaseous exchange

The exchange of oxygen between the environment and an animal always involves the diffusion of the gas through water, either the water of the environmental medium or the thin film that must cover the respiratory surface of all land animals. Whereas diffusion through a gas phase is a simple function of the molecular weight of the gas concerned, the diffusion of a gas through water must take into account the area over which diffusion takes place, and the distance over which the concentration gradient that is responsible for diffusion must operate. These factors are summed up in Fick's Law, which states that

$$\phi = D \cdot \frac{\mathrm{d}m}{\mathrm{d}x}$$

where ϕ is the rate of diffusion per unit area, and $\mathrm{d}m/\mathrm{d}x$ is a mathematical expression that relates the distance x over which diffusion must occur to the concentration gradient m over this distance. The *diffusion coefficient* D is a constant that varies according to the temperature and pressure. At 1 atm, D increases by about 3% for each rise of $1°C$.

Diffusion through liquids is a very slow process, perhaps exemplified by the fact that oxygen diffuses through air some three million times faster than it diffuses through water. The distance x is very small at the respiratory surfaces of animals, being represented by the thin water film and the thickness of the respiratory epithelium (the contents of which are aqueous), and therefore no problem is created at this level. It can be calculated that when $x = 10$ μm, the concentration of oxygen in the fluid exposed under the epithelium will reach 95% saturation in a second or less.

At the level of the whole organism, the operation of Fick's law is much more important. Using known values of oxygen need, it can be shown that for a spherical organism the diameter of the sphere must not exceed 1 mm if an adequate supply of oxygen is to reach its centre. These considerations apply to the resting state, and activity may reduce the figure by as much as one-half. Thus, unless some supplementary transporting system can be found, the acceptable limit of thickness of an organism is about 1 mm. In animals that seem to contradict this conclusion, such supplementary systems are in fact present. There is cytoplasmic streaming, aqueous channels through which water may pass or, as in sponges and sea anemones, a hollow body containing fluid that is either in continuous flow or is changed frequently. In large animals, circulatory systems are present that carry the blood into intimate contact with the metabolizing cells.

Small animals have a sufficiently large surface area in relation to their volume to be able to take in all the oxygen they need by diffusion across their body surface, provided that they are aquatic or live in moist conditions. In larger animals, the apparent respiratory area of the body surface becomes restricted to special regions, which are compensated for their reduced size by foldings or indentations of their surface. It must not be assumed that respiratory organs formed in this way are solely responsible for gaseous exchange. It has been calculated that cutaneous respiration in eels satisfies some 60% of their total oxygen need; and frogs use their lungs only when strenuous exercise demands it. The situation is rather different in truly terrestrial animals, in which it is impossible for the body surface to be kept moist, and it becomes virtually impermeable to oxygen, while the respiratory surface becomes sunken inside the body.

The supply of oxygen to the respiratory surface is often assisted in some way in the more complex animals. Among aquatic forms, such assistance is exemplified by the cilia of planarians, the branchial tufts of *Arenicola*, and the scaphognathite of decapod crustaceans. Land animals often possess pumping devices like the abdominal pumping rhythm of many insects, the air sacs of birds and the diaphragm of mammals.

4.3 Respiratory pigments

Circulatory systems aid gaseous exchange by removing the blood from the respiratory surface immediately it has been oxygenated. In addition, blood usually contains a *respiratory pigment* with a special affinity for oxygen, which increases the oxygen capacity of the blood many times.

In man, for example, the carrying capacity of the aqueous components of the blood is less than 0.3 vol per cent, whereas arterial blood actually contains some 19 vol per cent. The difference is due to the oxygen carried by the respiratory pigment. Without the activity of such pigments, the oxygen needs of many animals could not be satisfied.

Four respiratory pigments are found in animals: *haemoglobin, chlorocruorin, haemerythrin* and *haemocyanin*. Haemoglobin and chlorocruorin are rather similar in structure, since both are *haemochromogens*, with a molecular structure like that of chlorophyll and the cytochromes. Haemo-

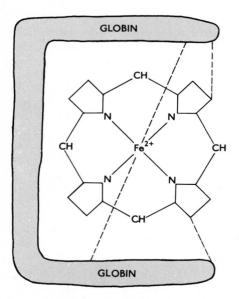

Fig. 4.1 Structure of haemoglobin. The porphyrin ring is linked to the globin through the ferrous iron, and probably also at the other points indicated by broken lines.

globin consists of a pigment *haem* combined with a protein *globin*. Haem is based on a structure known as a *porphyrin* ring (Fig. 4.1), in which the nitrogen atoms of the four *pyrrole* rings of which it is composed are joined to ferrous iron. The ferrous iron is linked in the haemoglobin molecule with the globin and the latter may also be linked to side groups on the individual pyrrole rings (Fig. 4.1). The presence of the globin molecule in a sense stabilizes this situation so that oxygen can combine loosely with the ferrous iron without actually oxidizing it. The combination with oxygen results in differences in the electron distribution between the iron and the nitrogen atoms. Thus, although it is possible to oxidize or reduce haemoglobin by the use of suitable chemicals, and in these cases the reaction is irreversible, oxidation and reduction are not involved when oxygen is taken up or given up. The iron remains in the ferrous state throughout.

Haemoglobin is capable of taking up one molecule of oxygen for each atom of iron in its molecule. The combined form, which is known as *oxyhaemoglobin*, is therefore often written as HbO_2. The oxygen is easily displaced from oxyhaemoglobin by other gases that do form stable chemical compounds with it. This is the reason why carbon monoxide in quantity is lethal to man; the gas combines chemically with the haemoglobin, and thereby prevents it from being available as an oxygen carrier.

Haemoglobin is by far the most widely distributed of all the respiratory pigments. It is found in all vertebrates apart from a handful of fishes; and in one hemichordate, *Discoglossus*; in many annelids, including *Nereis*, *Arenicola*, *Lumbricus*, *Tubifex* and *Hirudo*; in a few molluscs, such as *Planorbis* and *Limnaea*; in a few arthropods like *Daphnia*, and chironomid larvae; and in several sea cucumbers. Invertebrate haemoglobins have sometimes been separated from vertebrate ones under the separate name *erythrocruorin*, because the proteins of invertebrate haemoglobins are larger than those of vertebrates. However, vertebrate pigments are not at all homogeneous, and it seems better to unite all pigments with the same basic structure under the one term. In fact, the blood of many animals does not contain a single type of haemoglobin, but a mixture of two or three types. The differences between the types are due to their containing different globins, which confer on the molecules that contain them their distinctive properties, solubility, liability to denaturation and oxygen affinity among them.

Chlorocruorin is structurally very similar to haemoglobin, but there is a different side-chain on one of the pyrrole rings, and the pigment is green in colour and not red. It is a pigment that is very restricted in its distribution, being present only in some members of three families of polychaets, and especially in the Sabellidae and Serpulidae.

Despite their names, neither haemerythrin nor haemocyanin contain haem, and they both differ chemically from one another, as well as from haemoglobin. They do consist of a metal combined with a protein, the metal being iron in haemerythrin and copper in haemocyanin. The former pigment is brown in colour, the latter blue when oxygenated and colourless when evacuated of oxygen. Haemerythrin is found only in one polychaet worm, *Magelona*, in sipunculids, and in the brachiopod *Lingula*. Haemocyanin occurs in cephalopod molluscs and some gastropods, including *Helix*; in the decapod crustaceans; and in *Limulus* and one or two other arachnids.

4.4 Properties of respiratory pigments

Since haemoglobin is the vertebrate pigment and is more widespread than any other, it is not surprising that it is the one whose properties are best known, followed by haemocyanin. However, there is no reason to suppose that the other two pigments are fundamentally different in their properties.

When a solution of a respiratory pigment in the deoxygenated state is exposed to a gradually increasing oxygen tension, the gas is absorbed rapidly at first but after a time the rate of absorption becomes progressively slower, and approaches the level of 100% saturation asymptotically. This

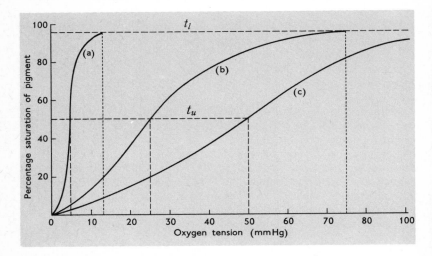

Fig. 4.2 Oxygen equilibrium curves of (a) the lug-worm *Arenicola*; (b) man; and (c) the pigeon. (Re-drawn from Prosser and Brown, 1961, *Comparative Animal Physiology*, 2nd ed., W. B. Saunders, Philadelphia and London)

relationship between oxygen tension and its absorption by the pigment produces a characteristically shaped graph known as the *oxygen dissociation* or *oxygen equilibrium* curve (Fig. 4.2). The latter term is to be preferred because the term dissociation refers only to the unloading of oxygen by the pigment, whereas the graph also describes how it takes up oxygen, or the process of *association*.

Because of the asymptotic nature of the top of the equilibrium curve, it is impractical to use the 100% saturation level as the maximum useful capacity of the pigment, and the *maximum loading tension* used is that observed at 95% saturation (t_l in Fig. 4.2). It is helpful to have a measure of the *unloading* capacity of the pigment as well, and this is taken as the point at which the quantities of oxygenated and deoxygenated pigment are equal, or 50% saturation (t_u in Fig. 4.2). The unloading tension provides a valuable reference figure to use in comparing the equilibrium curves of different pigments.

Three pigments with widely-differing equilibrium curves are shown in Fig. 4.2. They have been chosen to illustrate the extent to which the oxygen capacity of a blood pigment of the same basic type—in this case haemoglobin—may vary between species. The equilibrium curve of *Arenicola* haemoglobin rises rapidly to become 95% saturated at about 13 mmHg oxygen tension. The equilibrium curve for the pigeon represents the other extreme, since it never reaches 95% saturation at the highest level represented by the graph. The equilibrium curve for *Homo* falls between these two extremes. It should be noted that the limit of 100 mm Hg oxygen on the graph represents very nearly the maximum tension to which a respiratory pigment will be subjected in a terrestrial animal. The reason is that the placing of the respiratory organs inside the body creates a *dead space* through which the inspired air must pass before it reaches the actual respiratory surface. On expiration some of the expired gases, which are richer in carbon dioxide and depleted in oxygen content, remain in the dead space to mingle with the air taken in during the next inspiration. The result is that the oxygen tension in contact with the respiratory surface is about 100 mmHg at the most, in contrast to a pressure of 159 mmHg in the outside air. This is the situation in mammals such as man. The way in which the large air sacs of birds function is not properly understood, but their effect is also to lower the pO_2, although to a lesser extent than in mammals, so their curve is farther to the right, as shown in Fig. 4.2.

Comparison of the three curves given in Fig. 4.2 suggests that the shape of the equilibrium curve might have adaptive significance. At the t_u of pigeon haemoglobin the pigment will be unloading to a considerable extent, whereas the *Arenicola* haemoglobin is unloading hardly at all at t_u. Much of the oxygen contained in *Arenicola* haemoglobin is unloaded at saturation levels between 80% and 95%. The total amount which it will take up is small by comparison with the pigeon, but it is absorbed very rapidly. It is evident that a pigment like that of *Arenicola* is of value to an animal that lives in low oxygen tensions, whereas one like that of the pigeon is of value in high oxygen tensions.

The partial pressure of carbon dioxide present in the blood causes a shift in the oxygen equilibrium curve in the case of most haemoglobins and many varieties of haemocyanin. As the CO_2 tension is increased, the curve tends to shift from the left of the graph to the right (Fig. 4.3), a change known as the *Bohr effect*. In general, this would appear to be advantageous, since it means that more oxygen will be given up in tissues in which the carbon dioxide is rising due to high metabolic activity. However, in the case of a curve already well to the left, a shift to the right might be dangerously large in a pigment that must operate over a narrow range of oxygen tension, and it is significant that aquatic animals that live in stagnant water are less sensitive to CO_2 than those in moving water. In general, we should expect to find that the sensitivity of a pigment to CO_2 varied according to the initial shape of its equilibrium curve and the available oxygen in the environment.

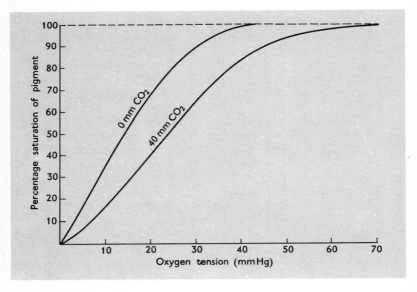

Fig. 4.3 Effect of carbon dioxide (in mmHg) on the oxygen equilibrium curve of haemocyanin in whole blood of the spider crab *Maia*.

We have already seen that more oxygen is present in air than in water, and is more readily available. For this reason there are differences in the shape of the equilibrium curve between aquatic and land animals. In general, the equilibrium curves of aquatic poikilothermic vertebrates are more to the left of the graph than those of terrestrial ones. There are also differences between animals in different aquatic habitats, which are exemplified in certain fishes found in rivers. The characteristic fish of the upper reaches of rivers is the trout, which lives in water that is fast-flowing, turbulent, and relatively shallow by comparison with lower reaches, all

factors which ensure a high oxygen content in the water. The characteristic fishes of the lowland reaches are the cyprinoid fishes belonging to
the carp family. The water in which they live is deeper and more sluggish
than that of the upper reaches, and hence is less well oxygenated. A simple
experiment demonstrates the adaptation of these fishes to their different
environments. If a trout is placed in a bottle of water from which further
oxygen is excluded it is soon obviously distressed, whereas a carp introduced at this stage shows no sign of distress. It is found that the equilibrium curve of trout haemoglobin is further to the right of the graph
than that of the carp, and there is also a difference in the sensitivity of the
two pigments to CO_2. An increase of 10 mmHg in the pressure of CO_2
to which the pigments are exposed results in a shift in the value of t_u of
the trout pigment that is three times greater than that of the carp
pigment.

There are a number of invertebrates which can live on river bottoms
covered by slowly decomposing mud, where the oxygen pressure must
be very low, such as chironomid larvae and tubificid worms. Other invertebrates live in stagnant ponds, like the snail *Planorbis*; or are subjected
to periods when the flushing of their burrows is no longer possible, like
Arenicola and *Nephtys*. In *Tubifex* and *Chironomus* larvae the value of
t_u is 0.6 mmHg at 17°C; in *Arenicola* and *Planorbis* about 2 mmHg.
These values may be compared with the t_u of human haemoglobin, which
is 27 mmHg at 37°C. The partial pressures of oxygen in the burrow of
Nephtys when the tide is out is about 7 mmHg, and that in the burrow
of *Arenicola* about twice this figure. Thus, all these forms possess a pigment (haemoglobin in the case of all the animals named) whose equilibrium curve is well to the left of the graph, and adapted to low oxygen
tensions. Carbon dioxide either has no effect on the oxygen capacity of
the pigment or only a very slight one. It has sometimes been suggested
that in animals like *Arenicola*, which is subjected alternately to high and
low oxygen tensions through the tidal rhythm, and *Planorbis*, which is
found in a range of freshwater habitats with a variety of oxygen concentrations, the haemoglobin might act as a kind of oxygen 'store' whose contents would be released in adverse conditions of low oxygen concentration.
However, it has been demonstrated that the oxygen contained in the
haemoglobin of *Arenicola* would be quite inadequate to sustain the animal
for more than a fraction of the time the tide is out, at the level of oxidative
metabolism maintained by the animal while the tide is in. It may be that
the pigment helps to extend the time over which the worm must adjust
itself to conditions of low oxygen supply, thereby ensuring a smoother
transition between the two states.

Respiratory pigments in tissues are often different in detailed molecular
structure and in molecular weight from the corresponding pigment in the

blood of the same animal, the mammalian *myoglobin* found in muscle being one example. Like, myoglobin, they usually have a higher affinity for oxygen than the blood pigment. Myoglobin is often thought of as an oxygen store, but it can be shown that this 'store' of oxygen would last for about one second of vigorous activity. It might possibly help an animal over the short time lag between the start of vigorous activity and the reflex opening of the capillary bed of the muscle, but it could do little more. Myoglobins also occur in a number of invertebrate tissues, and the suggestion has been made that here they facilitate the diffusion of oxygen through the tissue, but this has not yet been proved.

It is evident that the pigments of animals such as *Arenicola* and *Planorbis* can be non-functional, depending on the environmental conditions. Curiously, the pigments of some animals exhibit equilibrium curves which, when considered in relation to the oxygen tensions the animal is ever likely to encounter in the environment, show that they are always non-functional. Thus, the shed blood of *Limulus* is colourless, indicating that its haemocyanin contains no oxygen; and the amount of oxygen that the haemoglobin of certain nematode gut parasites could hold in the environmental circumstances in which they live is far below their level of oxygen utilization, and the worms are unaffected when it is inactivated. Mostly, however, all four respiratory pigments are functional, and their properties are apparently very similar.

Atmospheric pressure diminishes rapidly with altitude. At 1500 m the oxygen tension is down to almost one-half that at sea level. Since the output of CO_2 remains much the same, the net result is that more CO_2 remains in the blood, and the value of t_u shifts to the right, producing further oxygen lack. The increased CO_2 provides some compensation for its action by causing the respiratory centre in the medulla oblongata (§10.18) to increase the ventilation rate, and the resulting panting produces a quicker replenishing of the air in the lungs. In addition, the spleen contracts, increasing the number of red corpuscles, and hence of available haemoglobin, in the circulating blood. If the exposure to high altitudes is prolonged, the kidney threshold to bicarbonate changes, so that more is excreted to reduce the level of CO_2 in the blood.

4.5 Carbon dioxide transport

Carbon dioxide is not transported simply in physical solution in any animal which has been examined. In man, the quantity carried by the blood may be as high as 50 vol per cent, whereas only 2.5 vol per cent could be carried in physical solution. The corresponding figures for the skate are 11 and 4.8 vol per cent, and in *Astacus* about 30 and 6 vol per cent. There are one or two reported anomalies, like those ascidians which are said to have less CO_2 in the blood than occurs in the surrounding water; but these are rare.

Transport by physical solution will be automatic, but it is clear from the above figures that one or more other mechanisms must exist. Carbon dioxide can combine with the amino groups of amino acids and proteins, including haemoglobin, to form *carbamino-compounds*. These react reversibly with oxygen, so that CO_2 is liberated at the respiratory surface when oxygen is taken up.

About one-fifth of the total CO_2 carried by vertebrate blood is in the form of carbamino compounds, but most of it is transported as bicarbonate. The CO_2 first goes into physical solution, and being a highly diffusible gas, passes into the red blood corpuscle. In both the plasma and the

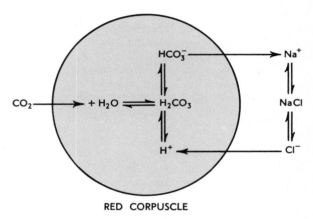

RED CORPUSCLE

Fig. 4.4 Diagram to illustrate the chloride shift. For explanation see text.

red corpuscles, a small amount of the gas combines with the water of the medium to form carbonic acid

$$CO_2 + H_2O \rightleftharpoons H_2CO_3$$

but this is a very slow reaction as well as being small in extent. Vertebrate blood corpuscles contain an enzyme *carbonic anhydrase* that catalyses the formation of carbonic acid, which in turn dissociates to form bicarbonate ions:

$$H_2CO_3 \rightleftharpoons H^+ + HCO_3{}^-$$

The bicarbonate ion is also highly diffusible, and since its concentration is now much higher within the corpuscle than outside, it readily diffuses into the plasma. This would cause a rise in the hydrogen ion concentration of the corpuscle, but the necessary balance is restored by the entry of chloride ions into the corpuscle from the plasma. The cations in the plasma, chiefly sodium, balance the bicarbonate ions leaving the corpuscle, and the net result is that carbon dioxide has been transformed into bicarbonate carried in the plasma in exchange for chloride ions which have entered the corpuscle (Fig. 4.4). The CO_2 tension at the respiratory surface is less than in the respiring tissues, and since the whole process depends on the initial concentration of CO_2, the reverse process takes place there.

Some of the hydrogen ions in the corpuscle do not take part in this process, which is known as the *chloride shift*, but instead are taken up by the protein of the haemoglobin. This has the effect of reducing the negative charge normally present on the haemoglobin, which in turn reduces its affinity for oxygen. This may be the reason for the Bohr effect.

Carbonic anhydrase is present in some invertebrates, in the blood or coelomic fluid of certain annelids and the gills of some arthropods and molluscs. It is also found in the respiratory trees of echinoderms. In all these cases it presumably facilitates the exchange of CO_2. Many invertebrates may be sufficiently small for carbon dioxide to be transported in physical solution, for it is a highly diffusible gas.

4.6 Air-breathing aquatic animals

In all predominantly terrestrial groups of animals there are some species which have returned to an aquatic life, and have thereby subjected themselves to an environment in which oxygen is much less readily available.

The resulting problems are not acute for very small animals even if they possess impermeable skeletons, such as small arthropods, nor for larger animals with moist and permeable skins. It is the larger arthropods and other large animals with impermeable coverings that face a more acute problem. Many of them must depend on oxygen which they replenish periodically at the surface, and in order to avoid over-frequent visits to the surface they must use their oxygen as economically as possible. There are a number of structural adaptations to this situation. Larvae of the mosquito *Culex* possess a circle of hydrophobic hairs around the functional spiracles at the posterior end of the body and these allow the spiracles to be exposed to the air at the surface of the water. They require to come to the surface every 10–15 minutes. Many aquatic insects have developed a *plastron*, a layer of hairs over the body surface that traps air beneath it when the animal surfaces. Some water beetles trap air under their elytra, and water-spiders actually convey a water bubble down to their aquatic webs. Certain insect larvae, such as dragonfly and mayfly larvae, have developed external gills that connect with the internal tracheal system. Others, larvae of some beetles and flies, can actually puncture aquatic plants with the specially developed sharp edges of their spiracles and extract the air from the large air spaces of these plants.

Most pulmonate snails continue to use their lung cavities for breathing air, but a few species fill them with water to produce a water lung. This system does not seem possible for terrestrial vertebrates that have returned to water, and they all depend on air for their oxygen. Despite this, they are able to remain submerged for what appear at first sight to be impossibly long periods. For example, whales can stay under water for up to $1\frac{1}{2}$ hours, the alligator for up to 2 hours. These times may be compared with the human record of about $2\frac{1}{2}$ minutes. The amount of air an alligator can take with it beneath the surface would be sufficient to maintain it for about 15 minutes of respiration at the same rate it exhibits in air, and the same is true of other air-breathing aquatic vertebrates. Clearly, special adaptations are involved.

In all cases examined, the heart rate of diving animals is markedly slowed (*bradycardia*) when they are beneath the surface compared with when they are at the surface. For example, the heart beat of the alligator slows from 41 beats/min at the surface to 2–3 beats/min when submerged. The corresponding reduction in the seal is from 80 beats/min to 10 beats/min. The slowing is due to vagal inhibition (§11.15). Cardiac output therefore drops, but the arterial pressure to essential organs is maintained by selective vasoconstriction. In birds and mammals the capillary beds of the skin and visceral organs (except liver, thyroid and adrenals) and even in some cases, including the alligator, of muscle are shut down, but the circulation to the heart and the brain are kept normal or even increased. The special capillary plexuses of diving mammals may in some cases be used as arterio-venous shunts to by-pass completely non-essential organs.

There is also an increased element of oxygen storage. The blood haemoglobin and muscle myoglobin are generally in higher concentration in diving mammals, and the blood volume is greater. In mammals, too, there are venous sinuses in the abdomen and an enlarged part of the posterior vena cava controlled by a special sphincter, which both increase blood volume and enable some extra control of blood flow.

The proportion of *tidal air*, i.e. the volume of air actually inspired during ventilation, to total lung capacity is much higher in diving vertebrates, e.g. in the porpoise it amounts to some 80% of the total compared with about 25% in man. Thus, for a comparable lung volume, much more of that elusive substance 'fresh air' can be taken in before a dive. Since lung volume is not significantly increased, there is no danger of the 'bends' as there is in human divers. The latter consume air by breathing all the time and the nitrogen in their blood increases as they continue to breath. Diving animals do not continue to breath and the quantity of nitrogen in their blood cannot increase beyond that taken in before diving in one breath. In the human diver a sudden reduction in pressure on surfacing causes the nitrogen to be released as bubbles of gas, and it is this that causes the 'bends'. The quantity of nitrogen in diving animals is insufficient to cause difficulty.

Diving animals exhibit a higher tolerance to carbon dioxide, so that the respiratory centre in the medulla oblongata (§10.19) is not very sensitive to it. During a dive this aids in preventing ventilation, and on re-surfacing it prevents undue panting. Although many diving animals build up a massive oxygen debt (§11.3), and are assisted in doing so by a reduced sensitivity to lactic acid, the large tidal volume, coupled with an enhanced Bohr effect, helps to abolish it without undue panting being required. Since violent movement is not normally undertaken while the animal is submerged, the net result of all the foregoing adaptations is to reduce oxygen consumption by the tissues and enable the animal to stay under for prolonged periods.

4.7 Respiratory quotient

The amount of oxygen consumed, and the amount of carbon dioxide given off during respiration, are often used as a measure of the extent of oxidation of fat, carbohydrate and protein by a particular animal. For example, in the oxidation of a hexose sugar the quantity of carbon dioxide formed is equal to the quantity of oxygen utilized:

$$C_6H_{12}O_6 + 6O_2 = 6CO_2 + 6H_2O$$

Here, the ratio $\dfrac{\text{volume } CO_2 \text{ formed}}{\text{volume } O_2 \text{ utilized}}$

is equal to unity. This ratio, which is known as the *respiratory quotient* (RQ), is different for each of the three major bulk foodstuffs. In the case of a fat such as triolein, the equation for its oxidation takes the form

$$\underset{\text{triolein}}{C_{57}H_{104}O_6} + 80\ O_2 \rightarrow 57\ CO_2 + 52\ H_2O$$

and the RQ is therefore equal to 57/80, or 0.71. Protein is not completely oxidized in the tissues, but its RQ can be determined by indirect means, and shown to be between 0.8 and 0.82.

It is evident from these figures that an omnivorous diet should be represented by an RQ of about 0.85, and a high carbohydrate diet by a figure in excess of 0.9. The RQ can only be a rough guide to the type of diet and needs to be used with caution, because other substances may be metabolized, and because of the possibility of interconversion of food substances within the body of the animal. Furthermore, there are circumstances in which some of the CO_2 contained in the bicarbonate reserve of the body may become 'washed out' of the animal during the period of the experiment. For this reason, measurements of RQ need to be made over a fair period of time. If the RQ when properly measured exceeds unity, it indicates that fat must have been formed from carbohydrate; and conversely, if the RQ falls below 0.7, it indicates that carbohydrate must have been formed from fat.

4.8 Metabolic rates

The amount of oxygen consumed is often expressed in terms of Q_{O_2}, which the number of ml. O_2 consumed/g dry body weight/h (sometimes wet weight is used). The value of Q_{O_2} can vary widely in the same animal due to the operation of a number of factors. As would be expected, it rises rapidly during activity. For example, the Q_{O_2} of the moth *Vanessa* when resting is about 0.6, but rises to about 100 in flight; and that of the resting mouse has a value of 2.5, which rises to 20 when the animal is running (both values determined at 20°C). Reproductive activity in the female is often accompanied by a rise in respiration; there may be changes during the life history, and there are often diurnal, lunar or seasonal fluctuations. It is obvious that great care is necessary in drawing conclusions from respiratory data, and that this should have been obtained over a reasonably long period in well-defined conditions.

Most of the factors referred to above as influencing the respiration rate, and hence the metabolism of an animal, are inherent in the animal's constitution. However, if care is taken to define these innate factors and take them into account, the outstanding variable is that due to activity. In order to eliminate it, measurements of metabolic rate are made when the animal is at complete rest. In this condition, the energy expended by the animal is assumed to represent that required to maintain the animal in proper condition. In homeotherms this level of metabolism is known as the *basal metabolic rate* (BMR), and in poikilotherms as the *standard metabolism*. In practice, 'complete rest' is an unobtainable ideal, since muscular activity may vary even in sleep, and there may be variation due to other activities within the body, notably digestion and hormonal effects. For these reasons the BMR is measured well after a meal has been taken, and if an animal normally exhibits it, during sleep.

Control of the external temperature is also important in measuring basal metabolism (the term will be used for all animals). Heat loss alters with external temperature, but the relationship is not a linear one, and it is therefore necessary to employ a standard temperature. The standard temperature should ideally be within what is called the 'zone of thermal neutrality', the temperature range over which heat is not being lost to combat cold conditions, nor gained from an environment that is too warm. Unfortunately, comparative studies have been complicated through the employment of the same temperature (16°C) for both homeotherms and poikilotherms in many cases. This temperature is satisfactory for many poikilotherms, but not for homeotherms, because the BMR of the latter at 16°C can appear to be twice what it is at 28°C, yet the latter is much more like a neutral temperature for them.

The commonest way of expressing basal metabolism is in terms of heat production, the units used being kJ (kcal). Heat production is normally calculated from the oxygen consumption. From the equations describing the oxidation of the major classes of foodstuffs, it is possible to calculate the amount of heat equivalent to the utilization of 1 l of oxygen for each of them. These figures amount to 21.21 kJ (5.05 kcal) for carbohydrate, 19.7 kJ (4.69 kcal) for fats, and 17.84 kJ (4.25 kcal) for protein. These figures are sufficiently close for a single figure to be used for an animal with an omnivorous diet, for example 20.16 (4.8) is used for man. If the food intake of an animal is known precisely, and particularly if it uses overwhelmingly one type of food, the accuracy is greater.

The disadvantage of this method, which is known as *indirect calorimetry*, is that it assumes that metabolism is completely aerobic. It is not known how far this may be true of individual animals. In frogs, for example, the correlation between heat production and oxygen consumption is known to be poor, and this has been presumed to be due to a degree of anaerobic respiration in this animal. For complete understanding of an animal's metabolism, there is clearly no substitute for a determination of the calorific value of food ingested, matter excreted or secreted (including eggs), and oxygen consumption. The direct measurement of heat production (*direct calorimetry* of the animal) would, of course, save a great deal of trouble, but it is fraught with technical difficulties.

The reader will by now be wondering, in view of all the reservations made, whether the figures given in the literature for heat production can be accepted at all. Fortunately, they are mostly sufficiently reliable for certain generalizations to be possible, even though many results might not be very useful for detailed analysis.

Heat production is normally related graphically to the body weight of an animal. However, when plotted in this way the heat production varies not only between species, but also among animals of different size belonging to the same species. Thus, a rat weighing 600 g has a heat production of 314 kJ (75 kcal)/kg body weight, but the corresponding figure for a 150 g rat is 473 kJ (113 kcal)/kg. This means that for comparative purposes a number of determinations must be made on different sizes of the same species, and the mean heat production determined. In measuring 'body weight', the 'inert' components have to be taken into account, structures such as fur, feathers, arthropod integuments and molluscan shells, which may amount to 10% or more of the apparent body weight. Gut contents in ruminants may equal 15% of the apparent body weight.

When all these factors have been allowed for, it is found that the relationship between body weight and heat production is not constant for all animal species. If metabolism is denoted by M and weight by W, the relationship between them can be expressed as

$$M = KW^b$$

where K and b are constants. The constant b gives the rate at which oxygen consumption or metabolism varies with size, and if metabolism were directly proportional to body weight, b would be equal to 1. In fact b normally varies between about 0.67 and 1, and the value in developing young is always lower than in adults. The value of K also varies considerably, being between 10 and 13 in mammals; but even here, it is found that the metabolism of small mammals is much greater than that of large mammals, and it can be calculated that it would be impossible for a mammal weighing less than 3.5 g to get sufficient food to keep itself active.

It might be thought that surface area and metabolism would exhibit a constant relationship among animals. In fact, it fluctuates much as the weight relationship does, and not necessarily to the same extent or in the same direction for any given species. This shows, incidentally, that the relationship between surface area and body weight is not precisely the same for all animals. Clearly, surface area must affect heat loss, but so will the insulation of the body surface.

We can therefore conclude that metabolism is not directly correlated with any body parameter, but that the trend is for larger animals to metabolize less per unit weight than small animals. In addition to this general trend, it is evident from metabolic studies that homeotherms have a much higher BMR than poikilotherms, their heat production being some five to eight times that of the latter. This is, of course, due to the ability to maintain the body temperature at a constant high level.

5

Excretion

When carbohydrates, fats and proteins are utilized as fuels for the liberation of energy, water and carbon dioxide are produced. The water enters the body fluids of the animal, and its fate becomes inseparable from the general problems of water balance, dealt with in Chapter 6. The way in which carbon dioxide is eliminated has been described in Chapter 4. There remains a third major excretory product, the nitrogen-containing group which is split off from amino acids during their metabolism. Some protein is used by animals, in the form of its constituent amino acids, as raw material from which other proteins may be synthesized (§2.12), and some is used as an energy-providing fuel. Protein cannot be stored in the bodies of animals, and any surplus must be eliminated from them, together with the remains of protein from dead and damaged cells, autophagy, and cellular turnover of protein. This must all follow the same pathway as amino acids used immediately for fuel. Some aquatic invertebrates do excrete a proportion of their waste and surplus nitrogen as amino acids, but this method is wasteful of carbon, and the vast majority of animals degrade their amino acids to keto acids and ammonia, as described in §3.19. The basic need in nitrogenous excretion is therefore the elimination of ammonia.

5.1 Deamination and transamination

The first stage in the breakdown of amino acids is the removal of their nitrogenous group as ammonia. Such *deamination* results from the action of a variety of enzymes, which are either oxidative or hydrolytic. Oxidative deamination is catalyzed by a group of enzymes known as *amino-acid oxidases*, and although the process actually involves several steps with flavine compounds acting as hydrogen acceptors, it may be summarized thus:

$$R \cdot C \overset{\cdots}{(H \cdot NH_2)} \cdot COOH + \tfrac{1}{2}O_2 \longrightarrow R \cdot C{=}O \cdot COOH + NH_3$$

amino acid   keto acid

Even more important than the amino-acid oxidases are a number of hydrolytic enzymes that deaminate specific amino acids, notably the sulphur-containing amino acids, and aspartic and glutamic acids. For example, in the non-oxidative deamination of glutamic acid, the enzyme *glutamic dehydrogenase* removes two hydrogen atoms from a combination of glutamic acid and water, with NAD as co-enzyme, and a keto acid is again formed:

$$COOH \cdot C \overset{\cdots}{(H(NH_2))} \cdot CH_2 \cdot CH_2 \cdot COOH + H_2O \longrightarrow$$

glutamic acid

$$COOH \cdot C{=}O \cdot CH_2 \cdot CH_2 \cdot COOH + NH_3 + 2H \text{ (to NAD)}$$

α-ketoglutaric acid

In most cases, both kinds of deamination lead eventually to either pyruvate or acetate being formed, and these can then enter the metabolic pathway for carbohydrate at the appropriate point (§3.19).

The ammonia is either eliminated unchanged, or in the form of a more complex nitrogenous substance which is normally either *urea* or *uric acid*. These two latter compounds are produced by synthetic mechanisms that represent relatively slight elaborations of normal cellular processes. Since these processes follow a definite pattern, they depend on a fairly small number of specific amino acids, into which the others must be converted from time to time. Such conversions, which are also necessary for the normal metabolism and transport of proteins and may be used in interconversions of protein with fat and carbohydrate, are effected by *transamination* which, as its name implies, involves the transfer of the amino group —NH$_2$ from one compound to another, by means of enzymes known as *transaminases*. An example of transamination is the formation of the amino acid alanine by the transfer of the amino group from glutamic acid to pyruvic acid, in the course of which α-ketoglutaric acid is produced:

$$COOH \cdot CH_2 \cdot CH_2 \cdot CH(NH_2) \cdot COOH + CH_3 \cdot C{=}O \cdot COOH \rightleftharpoons$$

glutamic acid   pyruvic acid

$$CH_3 \cdot CH(NH_2) \cdot COOH + COOH \cdot CH_2 \cdot CH_2 \cdot C{=}O \cdot COOH$$

alanine   α-ketoglutaric acid

Transaminations are, as shown, reversible, and the direction in which the reaction proceeds will be governed by the needs of the cell. Transamination to form glutamic acid, which is then deaminated by glutamic dehydrogenase, is one of the most important sources of ammonia in animal cells.

5.2 Synthesis and properties of excretory products

No animal is known which produces only one of the three excretory materials, ammonia, urea, uric acid, to the exclusion of the others. Usually, a certain amount of all three is excreted, but with one predominating over the others. This is not surprising in view of the fact that urea and uric acid are made by synthetic processes already existing in the cell for other purposes. To understand why one excretory product should be dominant over the other two, it is necessary to know something about the properties of the three substances.

Ammonia NH_3 is highly soluble in water, with which it forms ammonium hydroxide NH_4OH. Ammonium ions and ammonia diffuse readily through water, and pass rapidly across cell membranes. Thus, if the animal is surrounded by a large volume of water, they will be quickly dispersed in a concentration made negligible by the bulk of the water. Rapid elimination of ammonia is essential, as it is toxic to animals in not very high concentrations. Invertebrates are generally more tolerant to it than vertebrates, but even the most tolerant of them, such as *Sepia* or the earthworm, do not exhibit a blood concentration of ammonia higher than 5 mg/100 cm^3 blood. This concentration is fatal to mammals, in which the level in the blood does not normally exceed 0.1 mg/100 cm^3, a figure fairly typical of vertebrates as a whole.

Urea $CO(NH_2)_2$ is much less toxic and slightly more soluble in water than ammonia. Its high solubility means that water must be used for its elimination, but the amount required is much less than for an equivalent quantity of ammonia, because its relative lack of toxicity allows it to be concentrated to a much greater extent. The quantity of urea in the blood is small, but it can be concentrated by the excretory organ without damage to it, and stored safely in a bladder if necessary. In man, for example, the blood contains between 0.1015 and 0.04 g/100 cm^3 whereas the urea in the urine amounts to about 2.0 g/100 cm^3.

Urea is produced in animals by a cyclic process known as the *urea cycle* or *ornithine cycle* (Fig. 5.1). The urea is formed by the action of the enzyme *arginase*, which catalyzes the formation of urea from the amino acid *arginine*, and the purpose of the urea cycle is to regenerate the arginine using excretory ammonia. A molecule of arginine contains four nitrogen atoms, and in the formation of a molecule of urea two of these are split off, leaving the remaining two as part of the amino acid *ornithine*.

During the course of the urea cycle the ornithine is converted back into arginine, the two nitrogen atoms required for this purpose being obtained as ammonia by the deamination of amino acids. One molecule of ammonia enters the cycle in combination with carbon dioxide in the form of *carbamyl phosphate*, which becomes attached to the ornithine to produce *citrulline*. The second molecule of ammonia is then injected into the cycle in the form of *aspartic acid*, which joins on to the citrulline to form a complex compound with four nitrogen atoms. This complex loses part of its molecule as *fumaric acid*, to become arginine. The fumaric acid can theoretically be aminated through several steps to re-form aspartic acid, but since one of these steps leads to the production of oxaloacetic acid, the fumarate really enters the TCA cycle (§3.17), and the aspartic acid used in the urea cycle is probably obtained by transamination of oxaloacetic acid with glutamic acid, the former being derived from carbohydrate metabolism as much as from fumaric acid. Figure 5.1 shows that the cycle must be driven by an external source of energy, which is ATP. In all, four high-energy phosphate groups must be detached from ATP in one turn of the cycle, i.e. two moles of high-energy phosphate are used for each nitrogen atom excreted.

The urea cycle is best known from mammalian liver, but the necessary enzymes have been found in many other animals, both vertebrate and invertebrate. With the exception of an additional step in those animals not normally urea-secreting, the urea cycle is a normal pathway for the metabolism and transamination of certain amino acids, and it is therefore to be expected that it should be widespread in occurrence. Reports of its absence should be treated with caution, since all the steps of the cycle may not take place in the same organ within a given animal. Indeed, arginase is present only in the kidney of birds, whereas in other vertebrates it occurs in the liver. Nevertheless, the cycle may possibly be absent in some snakes and birds whose main excretory product is uric acid. Even in these cases, the enzyme arginase is still present, and so some urea is formed from arginine taken in with the food.

Uric acid is more complex in structure than urea (Fig. 5.2), being a cyclic compound containing four nitrogen atoms in its molecule. It is virtually insoluble in water and therefore non-toxic. In those animals in which it is the major excretory product, its lack of solubility allows it to be voided as a thick paste or a solid pellet, and it is accordingly especially suitable for production by animals living in conditions in which water is scarce.

The method by which uric acid is formed is even more complex than the urea cycle, but again it is a series of reactions that is normally found in the animal body, this time for the production and interconversion of nucleotides, and which has become especially active in animals that use

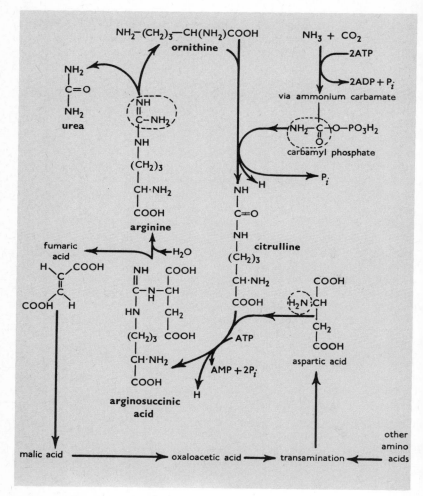

Fig. 5.1 The urea cycle (ornithine cycle).

uric acid as their main excretory product. Animals in which another product predominates possess an enzyme *uricase*, that converts the uric acid
they manufacture to urea (see above).

The essential features in the production of uric acid are the incorporation of the amino acids *glutamic acid* and *glycine* and the hexose sugar
ribose into a nucleotide, *inosine-5-phosphate*, from which the ribose is

eventually removed to leave, after a further dehydrogenation, uric acid (Fig. 5.2). Eight molecules of ATP are converted to ADP to provide the energy for these reactions, and since uric acid contains four atoms of nitrogen, the energy cost per atom is the same as for urea. Both urea and uric acid are therefore relatively costly to synthesize, in terms of energy loss per nitrogen atom excreted. Moreover, the synthesis of urea involves the loss of one carbon atom for every two nitrogen atoms excreted, and the synthesis of uric acid 5 carbon atoms for every 4 nitrogen atoms excreted. Any advantage the excretion of these two substances in place of ammonia may confer upon an animal is clearly not a metabolic one.

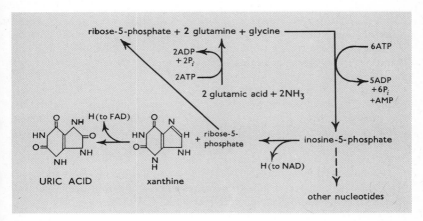

Fig. 5.2   The uric acid cycle in outline.

Apart from excretory products from protein sources, a relatively small percentage of nitrogenous waste is derived from the breakdown of nucleic acids, largely those taken in with the food. Breakdown of nucleic acids yields principally the purines adenine and guanine and the pyrimidines (§3.6). Some animals excrete these as such, and this may be the normal method of excretion of pyrimidines. But although some purines are excreted without change, principally by flatworms and annelids, the majority of animals possess enzymes that convert them into xanthine, from which uric acid is formed (Fig. 5.2). Animals that contain uricase degrade this acid further, first into *allantoin* and then into urea. If urea is not the major excretory product, the animal will possess the enzyme *urease*, and this converts it into ammonia. Thus, except in those cases in which the primary products of nucleic acid breakdown are excreted direct, the end-products of their metabolism will be the same as the end-products of protein metabolism.

5.3 Occurrence of excretory products

Although ammonia, urea and uric acid are the major nitrogenous excretory substances, others undoubtedly exist. Many invertebrates produce a range of unidentified substances, which may amount to 10%, 20% or even 30% of the total nitrogen excreted. A few extreme cases have been reported in which the quantity is even higher, as in the slug *Arion*, in which they may form as much as 60% of the total. Quite high amounts of undetermined products have also been found in some vertebrates from fishes to mammals, the quantities claimed being as great as those in invertebrates. Some other substances, whose identity is known, may be excreted in quantity by individual groups of animals. Spiders excrete almost exclusively the substance *guanine*, and since nucleic acid metabolism can account for only a small proportion of this, they must possess a synthetic method for its production, the details of which are unknown. Guanine is even less soluble than uric acid, and so requires no water for its elimination.

Marine teleosts excrete quite a large proportion of their nitrogen—as much as one third in some cases—as *trimethylamine oxide*, a soluble non-toxic substance whose mode of synthesis is unknown, and which increases in the flesh of dead fish to give them their characteristically 'high' smell. Many vertebrates excrete *creatine* or more usually its derivative *creatinine*, but except in the case of some marine fishes, this does not usually amount to more than 10% of the total.

In terms of their properties, trimethylamine oxide and creatine may be equated with urea, and guanine with uric acid. Thus, the identified excretory products fall into three groups with distinctive properties, exemplified by the commonest substances they contain, ammonia, urea and uric acid. Except in a few instances, found among both invertebrate and vertebrate groups, these identified substances form the greater part of the nitrogenous compounds excreted, and this permits us to discern an apparent pattern of nitrogenous excretion in which the properties of the predominant excretory product of an animal may be linked with its type of habitat, whether adult or embryo. It should be remembered, however, that no animal excretes one type of product exclusively, and this fact accounts for the considerable plasticity of excretory pattern in the course of evolution.

It will be evident that the excretion of ammonia, which may be presumed to be the primitive excretory material, is especially suited to an aquatic habitat. It is found as the chief excretory product in the majority of aquatic animals, not only in those which are primitively aquatic like marine invertebrates and the fishes, but also in many secondarily aquatic animals such as the larvae of the insects *Aeschna* and *Sialis*, and aquatic tortoises and turtles. Anuran tadpoles excrete mainly ammonia, whereas the adults produce urea, and the proportion of urea to ammonia increases the drier the conditions. Although many aquatic animals contain excretory organs, much of the ammonia they excrete is eliminated from their gills. This is to be expected, because the gill surface is highly vascular and in direct contact with the surrounding water, thus permitting simple diffusion to take place. However, ammonium ion is also exchanged for the sodium taken up by freshwater animals (§6.5) and this cannot be a simple diffusion process. In some fishes at least, the gills are themselves the site of production of much of the ammonia excreted. In the sculpin *Myoxocephalus* most of the ammonia excreted is passed out through the gills, but less than 15% of this represents blood ammonia. The remainder is formed in the gills from glutamic acid and glutamine (first transformed into glutamic acid by the enzyme *glutaminase*), deamination being due to the enzyme glutamic dehydrogenase referred to earlier.

Urea is clearly a useful excretory product for a semi-terrestrial animal that is not exposed to a scarcity of water, but is not necessarily surrounded by large quantities of it. It is accordingly the main product of adult terrestrial amphibians, shore gastropods, some earthworms, and the semi-terrestrial tortoises and turtles. In very dry conditions amphibians accumulate urea in their tissues and excrete it only when water is again available, and the governing factor in their retention of urea is water turnover. The special cases of elasmobranch fishes and mammals are dealt with below.

All genuinely terrestrial animals, including insects, terrestrial gastropods, terrestrial reptiles and birds excrete mainly uric acid, with exceptions to be mentioned presently. Insects and land snails excrete solid pellets of uric acid, the rectum or malpighian tubules of insects being especially efficient at resorbing water from both the excretory and faecal material. Reptiles and birds are not quite so efficient, and void a thick paste of uric acid crystals. In birds, the urine is already concentrated when it enters the ureters, which have to force it along by peristaltic action. It is probable that cloacal resorption of water takes place in birds and reptiles, causing the urine to become even more concentrated. A certain amount of urea is voided by birds and reptiles, but in desert reptiles, for which the conservation of water is of paramount importance, the urea that is formed by the liver is resorbed by the cloaca and re-processed into uric acid.

At least some snails lack the ornithine cycle, and hence must produce either ammonia or uric acid. Freshwater snails such as *Limnaea* and *Paludina*, which are believed to have returned to the water from a terrestrial habitat, excrete a good deal of uric acid; whereas the prosobranch *Potamopyrgus* (= *Hydrobia*) *jenkinsi*, which is known to have penetrated into fresh water from the sea relatively recently, excretes only minute quantities of uric acid, its major product being ammonia. Marine prosobranchs like *Buccinum* and *Gibbula* excrete ammonia, but the shore-living periwinkles excrete significant quantities of uric acid and urea.

Exceptions to the apparent correlation between a land habitat and the excretion of uric acid are found in the Crustacea. Both the shore crab *Carcinus* and the apparently terrestrial woodlice excrete most of their nitrogen as ammonia. Neither is genuinely terrestrial, as woodlice are unable to survive dry conditions and their behaviour is directed towards keeping them in a moist habitat, but it is perhaps surprising that ammonia has not been superseded by urea as their chief excretory product. Presumably their tolerance to ammonia is above average.

Although the excretion of ammonia appears to be the primitive condition, there is no fundamental reason why an aquatic group should not utilize urea as its excretory product. In fact, urea is the major product of elasmobranch fishes, but they are atypical in retaining it in their blood at the high level of 2.0–2.5 g/100 cm^3 blood. By this means, marine elasmobranchs are able to overcome their osmotic problems (§6.7), but urea production and retention appears to have become such an established feature of the group, that its elimination by freshwater elasmobranchs is difficult, and although they retain less than their marine relatives, sufficient remains in their blood to constitute an additional osmotic problem.

5.4 Excretion during embryonic development

It has been suggested that the evolution of uric acid as an excretory substance by terrestrial animals was a result of their form of embryonic development. Insects, reptiles and birds possess in common a *cleidoic* egg, a yolky egg enclosed in a hard shell that permits some gaseous exchange but prevents the passage of water between egg and environment. Because the young of terrestrial animals must hatch at an advanced stage of development by comparison with the more protected aquatic embryo, the egg must both contain ample supplies of food and provide a receptacle for the storage of the excretory matter that results from their consumption. The storage of a soluble substance such as urea would introduce osmotic complications, and the need is therefore for a product which is both nontoxic and insoluble. These requirements are met by uric acid, which is stored in solid form until the egg hatches.

Although mammals are a terrestrial group, they excrete urea. It is usually supposed that this feature is related to placentation. The embryonic urea, being soluble and non-toxic, can diffuse through the placenta into the maternal blood stream and be eliminated through the parental kidney. The mammalian excretory organs reflect the ureotelic habit in that the kidney possesses a collecting region, the pelvis, and there is a well-developed bladder and no cloaca.

Many animals alter their predominant excretory product at stages in their life history, like the amphibians already referred to. The embryos of insects, reptiles and birds, which show a tendency to excrete uric acid from the beginning of development, have been claimed to produce chiefly ammonia in the early stages followed later by urea and eventually uric acid; and this was suggested as an example of evolutionary recapitulation. However, the claims originally made in this respect concerning the chick embryo have not been confirmed, and it has always seemed unlikely that a wholly terrestrial animal would excrete much ammonia at any stage in its life history. It is certain that many members of these groups do not exhibit sequential changes of excretory products during their life histories, and it is probably better to treat all such claims with scepticism.

6

Osmotic and Ionic Regulation

It is generally agreed that life began in the sea. Although many living organisms have moved into other habitats in the course of evolution, water remains an essential constituent of all living matter. As we have seen in earlier chapters, it is the universal biological solvent in which all vital processes take place, and it is often a participant in those processes. In all groups of animals, except perhaps the insects, the embryo must be surrounded by watery fluid if it is to develop. The conservation of an adequate internal quantity of water is therefore one of the primary requirements of living organisms.

Cellular processes can only be executed in a salt solution of given composition. The exact composition varies between species but it is never the same as sea-water, even in animals that inhabit the sea. At one time it was thought that the composition of the sea when life arose was different from that of today, and that the ionic concentration of body fluids resembled that of the supposed 'archaeocean'. It is now known that the composition of the sea has varied relatively little during the history of the earth, and it appears that living organisms became sufficiently independent of their environment early in their evolution to evolve their own 'internal sea'. The result is that there is a universal need for living organisms to regulate the extent to which different ions are allowed to pass between themselves and their environment.

6.1 The passage of water and salts across cell membranes

Some of the properties of cell membranes have already been described (§3.3). They are often thought of as being analogous to a sieve through which water can pass but which will permit the passage of only small solute particles. In general, small hydrated ions such as those of potassium, chloride and ammonium pass more easily than larger ions such as sulphate or organic ions. The pore diameters of the boundaries of different cells can range from 0.4 nm, as in *Amoeba*, to 6–12 nm in the cells lining blood capillaries. Most cells are nearer the lower figure cited, but we should expect to find adaptive differences between cells in this respect, and to discover that they result in variations in the extent to which the passage of different ions across the membrane is regulated.

The direction in which an ion will move is determined by the potential difference and the concentration gradient across the membrane. These two opposing influences reach an equilibrium, the electrochemical potential, which is described more fully in §9.3. An ion will tend to diffuse from a higher to a lower electrochemical potential. If it moves in the other direction, the membrane must be performing active work in transporting it. Membranes may also actively speed the passage of ions in the direction of passive diffusion.

The restrictions placed on the passage of ions and other solute particles by the passive or active properties of cell membranes may result not only in the development of an internal composition different from that outside the cell, but also the setting up of a difference in total solute concentration between the two. This will lead to the development of an osmotic pressure which, unless the cell is actively able to prevent it, will cause water to flow across the membrane. This situation is found in a number of aquatic animals. In terrestrial animals the problem is usually one of conservation of water by preventing its loss through evaporation and excretion. In both cases, the burden of preserving normal water relations is largely borne by the external layer of cells, often aided by specializations of the gut and excretory organs, and the internal cells are not called upon to bear the strain to any extent. However, although cells may not normally be called upon to do so, they can usually tolerate some changes in the concentration of the extracellular fluid within the bodies that contain them, and we shall find that adaptation to changes in the salinity of the external medium may involve some concentration or dilution of the extracellular fluid.

6.2 Osmotic pressure of body fluids

The osmotic concentration of a fluid can be expressed in a variety of ways. Probably the most widespread method of measuring it is to determine the temperature at which the fluid freezes. The presence of particles in water—in the case of biological fluids the particles are mainly salts or their ions or nitrogenous compounds—lowers its freezing point by an amount that is proportional to the concentration of the particles. The depression of the freezing point is usually denoted by the Greek capital Δ. Alternatively, osmotic concentration can be described in terms of the molality of a fluid, which is the total number of moles per litre, usually expressed in milliosmoles for biological fluids; or in equivalent concentration of sodium chloride, as millimoles per litre; or as a percentage of the concentration of sea-water.

An *isotonic* solution is one that does not cause a cell immersed in it to change in volume. Since change in volume may be resisted by the rigidity of the cell wall, a better term to describe solutions which are equal in osmotic concentration is *isosmotic*. A solution which is more dilute than the reference solution is termed *hypoosmotic*, and one which is stronger *hyperosmotic*.

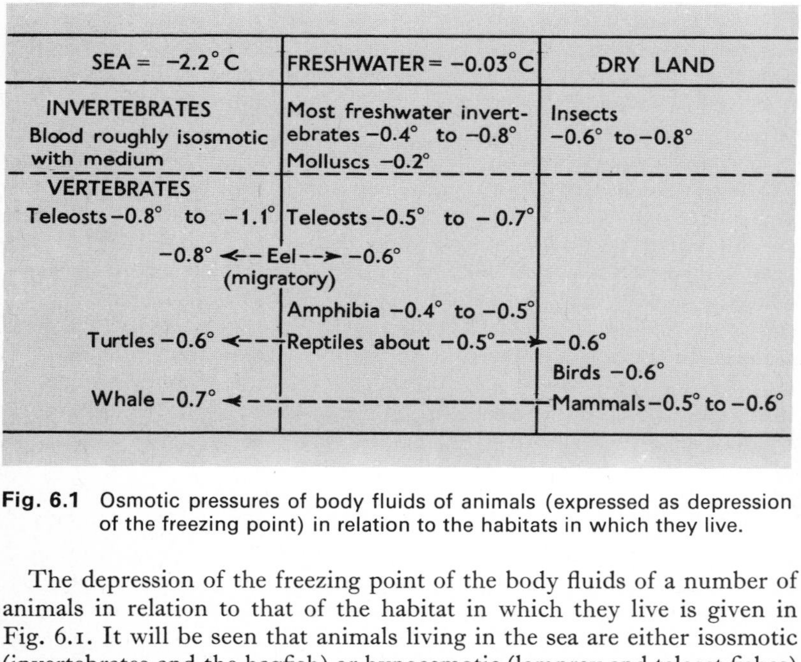

SEA = −2.2°C	FRESHWATER = −0.03°C	DRY LAND
INVERTEBRATES Blood roughly isosmotic with medium	Most freshwater invertebrates −0.4° to −0.8° Molluscs −0.2°	Insects −0.6° to −0.8°
VERTEBRATES Teleosts −0.8° to −1.1°	Teleosts −0.5° to −0.7°	
−0.8° ◄─ Eel ─► −0.6° (migratory)		
	Amphibia −0.4° to −0.5°	
Turtles −0.6° ◄─	Reptiles about −0.5° ─►	−0.6°
		Birds −0.6°
Whale −0.7° ◄ ─ ─	─ ─ ─ ─ ─ ─ ─ ─ ─ ─	Mammals −0.5° to −0.6°

Fig. 6.1 Osmotic pressures of body fluids of animals (expressed as depression of the freezing point) in relation to the habitats in which they live.

The depression of the freezing point of the body fluids of a number of animals in relation to that of the habitat in which they live is given in Fig. 6.1. It will be seen that animals living in the sea are either isosmotic (invertebrates and the hagfish) or hypoosmotic (lamprey and teleost fishes) with the medium. In hypoosmotic forms there will be a tendency for water to pass out of the body and for their tissues to be dehydrated. Freshwater animals, however, are always hyperosmotic to the medium and will therefore tend to gain water. The concentration of the body fluids of land animals is very similar to that of freshwater animals, and is often cited as evidence for their freshwater origin. The water problems of terrestrial animals are, of course, quite different from those of aquatic forms. They tend to lose water to the atmosphere, and the factors which govern this loss are, apart from structural adaptations, temperature and humidity. Although this does not strictly constitute osmoregulation, it is usually considered and compared with osmoregulation in aquatic forms.

6.3 Aquatic animals

Although marine invertebrates and the hagfish are isosmotic with sea-water, their ionic composition is different and is regulated by the body surface. In some groups of animals, such as coelenterates and echinoderms, the degree of regulation is not very great; but the majority of the marine members of the other invertebrate groups, and also the hagfishes, concentrate potassium and reduce sulphate and often magnesium by comparison with sea-water. A few molluscs actually concentrate magnesium, and in some animals it is not much altered. On the whole the calcium level remains more or less equal to that of the external medium. It is possible that some accounts of high blood calcium in decapod crustaceans have resulted from experiments which failed to allow for the fact that the blood calcium of these animals fluctuates with the moulting cycle (§13.13).

Ionic regulation depends on the properties of the external surface in contact with the medium, and on the differential absorption or extrusion of ions by that surface, using active ion pumps. Osmotic regulation may be achieved in a number of ways: by limiting the permeability of the outer covering to salts and water, or both; by altering the concentration and volume of any urine that may be excreted; and by active excretion or absorption of salts and water against a concentration gradient. It will be seen that osmotic and ionic regulation share the same regulatory mechanisms, and are inextricably interrelated. For this reason it is convenient to consider them together.

Marine teleosts and lampreys are hypoosmotic to the sea, and tend to become dehydrated through the osmotic loss of water from their bodies. Their general body surface is not very permeable to water, and the main areas from which water is lost are the epithelia lining the mouth, pharynx and gills. Despite this reduction in the possible area of water loss, the permeable surfaces are continually subject to the action of sea-water and a substantial problem remains. One way of gaining water is to drink it, and this is undoubtedly one mechanism adopted by teleosts. A fish like the eel may drink up to 200 cm³ of sea-water daily, and some 60–80% of the water will be extracted from it. But drinking sea-water creates a new problem, that of eliminating the salts without using the water gained to do so. The problem is aggravated by the fact that the body surface, though relatively impermeable to water, is permeable to salts, which tend to diffuse into the fish along their concentration gradient; and in fact some four-fifths of the total salt influx occurs by this route. The kidney might provide some help, but even if the urine was highly concentrated some water would be lost. In practice, the urine of marine teleosts is often slightly hypoosmotic even to their blood, and it is essential for them to keep down their production of urine to the bare minimum.

The way in which excess salts are eliminated was first demonstrated by what is known as a heart-gill preparation. The auricle of the heart was fed with a suitable saline, which was pumped by the ventricle through the gills, and then led off by tapping the dorsal aorta. The salinity and composition of the perfusing fluid and of the medium bathing the gills could be adjusted initially, and any changes in them detected by chemical analysis. This experiment, and more recent ones using isotopes, demonstrated that the gills of teleost fishes are able to secrete salts outward into sea-water against the concentration gradient. The amount secreted increases in proportion to the salinity of the perfusing fluid. The outward secretion of sodium and chloride is likely to be a property of all the cells of the gill epithelium, and the so-called 'chloride-secreting' cells are probably specialized for some other purpose, perhaps the secretion of mucus.

The turnover of salts in teleost fishes is very high, some 10–20% of the total sodium chloride of the fish being turned over each hour. This is in marked contrast to fresh water teleosts, in which the turnover is only $\frac{1}{2}$–1% per hour.

6.4 Penetration into fresh water

The successful penetration of marine animals into fresh water required the conquest of an osmotic problem which was the precise opposite of any which might exist in the sea. There is a certain minimum blood concentration that will support life indefinitely. The precise minimum varies between animals, tending to be less in brackish and freshwater animals than in marine animals (Fig. 6.1). But despite these variations, all animals seem to need to keep their blood concentration hyperosmotic to fresh water. The resulting osmotic pressure causes water to move into them across their permeable surfaces. Because of the need to preserve a minimum blood concentration, adjustment to this situation by allowing the blood concentration to fall in conformity with the osmotic pressure of the external medium can be permitted only to a limited extent, and other mechanisms must be employed.

We may expect to find such mechanisms in brackish-water animals as well as in truly freshwater animals. Brackish water is rather arbitrarily defined as diluted sea-water that ranges in concentration from 1.5% to 90% of that of pure sea-water. The upper end of this range represents the concentration at which the characteristically marine fauna begins to be markedly impoverished, and therefore the limit of dilution of the medium below which many marine animals are unable to cope. The fauna found in media representing the lower end of the brackish-water scale merges into that of fresh water, since at this dilution the osmotic conditions are not very different from those of true fresh water. The definition of brackish water embraces the estuaries of rivers, salt marshes, and the larger land-locked seas such as the Baltic and Caspian seas.

Most brackish waters will contain both marine animals which can tolerate low salinity, and true brackish-water animals which are not found in the sea. Characteristically marine animals in brackish water include the shore crab *Carcinus maenas* and the mussel *Mytilus edulis*. These marine forms tend to disappear from the brackish-water fauna when the water becomes equivalent to about 15% sea-water concentration. An example of a true brackish-water animal is the prawn *Palaeomonetes varians*, which can live in media ranging from 0.5% to 100% sea-water, but does not in fact live in genuinely marine habitats.

A common method of studying the effect of changes in the concentration of the external medium is to measure the osmotic pressure of the blood, usually by determining its depression of the freezing point, and this is plotted against the osmotic pressure of the medium. The resulting graph gives an indication of the ability or otherwise of the animal to regulate its salt and water content in media of different salinities. An animal that is incapable of any degree of regulation will give a curve in which the blood concentration is directly proportional to that of the external medium, and which, if it is extrapolated back, will pass through zero, although the animal will normally die before this point is reached experimentally. The greater the deviation of the curve from a linear curve that passes back through zero, the greater is the ability of the animal concerned to regulate its blood concentration despite the osmotic effects of the diluted medium.

Figure 6.2 shows a number of such graphs, from animals with differing powers of regulation. Linear curves are characteristic of the lug-worm *Arenicola marina* and the spider crab *Maia*. In the other cases shown, except for *Palaeomonetes*, the osmotic pressure of the blood is at first linear, but as the dilution of the medium is made progressively greater the internal osmotic pressure follows the external osmotic pressure to a decreasing extent, the blood becoming increasingly hyperosmotic to the medium. The ability to regulate is seen to be greater in *Carcinus* than it is in *Nereis diversicolor*, and it is accordingly not surprising to find that *Carcinus* can penetrate further into brackish waters than *Nereis*. The curve given by *Palaeomonetes* is rather different from the others, and reflects the true brackish-water nature of this prawn. Its blood concentration even in 100% sea-water is equivalent to only 70% sea-water. But it can maintain a blood concentration equal to 50% sea-water when the external medium is diluted to 5% sea-water. Below this level some further dilution of the blood occurs, and the animal dies in media more dilute than 0.5% sea-water. *Carcinus* is typical of marine animals that possess good regulatory ability. In 100% sea-water its blood concentration is equal to the surrounding medium. The blood concentration falls fairly rapidly at first in an increasingly dilute medium, until a level equal to

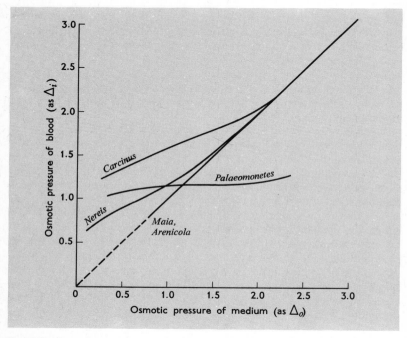

Fig. 6.2 Osmotic pressures of the body fluids of a number of marine or brackish animals (Δ_i) plotted against the osmotic pressure of the external medium (Δ_o). (Based on Beadle, 1943, *Biol. Rev.*, **18**, 172, and Potts and Parry 1964, *Osmotic and Ionic Regulation in Animals*, Pergamon, Oxford, from *Annls. Inst. océanogr.*)

60% sea-water is reached. *Carcinus* can maintain this concentration in media down to 15% sea-water, below which it dies. Since its minimum permissible blood concentration is higher than that of *Palaeomonetes*, *Carcinus* has to expend more energy in regulating in brackish water, and cannot stand the extremes of dilution tolerated by the former.

Graphs of the kind shown in Fig. 6.2 illustrate the fact that animals tend to counteract the dilution of the medium by reducing their blood concentration as far as this can safely be done, thereby reducing the amount of active work required to maintain ionic and osmotic regulation. It is evident from Fig. 6.1 that this has been a general feature of penetration into fresh water, for all freshwater animals possess a lower blood-concentration than their nearest marine relatives. It is interesting to note that teleosts, which are thought to have originated in fresh water, appear to have been unable to reverse this characteristic to any extent on their return to the sea.

A brackish-water animal from the sea like *Carcinus* uses all the methods of regulation outlined earlier in the chapter. It is found to be much less permeable to both water and salts than strictly marine relatives like *Cancer* and *Hyas*. In hypoosmotic media the production of urine by *Carcinus* increases, and some water is also secreted into the gut to be passed to the outside. The urine produced is hypoosmotic to the blood but hyperosmotic to the medium and therefore elimination of water by this method also entails some salt loss. Indeed, the antennary organ of crustaceans is now thought to be primarily an organ for ionic regulation rather than osmotic regulation. It produces a fluid that is high in magnesium and low in sulphate and potassium by comparison with the medium. Salt loss can be compensated for by the active uptake of salts from the medium by the gills. For example, in 40% sea-water *Carcinus* loses 4.4 mM/kg body-weight/h of sodium ions through diffusion from its permeable surfaces, and 3.8 mM/kg/h in its urine. This total net loss of 8.2 mM/kg/h is recouped entirely through the active uptake of sodium ions from the medium against the concentration gradient.

The pattern of osmoregulation exhibited by *Carcinus* is probably followed to varying extents by the majority of brackish- and freshwater animals. A rather different pattern is that observed in the planarian *Procerodes* (= *Gunda*) *ulvae*, which lives in tidal estuaries and areas of the shore in which fresh water enters the sea. It is thus subjected to alternate bathing by fresh water and sea-water. If *Procerodes* is placed in sea-water that is progressively diluted the animal swells, showing that it is fairly permeable to water. But eventually, provided a very small amount of calcium is present in the water, it adjusts its body volume to its previous level. By chemical analysis of the animal, the adjustment of its body volume can be shown to be achieved by the simple expedient of actively secreting ions outwards, which is done to a striking extent. In fresh water, *Procerodes* is found to possess an internal salt concentration that is only one-tenth of that in sea-water. The limitations of such a method are obvious, and the fact that the blood concentration reached in fresh water is dangerously low is illustrated by the observation that *Procerodes* cannot live indefinitely in fresh water. Such a method of adjustment is only suitable for an animal living in tidal conditions.

6.5 Freshwater animals

The osmotic problems of freshwater animals are the same as those of brackish-water animals except that the conditions to which they must be adjusted are more extreme. They meet these problems by increasing the extent of some of the osmoregulatory mechanisms employed by brackish-water animals. The permeability of the outer surface of fresh water invertebrates is less than in related brackish-water animals, although, oddly enough, the permeability of fresh water teleosts to water (but not salts) is higher than in their marine counterparts. The urine is generally hypoosmotic or in a few cases isosmotic to the blood, and in the great majority of instances hyperosmotic to the medium, as in brackish-water animals, and this is the main avenue through which salts are lost. However, the blood itself is less concentrated than that of brackish or marine animals, and the

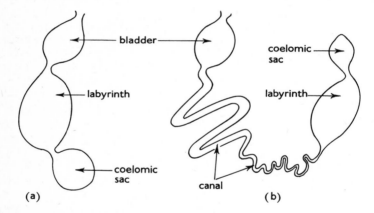

Fig. 6.3 Antennary organs of (a) *Carcinus* and (b) *Astacus*. (After Picken, 1936, *J. exp. Biol.*, **13**, 309)

concentration of the urine is less than in these latter forms. The secretion of more urine is therefore one way in which freshwater animals can get rid of their excess water, provided that the salts lost can be replaced. The fresh-water crayfish *Astacus* secretes urine equal to some 4% of its body weight in 24 h, the osmotic concentration of which is about one-fifth that of its blood. The antennary organ of this animal is seen to be more elaborate in structure than that of *Carcinus*, due to the interpolation of a quite complex canal segment between the labyrinth and the bladder (Fig. 6.3). The most likely explanation for the presence of this canal, which has received some experimental backing, is that it functions as a water-secreting segment.

Salt loss is again compensated for by the active uptake of salts from the medium, even when the external salt concentration is very low, chloride, sodium and potassium all being absorbed by the gills. *Astacus* can extract sodium from water containing as little as 0.02–0.09 mmol/l, and chloride from a medium hardly less dilute, while the common frog can extract chloride from a medium containing only 0.01 mmol/l. Some of the chloride is exchanged for bicarbonate produced through respiration, and some of the sodium for ammonium ions excreted following protein metabolism. Above an external concentration of 1.0 mmol/l the influx of chloride is almost independent of the external concentration, and in general the animal is much more sensitive to internal changes of chloride than to external changes.

Aquatic insect larvae conserve or extract salts not with their gills but with their guts. Either the rectal cuticle or the special anal papillae developed by some larvae, such as the yellow-fever mosquito *Aëdes aegypti*, can absorb salts and the gut wall is seen to be histologically differentiated in these areas by contrast with the non-absorbing areas. In *A. aegypti* the anal papillae are responsible for about 90% of the salt conservation, and the rectal wall accounts for virtually all the rest.

6.6 Terrestrial animals

Life on dry land poses water problems that result from the relative dryness of air and scarcity of water. The dryness of air introduces the risk of desiccation of all stages in the life history of the terrestrial animal, including the egg and the spermatozoon, the watery environment in which the latter swims to reach the ovum having gone.

The lack of a watery medium for fertilization has been overcome by the development of internal fertilization. Amphibians, which have not developed internal fertilization, are obliged to return to the water to breed. In spite of this disadvantage, some amphibians are able to live in unexpectedly dry habitats. The best example is perhaps the African desert toad *Bufo regularis*, which uses temporary pools for breeding, and lays eggs that develop with unusual rapidity.

Resistance to desiccation is achieved by the same basic mechanisms in all terrestrial animals, but these may be exaggerated in the most unfavourable habitats, such as deserts, and may then be accompanied by special adaptations, including behavioural ones. A common behavioural adaptation in deserts is the tendency to burrow into the ground when conditions are unbearably hot or dry. The interstitial spaces of even apparently dry soil are usually saturated with water, and burrowing therefore enables an animal to find a higher humidity than obtains above ground, thereby reducing water loss. Burrowing is the most important adaptation of many arthropods to desert conditions, and some possess specially shaped appendages for this purpose. By this means the desert woodlouse *Hemilepistus* is able to dig downwards for tens of centimetres. Many anuran amphibians also burrow into the sand, as also do many of the smaller mammals such as the kangaroo rat *Dipodomys*. Larger desert mammals, which include camels, donkeys and antelopes, cannot adopt this course but are able to survive by a combination of physiological adaptations, coupled with the advantage gained by a larger bulk in the reduction of the surface area/ volume ratio, and the correspondingly smaller degree of desiccation in relation to body weight. Birds clearly cannot burrow, and in dry conditions most of them cannot move very far from the available surface water.

Another behavioural adaptation of many desert animals is the restriction of activity to those periods when conditions are least damaging. Hibernation and diapause are often adaptations of temperate animals to adverse temperatures, but they are not osmoregulatory. Desert animals, however, may restrict their activity for osmoregulatory purposes, although they do escape the effects of the fiercest heat at the same time. In most desert species which live as adults for the greater part of the year there is a concentration of activity at dawn and dusk in summer; and then, during winter, when the need to gain heat is more important, they are at their most active around noon. Many desert insects are also believed to overweather the unfavourable dry season by egg or pupal diapause.

Those animals which are best able to withstand desiccation are often found to be able to tolerate changes in their blood concentration. For example, the blood of *Helix aspersa* may vary by over 300% in dry weather, and some insect adults and larvae seem to survive even more extreme changes. The larvae of the chironomid *Polypedilum vanderplanki*, which live in small temporary pools in Africa, are able to survive in a partially dehydrated state between the drying up and refilling of the pools. In the laboratory these larvae survived for 39 months after being dried to a point at which their water content was in the region of 8% of the body weight, a reduction of nearly ten times. In this condition they could also withstand extremes of heat. Larger animals are generally unable to tolerate such variations in blood concentration, but some desert mammals are able to survive considerable water loss, especially the camel and the donkey (see below).

Behavioural adaptations and unusual ability to withstand increased blood concentration are individual specializations that have arisen during evolution in addition to a general pattern of regulation. The major adaptations that enable animals to avoid desiccation in terrestrial habitats are a reduction in the permeability of the integument to water, and reductions in water loss due to excretion and respiratory ventilation. In other words, the ability to economize with water and conserve it are paramount among terrestrial osmoregulatory mechanisms.

There is in general a direct relationship between the dryness of an animal's habitat and the permeability of its integument. Among the vertebrates, there also seems to be a progressive increase in the impermeability of the skin from the amphibians upwards. For example, in desert amphibians, reptiles and rodents the total water loss is in the ratio 40:10:1 respectively. The impermeable surface in vertebrates takes the form of a horny layer produced by the epidermal cells, which may be continuously sloughed off or semi-permanent (Fig. 6.4). The outer surface is rendered more impermeable by a thin layer of wax. The integument of insects, and presumably also of arachnids, is rendered impermeable partly by the tanning of the outer part of the chitinous cuticle, the *exocuticle*, but principally again by wax that is made by the hypodermal cells and passed from them through pore canals in the cuticle to form a layer on its surface (Fig. 6.4). The wax layer is remarkably thin, of the order of 0.25 μm in depth, and is usually overlaid by a thin layer of cement, a hard substance which some workers believe to be shellac.

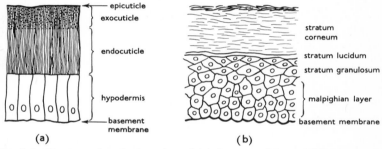

Fig. 6.4   Diagram of vertical section of (a) insect cuticle and (b) human skin. ((a)   Based on Wigglesworth, 1965, *The Principles of Insect Physiology*, Methuen, London)

The respiratory surface of animals must be kept moist for gaseous exchange, and is therefore both a potential water-losing structure and liable to desiccation. The first danger is minimized and the second abolished by the placing of the respiratory organ inside the body, as with the tracheae of insects, the lung-books of some spiders and the lungs of tetrapod vertebrates. Because moisture evaporates from the respiratory surface to saturate the outgoing gases, there is still a significant water loss from this site in tetrapod vertebrates. In man the loss may amount to 350–400 cm³ in a day. The ability to develop a large lung capacity and thus reduce the frequency of ventilation leads to some reduction in water loss through this route. It is a tendency found in a number of tropical mammals, such as the sheep and goats of East Africa and Australia.

Excretory water loss in terrestrial animals can be cut down by the excretion of a nitrogenous product that is solid, or nearly so. Insects, reptiles and birds excrete uric acid, which is very insoluble and may therefore be easily concentrated without toxic effects. The excretory organs of insects, the *malpighian tubules*, can resorb much of the water from the urates from which they form uric acid, and the rectum can absorb further water from the faeces, with the result that the faecal/excretory material is voided in what is virtually a solid form. In birds and reptiles the kidney and cloaca carry out similar functions in relation to the excretory material, and their uric acid is passed out in a semi-solid state. In these groups the loss of water through excretion is very small. For example, about two-thirds of the water lost by insects is lost from the spiracles, and about a third from the articular membranes, the loss through excretion being negligible. Mammals, however, excrete mainly urea, which is soluble and requires a fair amount of water for its elimination (§5.2). In man in cool conditions the water lost through excretion, respiration and sweating (an adaptation peculiar to some mammals) is of the same order in each case. In hot conditions the loss through sweating becomes overwhelmingly important. For this reason, some desert mammals either do not sweat, like *Dipodomys*, or restrict their sweating to cooler periods (see below).

Apart perhaps from desert insects, one of the best-adapted animals to dry conditions is the camel. The camel does *not* store water, either in its stomach or in its hump. The hump represents a unique case of a tendency common in desert mammals, to restrict fat stores to one large depot, thereby allowing the rest of the body to act as a more efficient radiator. It is not a potential source of water, since the extra oxygen needed to oxidize the fat for the production of metabolic water involves a loss of water due to the increased pulmonary ventilation that more or less cancels out the metabolic water gained.

In summer, the camel varies its thermoregulation to a greater extent than in winter, and may exhibit a morning temperature of 34°C and an afternoon maximum of 41°C, a range of 7°C. Since an adult camel weighs in the region of 450 kg, a considerable amount of heat must be absorbed for a gain of 7°C in body temperature, and since the temperature is allowed to drop at night, the heat gained during the day is dissipated. Thus, within limits, the camel warms up during the day and then dissipates the heat so gained during the night. Sweating is restricted, and begins only when the maximum temperature of 41°C is approached, to ensure that this temperature is not exceeded. In many mammals subjected to high temperatures, water loss from the blood makes it so concentrated that explosive heat death (§8.3) results. But in the camel water is withdrawn from the tissues, and the blood volume and concentration remain practically constant. The toleration of tissue water depletion in this animal is some three times greater than that of most other mammals, and unlike these it can eat normally until desiccation becomes very severe indeed. The output of urine is also kept low, and there is a high degree of water resorption from the faeces by the colon. By means of these adaptations the camel can survive at least 17 days in summer on dry food without drinking, and without using its fat to produce metabolic water. A camel kept for this length of time without water may take in one-third of its total body weight in water in 10 min when it is finally allowed to drink. Its ability to do this is actually one further adaptation, for its blood is much diluted at first, to an extent that would cause water intoxication in most animals. The donkey exhibits similar mechanisms to the camel, but to a lesser extent. But it, too, can take up some 20% of its body weight in water in less than 2 min!

The most obvious way a terrestrial animal can gain water is by drinking. Some animals can conserve their water so well that they do not need to drink or, like *Dipodomys* and *Zenaidura*, drink very rarely. Frogs are able to take up water actively through the skin, but this source is normally denied to animals whose skin is impermeable to water through the need to avoid its loss. Some insects, however, can take up water through their cuticle from air that is not even saturated. Insect cuticle does pass water more easily from outside to inside than in the reverse direction, but the main mechanism in at least some insects involves the uptake of water from saturated air sacs that form in the rectum. The most remarkable example is that of the rat flea *Xenopsylla brasiliensis*, which can extract water from air of only 50% relative humidity. Spiders and woodlice seem unable to gain water by this means, but in some way that is an elaboration of the normal drinking process they can draw water away from a moist surface against its capillary attraction.

In the metabolism of foodstuffs, water is produced. The oxidation of 1 g of fat yields 1.07 g of water, compared with 0.6 g from the oxidation of 1 g of carbohydrate. Utilization of protein to provide water is wasteful because it results in nitrogenous wastes that often require some water for their elimination. Fat is therefore preferred. In addition, food always contains some water, and even apparently dry food such as grain may contain 20–30% water by weight. These two sources, metabolic water and food, are sufficient for a number of animals or very nearly so, in the absence of any other water supply. Larvae of the mealworm *Tenebrio molitor* maintain a constant percentage of water in their tissues even when starved in unsaturated air, and it has been shown that under these conditions they convert storage food into metabolic water. At nearly 100% relative humidity so little water is lost that more is produced in metabolism than can be eliminated from the body. A variety of other insects possess similar abilities, including tsetse fly larvae, clothes moths and grain beetles. The kangaroo rat *Dipodomys* is virtually in this situation, for it drinks very little water.

6.7 Return to the sea

We have seen already that the penetration of marine animals into fresh water appears to have resulted in the evolutionary development of a blood concentration lower than that of the sea. The fact that marine teleosts have a blood osmotic concentration only slightly higher than their fresh water relatives has been taken as support for the view, which receives more support from fossil evidence, that teleosts were originally a freshwater group, some members of which have returned secondarily to the sea.

The concentration of mineral salts in the blood of elasmobranchs is similar to that of marine teleosts. Yet they maintain their total blood concentration at a level that is isosmotic with sea-water. This is achieved by retaining the nitrogenous excretory products urea and trimethylamine oxide in the blood instead of excreting them. The skin and gills are relatively impermeable to urea, and the kidney has a specially high threshold for its excretion. Even the young elasmobranch is provided with its store of urea, either directly from the parent as in the viviparous dogfish *Squalus*, or by its being placed in an egg-case impermeable to urea. By the time they hatch or are born, the young animals have developed their own urea-conserving mechanism.

The maintenance of a high level of urea in the blood is clearly an adaptation to marine conditions, and is presumed by most authorities to indicate that the elasmobranchs too were originally a freshwater group, but that all returned to the sea at some stage and were enabled to live there more easily through the development of urea conservation. On this view, freshwater elasmobranchs have returned once more to their habitat of origin. Since they possess a blood urea concentration only slightly below that of their marine relatives, it seems that they have been able to reverse this evolutionary change to only a limited extent (Fig. 6.5), and this provides them with an even greater osmotic problem than that which confronts freshwater teleosts. The presence of this amount of urea in the blood is additional evidence for a return to fresh water from the sea, since it would not have arisen in fresh water.

Some inland waters have a higher salinity than that of the sea, and certain animals have been able to evolve mechanisms that have allowed them to invade such waters. The osmotic problems involved are essentially similar to those facing marine teleosts, and like them the brine shrimp *Artemia* drinks the brine in which it lives and secretes salts into the surrounding water. The gut of this animal is an important osmoregulatory organ, having the ability to take up water against a concentration gradient. The first ten pairs of branchiae bear localized areas through which sodium and chloride are secreted in an outward direction.

A number of reptiles and mammals have returned to the sea, and some birds of the shore may also be considered as marine animals in an osmo-regulatory sense. They are all able to get rid of salt taken in through drinking salt water and eating salty food by secreting concentrated salt solutions, although this does involve the loss of a small but significant quantity of water. The excess salt is voided from special salt glands in

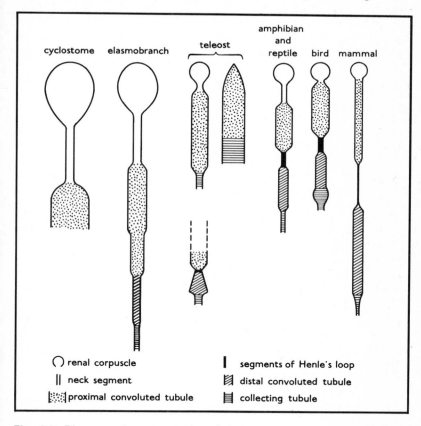

Fig. 6.5 Diagrammatic representation of the structure of the kidney tubules of various vertebrates. (After Marshall, 1934, *Physiol. Rev.*, **14**, 133)

marine birds and reptiles that probably represent modified mucus-secret-ing nasal glands in birds and modified lachrymal glands in reptiles. In the herring gull the secretion is largely sodium chloride at a concentration of about 718 mmol/l, some five times the sodium concentration of the blood. Although the potassium eliminated from this gland is of the order of only 24 mmol/l, this also represents a similar degree of concentration compared with the blood potassium level. The most concentrated nasal gland secretion known is that of Leach's petrel, which may reach a level of 1100 mmol/l sodium.

Marine mammals probably do not drink sea-water to any extent, and seem to rely instead almost entirely on their food for their source of water. Whales are a good example. The contents of their stomachs are almost dry, and the loss of water from their lungs is smaller than in land mammals because they have a higher rate of extraction of oxygen from the air. They do not need to thermoregulate by evaporation of water from their skin, and so avoid loss of water for this purpose. Mammals that eat fish, such as seals, already have some of their osmotic work done for them, since fish have a lower osmotic pressure than the sea.

6.8 Renal organs

Most metazoan groups possess structures that eliminate a watery fluid, usually termed 'urine'. These structures, which include the nephridial organs of flatworms, annelids and Amphioxus, the malpighian tubules of insects, the antennary glands of crustaceans and the kidneys of molluscs and vertebrates are often referred to as 'excretory organs'. They may indeed eliminate nitrogenous wastes in many animals, but this does not appear to have been their initial function, and it seems better to describe them by the neutral term *renal organs*.

In most, perhaps all cases, renal organs seem to have evolved as controllers of the body volume, although as we have seen, osmotic and ionic regulation are invariably linked, and it seems likely that both functions developed together as renal organs were elaborated. Our knowledge of these organs is decidedly scanty in some groups of animals, but in all the cases that have been studied an excess load of body fluid results in an increased flow of urine (*diuresis*). The force responsible for this increased flow is probably always an increase in hydrostatic pressure in the blood or body fluid, although this is inferred rather than confirmed in many animals. The fluid that is thereby forced into the renal organ is initially an ultrafiltrate of the blood or body fluid, identical in composition to them but lacking the larger molecules. The degree of ultrafiltration varies between different animals, and no doubt reflects the structure of the epithelium through which the fluid passes, and particularly its basement membrane. In some cases, a proportion of some of the smaller proteins may be present. In the open nephridia of earthworms and leeches, the nephridial fluid initially has the same composition as the coelomic fluid.

Since the ultrafiltrate contains many useful solutes, and since all animals undertake some degree of ionic regulation, the ultrafiltrate is subject to alterations in composition as it passes through the rest of the renal organ.

In general, these changes result from the absorption of salts and other solutes, either passively or actively, the osmotic absorption of water, and the secretion of unwanted substances into the lumen of the organ. The details are often imperfectly known or understood, but the mammalian

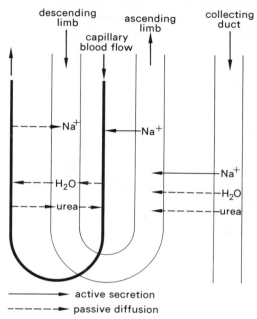

Fig. 6.6   Countercurrent exchange of Na^+, H_2O and urea in the mammalian kidney (inner medulla only). The force for the whole system is the active secretion of Na^+ from the ascending limb into the adjacent capillary bed. The net result is *dilution* of the urinary fluid. As a result, passive countercurrent exchange of Na^+ and H_2O occurs in the descending limb, resulting in *concentration* of the fluid. Further active secretion of Na^+ out of the collecting duct produces a diminution of volume and some transfer of water and urea into the ascending limb via the adjacent interstitial tissue. Urea is actively secreted early into the proximal tubule, but its subsequent movements and those of water, appear to be governed by the concentration changes brought about by the active secretion of sodium. The resorbed water passes into lymphatic vessels. (After Riegel, 1972, *The Comparative Physiology of Renal Excretion*, Oliver & Boyd, Edinburgh)

kidney has been more investigated than most, and a brief account of its working, as far as this is known, will serve to make clear the general principles on which all such organs must depend.

The vertebrate kidney is composed of many tubular structures known as *nephrons*. The morphology of nephrons varies, as shown in Fig. 6.5, but basic to all groups are the glomerulus, which produces the ultrafiltrate, and the proximal and distal convoluted tubules. Cyclostomes, however, lack the distal convoluted tubule. Sodium, potassium and chloride ions are resorbed by both convoluted tubules, and the proximal tubule also resorbs glucose and, where appropriate, water. In addition, the collecting duct that joins up the tubules and ultimately allows the kidney to be drained, can resorb sodium and chloride. A wide variety of organic substances, including urea in mammals may, be secreted into the nephron at a number of places.

In some fishes, and in all the higher vertebrate groups, an intermediate segment is interpolated between the convoluted tubules, which has become elaborated into a structure known as the *loop of Henle* in the truly terrestrial groups of reptiles, birds and mammals. The function of this additional segment is essentially the resorption of further water, but this is achieved by the active secretion of sodium ions in the *ascending* limb of the loop of Henle. The capillary bed in this region is so arranged that the blood flow is in the counter-current direction to the flow down the descending limb and the net result is a cycling of sodium from the ascending to the descending limb, but in the process water is withdrawn into the capillary bed, ultimately to leave the kidney via its lymphatic duct (Fig. 6.6). The only active work done by the kidney in producing the urine appears to be that needed for the secretion of sodium, as far as alterations in concentration are concerned. The osmotic changes and changes in concentration gradients produced by the active secretion of sodium then cause the movement of the other ions and of water. Urea and organic substances must, however, be actively transported into the tubule, and glucose is actively transported out of it.

The vertebrate kidney arose primarily to control the body fluid volume and the concentrations of certain ions. In fishes it may function as an excretory organ to only a very limited extent, since most of the excretory ammonia is liberated through the gills. When, however, the vertebrates emerged onto land, the type of excretory substance produced itself assumed osmoregulatory significance, and the kidney was the obvious organ to take on a major excretory role.

It may be noted that a typical, if primitive kidney is found in the hagfishes, a marine group that possesses a pattern of ionic and osmotic regulation similar to that of marine invertebrates and tunicates. Thus, although the major groups of fishes may be considered, on osmotic grounds, to have evolved in fresh water, it is no longer generally supposed that the vertebrate kidney developed in a fresh water protovertebrate as an osmoregulatory organ. It is probable that it evolved as an ion-regulating organ in a marine protovertebrate, and became important as an osmoregulatory organ when the early vertebrates penetrated into fresh water.

7

Hydrogen Ion Concentration (pH)

7.1 The nature of pH

It was explained in §2.8 that the removal or addition of electrons from or to atoms turned them into charged particles, or *ions*. Some substances, such as sucrose, do not form ions; others, such as hydrochloric acid, do:

$$HCl \rightleftharpoons H^+ + Cl^-$$

Ionization of hydrochloric acid is possible because an electron is removed from the hydrogen atom, giving it a positive charge, and donated to the chloride atom, giving it a negative charge. Some substances, like sodium chloride, are already ionized in the solid state, but in most cases ionization takes place when the substance goes into solution in water. Many biologically important compounds ionize, including water itself, the ionization of which is usually represented as

$$H_2O \rightleftharpoons H^+ + OH^-$$

equal quantities of hydrogen and hydroxyl ions being formed. In fact, water aggregates into larger molecules and as a result forms *hydronium* ions H_3O^+ as well as hydrogen ions, but the extent of this aggregation varies with temperature, and it is sufficient for our purposes to represent the ionization of water in the simple form given above.

Since ions are charged particles, their presence affects the electrical conductivity of water, and this fact can be used to measure the extent of ionization (or *dissociation*, a term used to describe the separation of the atoms as ions). In this way it can be shown that the hydrogen ion concentration of pure water at 22°C is 10^{-7} g/l. Put another way, one molecule in every 550 million molecules of water is dissociated into hydrogen and hydroxyl ions at this temperature. We can express this fact in the form

$$cH = \frac{1}{10 \text{ million}} \; N = 10^{-7} \text{ g/l}$$

where cH is the hydrogen ion concentration, and N refers to its normality. It is a simple matter to determine the cH of a substance such as hydrochloric acid, which dissociates almost completely in solution, if its normality is known. Thus, a solution of N/1000 HCl will possess a cH of 10^{-3}.

The ionization of water could equally well be described in terms of its hydroxyl ion concentration, cOH. Since the ionization of water results in equal concentrations of hydrogen and hydroxyl ions, cOH must also have a value of 10^{-7} g/l. In practice, hydrogen ion concentration is preferred, but since it is cumbersome to deal in negative powers of ten, the negative logarithm of this value is used instead, and this is known as the pH of the solution. In other words,

$$pH = -\log_{10} cH = \log_{10} \frac{1}{cH} \qquad (7.1)$$

The N/1000 HCl referred to above possessed a cH of 10^{-3}, and therefore

$$pH = -\log_{10} \cdot 10^{-3} = 3$$

Again, it would be possible to think of the dissociation of water in terms of pOH as well as pH; and clearly for water

$$pH + pOH = 7 + 7 = 14$$

The pH scale thus runs from 1 to 14, neutrality being at pH 7.0. Numbers *lower* than 7.0 denote an excess of hydrogen ions and an *acid* solution; numbers *higher* than 7.0 denote an excess of hydroxyl ions and an *alkaline* solution. It should be realized that the pH scale is a logarithmic one, and this means that a difference in pH of 1.0 represents a tenfold shift in hydrogen ion concentration. For example, a solution with a pH of 3.0 has a hydrogen ion concentration which is ten times greater than that of a solution of pH 4.0.

7.2 Effects of pH on animals

It is found that living cells and organisms can withstand large variations in the pH of the external medium. For example, eggs of the sea urchin *Arbacia* can develop in a medium of pH range 5.2 to 9.4, and the flagellate protozoan *Euglena* can live in media ranging in pH from 2.3 to 11.0. The crustacean *Chydorus* can live in a pH of 3.2 to 10.6, and isolated vertebrate and crustacean hearts and other tissues can tolerate a similarly wide pH range.

The considerable tolerance to pH exhibited by animals does not mean that cellular processes are not affected by the pH at which they work. It is found that the body fluids of animals, and as far as is known the intracellular fluid, tend to maintain a pH in the region of neutrality, and a figure outside the range pH 5–8 is rare. The gastric juice of mammals is quite exceptional in having a pH of about 0.9. Cell membranes are very impermeable to hydrogen ions because the positive charge on their external surface repels the positively charged ions. But the chief cause of the maintenance of body fluids in the region of neutrality, and of the ability of cells and organisms to tolerate changes in the external pH, is the possession by them of chemical *buffer* systems, which resist changes in pH when acid or alkali is added to them.

7.3 Buffer systems

The mechanisms by which animals maintain a constant pH are varied in detail but similar in principle, and are best introduced by a general example. Suppose we were to take three aqueous solutions, one of them pure water, the second a solution of sodium chloride, and the third a mixture of sodium hydroxide and potassium dihydrogen phosphate. If we were to add to 100 ml. of each of these 1 cm^3 of N/100 HCl, we should find that the first two of our three solutions were now markedly acid, but that our third solution would still be at pH 7.0. If we continued to add acid to the third solution we should find that it would continue to resist a change in its pH until an appreciable amount of acid had been added (Fig. 7.1). This solution clearly constitutes a buffer, and its *titration curve* in Fig. 7.1 is typical of other buffers.

Substances that dissociate into ions almost completely in solution are termed *strong*, and those that dissociate only slightly are termed *weak*. A buffer system is always composed of either a weak acid mixed with the salt it forms with a strong base, or a weak base mixed with the salt it forms with a strong acid. Biological buffers and those used in experimental work are mostly of the first type. In the example given above, the strong base is sodium hydroxide, and the potassium dihydrogen phosphate acts as a weak acid (see §7.4). Let us examine this kind of buffer system in general terms, denoting the weak acid as HA and its salt with a strong base as BA. If we add alkali to a mixture of these two in solution, the base will not be affected and the alkali will react with the acid:

$$HA + NaOH \rightleftharpoons NaA + H_2O$$

As we have seen, the ionization of water is very slight, and so the added OH$^-$ ions have been effectively removed from solution. If acid is added to the mixture its acid component is not affected, and it reacts with the base:

$$HCl + BA \rightleftharpoons HA + BCl$$

and because HA is a weak acid it ionizes only very slightly and has no significant effect on the pH of the system. Its formation means that the H^+ ions have been virtually removed from solution. Thus, a buffer system of this type has a kind of see-saw action, water being produced when alkali

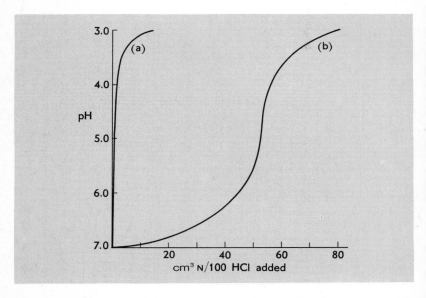

Fig. 7.1 The change in pH in (a) 100 cm³ each of distilled water and N/100 sodium chloride; and (b) 100 cm³ M/100 KH₂PO₄ brought to an initial pH of 7.0 by addition of NaOH, on progressive titration with N/100 HCl.

is added, and salt being produced when acid is added. The limit of the ability of the system to act as a buffer will obviously depend on both the absolute quantity of the buffering substances present, and on the ratio of one to the other, considerations to which we shall return in §7.4.

It is easy to accept that the production of water has no effect on the total ions in solution, in view of the figures given at the beginning of this chapter. It is perhaps less obvious that the production of weak acid also has no effect, and so it helps to consider a definite example of a weak acid. *Acetic acid* is sometimes used as a component of artificial buffer systems, and in solution it ionizes thus:

$$CH_3COOH \rightleftharpoons H^+ + CH_3COO^-$$

A $1M$ solution of acetic acid (its gram-molecular weight in 1 l of water) can be shown to dissociate at $25°C$ to the extent of only 0.4%. The pH of this solution can be calculated in two ways, by allowing for this degree of dissociation or by neglecting it. The calculated pH in both cases is 2.38, and a difference in answers is found only if further decimal places are considered, which is clearly pointless.

7.4 Natural buffer systems

Three major buffer systems are found in animals, and these involve phosphate, bicarbonate, or amino acids. A certain amount of all three is normally present in body and cellular fluids, and so the maintenance of a constant pH depends on a complex system. However, one of them is usually predominant in its buffering effects, and this is most often bicarbonate. Insects have an abnormally high free amino-acid level in their haemolymph, and some at least of them seem to use phosphate as their main additional buffering system. Since the phosphate system is most obviously like the generalized scheme given above, we shall deal with it first.

A solution of sodium dihydrogen phosphate ionizes in solution in the following manner:

$$NaH_2PO_4 \rightleftharpoons Na^+ + H_2PO_4^- \rightleftharpoons H^+ + HPO_4^{2-} \rightleftharpoons H^+ + PO_4^{3-}$$

The final ionization in this series is only half complete at a pH of nearly 13.0, and may be ignored from a physiological viewpoint. The predominant phosphate ions in solution at a physiological pH will therefore be $H_2PO_4^-$ and HPO_4^{2-}. Phosphate buffers are, in fact, usually made up as a mixture of the two salts, sodium dihydrogen phosphate NaH_2PO_4 and disodium hydrogen phosphate Na_2HPO_4. It will be evident from the scheme of ionization given above that the second of these two salts acts as a weak acid, since it produces some hydrogen ions in solution, whereas the first salt represents the salt of phosphoric acid H_3PO_4 with the strong base NaOH. The mixture of the two salts will behave like the generalized buffer described in §7.2. Thus, if hydroxyl ions are added, water is formed:

$$NaH_2PO_4 + NaOH \rightarrow Na_2HPO_4 + H_2O$$

The addition of acid results in the formation of the basic salt:

$$Na_2HPO_4 + HCl \rightarrow NaCl + NaH_2PO_4$$

Acids that ionize to produce more than one kind of ion, like the phosphoric acid H_3PO_4 from which these two salts are derived, are known as *polybasic* acids. Many common organic acids are polybasic, including the majority of those involved in metabolism. This is why substances concerned in metabolism are often written as ions or as salts. It is perhaps simpler to consider them as undissociated acids for the purpose of following their metabolic functions, and this is the procedure adopted in this book. But it should always be remembered that this is a convenient simplification.

The bicarbonate ion HCO_3^- can take up hydrogen ions from water to form carbonic acid:

$$HCO_3^- + H^+ \rightleftharpoons H_2CO_3$$

and although little carbonic acid is formed under normal conditions, the addition of alkali to this solution will cause it to react with the carbonic acid, remove it through the formation of water, and so facilitate the production of more carbonic acid:

$$H_2CO_3 + OH^- \rightarrow H_2O + HCO_3^-$$

and the net effect is that the added hydroxyl ions have been removed from solution, with the formation of more bicarbonate ion. If acid is added, it will react with the bicarbonate ion to produce carbon dioxide and water:

$$H^+ + HCO_3 \rightarrow H_2O + CO_2 \uparrow$$

Some of the carbon dioxide will go into solution and form carbonic acid, but since this is a weak acid its effect is negligible. Thus, bicarbonate can act as a buffer because it forms a weak acid in solution, and its action conforms to the generalized scheme previously described.

Amino acids contain in their molecule both a basic —NH_2 group and an acidic —COOH group (§2.13). They can be visualized as existing in the form of a neutral *zwitterion* in which a hydrogen atom can pass between the —NH_2 and —COOH groups, for example in the case of *glycine* the molecule can be pictured thus:

$$
\begin{array}{c}
NH_3^+ \\
| \\
H-C-COO^- \\
| \\
H
\end{array}
$$

By the addition or subtraction of a hydrogen ion to or from the zwitterion, either the cation form or the anion form will be produced:

$$\underset{\text{cation form}}{H_3N^+ - CH_2 - COOH} \underset{}{\overset{+H^+}{\rightleftharpoons}} \underset{\text{zwitterion}}{H_3N^+ - CH_2 - COO^-} \underset{-H^+}{\rightleftharpoons}$$

$$\underset{\text{anion form}}{H_2N - CH_2 - COO^-}$$

Thus, when OH^- ions are added to the solution of amino acid they take

H^+ ions from it to form water, and the anion is produced. If H^+ ions are added, they are taken up by the zwitterion to produce the cation form. In practice, if NaOH was added, the salt $H_2N—CH_2—COONa$ would be formed, and the addition of HCl would result in the formation of the hydrochloride $Cl—H_3N—CH_2—COOH$, but these substances would ionize in solution to some extent, to form their corresponding ions.

Amino acids differ in the degree to which they will produce the cation or anion form. In other words, a solution of an amino acid is not neutral, but is either predominantly basic or acidic, depending on which form is present in greater quantity. For this reason, different amino acids may be used as buffers for different pH values, and a mixture of them possesses a wide buffer range.

7.5 Dissociation constants

It will be apparent that the pH of a buffer system depends on the balance between its acid and alkaline components, and that the effect exerted by these is governed by the extent to which they will ionize in solution. We can express the hydrogen ion concentration in such a system in the form of the equation

$$cH = K_a . \frac{acid}{salt}$$

where cH has the meaning ascribed to it previously. The term K_a represents the dissociation constant of the acid from which the hydrogen ions are derived. The equation can also be derived more rigorously from the Law of Mass Action, and the student familiar with this Law should attempt the exercise.

The above equation is expressed in terms of the hydrogen ion concentration, cH. If we now convert it into an expression for pH, using equation 7.1, it becomes

$$pH = \log K_a + \log \frac{salt}{acid}$$

and if, by analogy with the term pH, we denote $\log K_a$ by pK_a this further becomes

$$pH = pK_a + \log \frac{salt}{acid} \tag{7.2}$$

Equation 7.2 is known as the *Henderson–Hasselbach* equation. It enables the pH of a buffer solution to be calculated for any ratio of salt to acid, provided that the pK_a is known. From equation 7.2 it will be seen that when the ratio of salt to acid is unity, $pH = pK_a$, and this fact allows the pK_a for a system to be determined, by making up a mixture of equimolar concentrations of salt and acid and measuring its pH.

The Henderson–Hasselbalch equation applies to dilute solutions of weak acids. It does not apply quite so closely to strong acids, nor to situations in which the proportion of acid or salt becomes grossly different. Fortunately, the conditions which it does describe are those generally found in biological situations. Let us take as an example of its application the situation in human blood. The chief buffer system in human blood appears to be bicarbonate, and we can check this quite simply by determining the ratio of HCO_3^- to H_2CO_3 in human blood, which is found to be 20:1, and finding the pK for H_2CO_3, which is 6.1. The pH of the blood should then be given by

$$pH = 6.1 + \log 20 = 6.1 + 1.3 = 7.4$$

8

Temperature Relations

8.1 Temperature and survival

Living organisms do not utilize the whole of the available range of environmental temperatures. Some primitive plants can withstand temperatures as high as 90°C, but animals are less hardy than this and even hot-spring protozoans cannot tolerate temperatures much in excess of 50°C. As shown in Fig. 8.1, most animals live at temperatures between 0°C and 50°C, although relatively few of them actually live in habitats with such extremes. Some animals can live at below freezing point, as in Arctic conditions, but we shall find that the ability to live at extremes of cold or heat is the result of resistance to such temperatures rather than tolerance of them by the body. Homeostatic mechanisms have been evolved to prevent the body temperature from reaching extreme environmental temperatures or, if these mechanisms are inadequate or impossible, there may be behavioural adaptations designed to avoid exposure to them.

Although animal life may exploit the full range from 0°–50°C, most species normally live and survive over a very much narrower range of temperature within this wider one. A limited degree of adjustment or *acclimatization** demonstrates that the ability to live in a particular range of environmental temperatures is a result of genetic selection.

An aquatic environment poses fewer problems than a terrestrial one, given a minimum quantity and depth of water. Water has a relatively low heat conductivity and a high specific heat, and therefore takes a long time both to heat up and cool down. Because of this, large and rapid

* The student may find this word restricted in the literature to natural adjustments in the field, and the word *acclimation* applied to adjustment in laboratory experiments.

fluctuations in the temperature of the air outside the water are transformed into smaller and slower heat changes within it. Air is quickly
warmed and cooled, and consequently it is among terrestrial animals

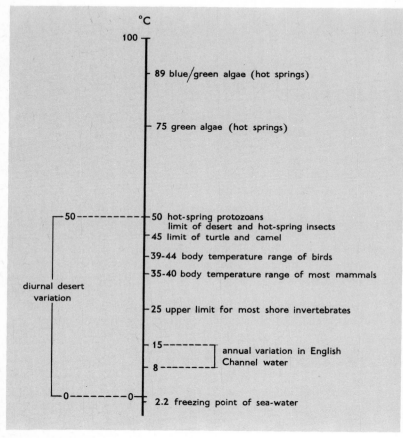

Fig. 8.1 Temperature ranges of animals and other important temperatures.

that the most far-reaching adaptations to temperature extremes have been
evolved.

8.2 Heat balance in animals

Heat can be gained by animals in two ways: by the metabolism of foodstuffs, and by the direct absorption of solar radiation. Loss of heat occurs
mainly through *convection* and *evaporation*. Liquids and gases do not

conduct heat very well, but they do exhibit convection, which is the term used to describe the familiar fact that warm liquids or gases rise and cold ones sink. Convection therefore causes heat, in effect, to flow and it is this movement that carries heat from the interior of an animal's body to its exterior surface and thence away into the surrounding air. *Radiation*, the carrying-away of heat in wave-form, is probably less important than convection and evaporation in most cases, but it is undoubtedly a significant factor. *Conduction* of heat is more important between solids, and comes into play when an animal is in contact with the substrate, when heat may move either way according to whether the animal or the substrate is hotter.

Evaporation is a major cause of heat loss in the majority of terrestrial animals. Unless the surrounding air is fully saturated, such animals must lose water by evaporation and the energy lost results in a cooling of the body surface. Evaporation is clearly greatest when the surface of the animal is moist, but even apparently hard-bodied and impermeable animals like insects suffer an appreciable water loss from their cuticle.

8.3 Poikilotherms

The majority of animals are *poikilotherms*, a term used to denote animals in which heat loss and heat gain tend to cancel one another out and the body temperature approaches that of the environment, although it may not be precisely equal to it. There may be areas of high metabolism in a poikilotherm that are at a slightly higher temperature than the rest of the animal, even in aquatic organisms. An example would be the myotomes of a rapidly swimming fish.

It is usual for moist-bodied terrestrial poikilotherms to exhibit a slightly lower temperature than that of the surrounding air, due to evaporation of water from the body surface. However, such deviations from the environmental temperature are small, and a poikilotherm is virtually at the mercy of its environment unless it can mitigate the effects of the external temperature or move to a more suitable one. Many fishes, such as the catfish and the herring, possess special heat receptors that enable them to seek out favourable temperatures. Desert animals tend to be active only at those times of the day when the temperature is reasonable. The horned toad *Phrynosoma* burrows into the ground at temperatures below 20°C or above 40°C, and desert iguanas seek shade when their body temperature approaches about 43°C. Desert grasshoppers rise up on their legs when the sand becomes very hot, in order to avoid overheating through conduction; and insects generally will not remain on sand where its temperature approaches 50°C, but instead fly or burrow.

A number of poikilotherms increase their body temperature by various methods. Many snakes and lizards bask in the sun and increase their body temperature by absorbing solar radiation and by conduction from the substrate. They may achieve a rise in body temperature of 20°C or more by this process. Other animals may use metabolic heat, generated quickly by means of enforced muscular activity. By fluttering the wings, some flying insects can raise the temperature of their flight muscles to a level at which they are efficient enough for take-off to be possible. The generation of metabolic heat in this way can be considerable; for example, bees can increase the temperature of their flight muscles by over 12°C.

Social methods of overcoming the disadvantages of poikilothermy are also found. Larvae of the butterfly *Vanessa* cluster together in the cold, and the group often has a temperature that is 1.5–2.0°C higher than the air temperature. Ants site their nests under stones or earth that are directly in the path of the sun's rays, and open or close their nests according to the temperature of the surrounding air. Workers of the social wasp *Polites* fan the comb with their wings when it is hot, and in extreme heat they apply water to the combs to induce them to cool through its evaporation. Some pythons coil themselves around their eggs, and thereby keep them at a temperature some 2°C above that of the environment.

8.4 Heterotherms and homeotherms

In contrast to poikilotherms, homeotherms maintain their body temperature at a more or less constant level, irrespective of the environmental temperature. The ability to become, within limits, independent of the environment in this way has obvious survival value, a fact which is plain when poikilotherms and homeotherms are observed side by side in the same cold conditions. The heat required to maintain the body temperature of homeotherms is derived from metabolism, and these animals therefore metabolize at a much higher rate than poikilotherms. In general, homeotherms metabolize some five to six times more intensely than any poikilotherm kept at the same body temperature. Although temperature is not the only factor involved, an increase in temperature does speed up most body processes, and it is found that these proceed more quickly in homeotherms. For example, the nerves of a homeotherm conduct faster than those of a poikilotherm, its muscles contract faster, and so on. Of course, in hot conditions the homeotherm will produce more heat than it needs, and must then get rid of the excess, but this seems to be a relatively slight disadvantage.

Homeotherms usually maintain a body temperature of about 37°C. Some mammals have a lower body temperature than this; most birds a higher one. This temperature is kept very constant in the majority of cases, the only variation in normal circumstances being a diurnal one which does not usually amount to more than about 2°C. There are slight differences between various parts of the body, and in particular the outside of a homeotherm is slightly cooler than the centre. For this reason, a definite standard, the colonic temperature, is used for critical work.

Young homeotherms are virtually poikilothermic, and the ability to regulate body temperature develops slowly after birth. A new-born rat has a body temperature less than 2°C above that of the environment, when the temperature of the latter is between 24 and 37°C, and the difference drops to less than 1°C at external temperatures below 20°C. As it matures, the regulation of the young rat improves, but it does not approach the adult condition until 20 days after birth.

Primitive mammals such as the spiny ant-eater *Echidna* and the duck-billed platypus *Ornithorhynchus* possess poorer powers of temperature regulation than birds and higher mammals, and for this reason they are sometimes regarded as intermediate between poikilotherms and homeo-therms, and termed *heterotherms*. *Echidna* can maintain a body tempera-ture of 34°C in an environmental temperature of 35°C, but as the external temperature is lowered its body temperature tends to drop, so that at an external temperature of 5°C its internal temperature is only 26°C. The platypus is similar, and some other relatively primitive mammals, such as opossums and sloths, are little better. Some homeotherms exhibit a greater than average diurnal fluctuation in body temperature. Bats cool down to air temperature during the day, presumably for the sake of metabolic economy, and since they must be at an internal temperature of over 30°C before their flight muscles can function properly, they adopt the pro-cedure of some poikilotherms of shivering and stretching their muscles in order to warm up. Some small birds like sparrows have a diurnal fluctuation of 37–43°C, which is correlated with the degree of activity they undertake by day compared with that at night.

8.5 Homeothermic mechanisms

The skin of reptiles is heat-absorbing, enabling them to obtain heat from solar radiation, but homeotherms use their skin for the dissipation of heat, chiefly by convection and evaporation. The heat balance of homeotherms is thus between the heat loss due to these physical factors and the metabolic heat production. The maintenance of a constant tem-perature likewise depends on two sorts of mechanism, *physical* and *chemical*.

Loss of heat due to convection and radiation can be reduced by lowering the temperature of the body surface so that there is as small a heat gradi-ent as possible between it and the surrounding air. The conduction of heat from the warm body to the colder exterior surface, which is much greater in homeotherms on account of the heat-dissipating function of their skin, may be reduced by the presence of an insulating layer between the body surface and the rest of the tissues. Both methods of lessening the heat loss are present in homeotherms, through control of the blood circulating in the skin capillaries and the existence of insulating layers of fat beneath the skin and fur or feathers on its surface.

In the skin of mammals there are a number of temperature receptors which probably take the form of free nerve endings. When the skin becomes cooler, the cold receptors among these increase their discharge of impulses into the central nervous system in proportion to the drop in the external temperature. As a result, reflex constriction of the skin capillaries occurs, and blood is thereby withdrawn from the skin to the interior of the animal. The inside of the animal becomes a little warmer in consequence, and this causes the stimulation of further temperature receptors in the brain, which act in such a way that the local skin reflexes are reinforced. The fur or hair is also erected through the reflex stimulation of their individual *pilomotor* muscles, leading to an increase in the amount of air trapped between the hairs or feathers, and hence in the depth of the insulating layer surrounding the skin. Further cooling brings other central reflexes into play, resulting in postural changes such as curling up of the body, and in shivering as a means of greater heat production by the muscles.

Heating of the skin produces the reverse effects. The skin blood vessels open up, and the skin becomes red and flushed in consequence, thereby permitting the maximum loss of heat to the air. The heat itself reinforces this reflex activity through its relaxing effect on the capillaries. Central reflex activity results in an increase in the volume of blood circulating through the skin, by means of withdrawal of tissue fluid from the internal organs, and this allows the skin to employ its maximum heat-losing capacity, when coupled with an increase in the circulation rate. The fur is lowered to reduce the size of the insulating layer of air.

The control of the evaporation of water from the body surface is an important method of physical regulation of body temperature. Nearly 2·5 kJ (0·6 kcal) is used in the evaporation of 1 cm^3 of water, showing that if the water can be easily replaced, evaporation represents a useful method of losing heat. Man is perhaps a special case since he lacks fur, but it is interesting to note that in a really hot temperature he can produce over 1 l of sweat an hour, equivalent to a heat loss of nearly 2500 kJ (600 kcal) in that time. Evaporation is used to aid in heat loss by all the higher mammals, although few are as efficient in controlling it as man. Birds, marsupials, rodents and lagomorphs lack sweat glands and therefore must use other methods of evaporating water. Rodents and many marsupials salivate copiously and lick their fur, and birds pant and thereby lose water vapour from their air sacs. Loss of water by panting is also resorted to by other mammals, such as dogs and cattle.

As in the case of poikilotherms, homeotherms may exhibit behavioural adaptations to cold and heat, and this is especially true of their young during the period when they are unable to regulate. The building of nests with suitable insulating materials is found in a number of mammals and most birds, and the young huddle together inside them. Many birds and mammals seek the shade, or burrow into the ground in both hot and cold weather, or wallow in shallow water, or take dust baths to lose heat by conduction. The seasonal migration of birds and fishes may also be regarded as a behavioural mechanism for avoiding unfavourable temperatures.

Insulation of the body surface by fat, fur or feathers varies according to habitat and season. Permanent residents in cold habitats have thicker coats of fur than related species from warmer habitats, and winter coats of animals subjected to seasonal changes in temperature are thicker than summer coats. The differences may be considerable, and it has been estimated that on average the insulation of Arctic animals is about nine times greater than in comparable tropical species. Dermal fat is commonly restricted to one area of the body in desert mammals, the most famous example being the camel's hump. This permits the rest of the body surface to act more effectively as a radiator for heat dissipation during the cool parts of the daily cycle. In cold habitats the dermal fat is usually increased, the most striking examples being whales and seals. Seal blubber is so effective in insulating the skin from the rest of the body that the skin temperature is only slightly higher than that of the surrounding water. Sometimes the extremities of animals cannot be insulated in this way and become very cold, as in the feet of birds or the flippers of seals. In such cases the arteries and veins of these extremities often intertwine in a *rete*, so that heat transfer from artery to vein is rapid and heat loss is minimal, but the extremity still receives an adequate supply of oxygen.

By such means Arctic animals like the husky and the lemming are actually able to maintain their metabolism at a lower level in winter than in summer, and others exhibit no seasonal variation in metabolism. These results are achieved by physical means, and it is generally true that temperature regulation depends primarily on physical rather than chemical mechanisms in all homeotherms. This is to be expected, because the conditions in which chemical regulation would be used would be those in which food was not likely to be plentiful. Vital metabolic resources could not be spared for temperature regulation in such circumstances. Chemical regulation is thus a short-term supplement to physical regulation. It is probable that heterotherms have relatively poorly developed physical mechanisms for regulation, and since they can only expend a limited amount of metabolic energy for this purpose, they are forced to be poor regulators.

When it becomes necessary for physical methods to be supplemented by chemical ones, it is found that the major sites of increased heat production are the muscles and the liver. Muscle metabolism rises through asynchronous twitches of groups of fibres, and if this is insufficient shivering develops. The adrenal cortex and thyroid gland increase their activity and the secretion of adrenaline by the adrenal medulla rises. As a result, there is a general increase in metabolism (see Chapter 13), but if this is insufficient to bolster up the physical methods the animal sinks into a coma, and eventually it stops breathing and dies.

8.6 Central control of homeothermy

By appropriate lesions of the brain, it is a relatively easy matter to show that the co-ordination of temperature regulation is carried out by the hypothalamus in the brain. When the hypothalamus is subjected to highly localized electrical stimulation, it is found that there are a number of areas within it which control different parts of the regulatory system. These consist of a *hot* centre and three *cold* centres. When the hot centre is stimulated, it results in the dilation of the skin vessels of the animal, sweating and panting. The cold centres are concerned respectively with (i) constriction of the skin vessels, blood volume changes and the cessation of sweating, (ii) muscular contraction and shivering, and (iii) metabolic changes. Hormonal effects are presumably mediated through the activity of the neighbouring pituitary gland. The efferent pathways for the nervous effects will be through the motor neurones of the spinal cord in the case of shivering, and through the autonomic nervous system in the case of sweat glands and blood vessels.

A clue to the evolution of this elaborate control system is to be found, as might be expected, in amphibians and reptiles. In turtles and frogs there is a direct relationship between arterial pressure and the temperature of the blood. The co-ordinating centre for this relationship is the hypothalamus, and it is not improbable that the development of this part of the brain as the thermoregulatory centre took place as the result of this relationship, coupled with its proximity to and influence upon the pituitary gland (§13.1) and its undoubted association with the phenomenon of sleep (see below). The other major change required was a switch from the heat-absorbing skin of reptiles to the heat-dissipating skin of homeotherms. A heat-absorbing skin needs a good blood supply to carry the heat gained away into the interior of the animal, and presumably some vasomotor control to take maximum advantage of the prevailing situation. Hence, the skin of reptiles possesses properties that must have facilitated its conversion to a heat-dissipating function.

8.7 Cold death and cold resistance

The exposure of animals to temperatures outside their normal environmental range leads to their death if it is continued for a certain minimum time. Since the temperature range over which one species of animal can survive may be very different from that of another species, the lethal temperature for one animal may be one that can be tolerated by another animal without distress. For this reason, the concept of temperature extremes, and the terms *heat death* and *cold death* are relative, and may refer to very different temperatures, depending on the species under consideration.

Most animals are killed by external temperatures even a few degrees below freezing point, but some can withstand freezing and subsequent thawing. It has been suggested that the mechanical disruption of cells by the formation of ice crystals is often responsible for death at temperatures below freezing. For example, when hibernating moth prepupae were put into a temperature of $-10°C$ to $-20°C$, ice crystals were seen to develop inside their bodies. Most of the prepupae recovered when they were allowed to thaw, but in those cases in which ice crystals had been formed in the heart mechanical disruption of the heart tissues occurred on thawing, and the affected prepupae died.

This cannot be the only explanation for death at temperatures below freezing point, since many frozen animals die when there is no disruption of tissues by ice crystals. The slowing of metabolism by the cold may be of some importance in conjunction with other factors, but it is probably not itself a critical factor except possibly in homeotherms. Perhaps the most important effect produced by sub-zero temperatures is upon the intracellular salt concentration. When a saline freezes, water crystallizes out and the salt left in solution becomes more concentrated because less water is available as a solute for it. Since the freezing of an animal is usually from the outside in, the extracellular fluid tends to freeze first, with the result that the salts in it become more concentrated and this causes water to be drawn osmotically out of the cells. Although the intracellular fluid will now freeze at a lower temperature than before, if the external temperature continues to fall this fluid will eventually freeze and a point will be reached at which the salt concentration within the cell becomes too great and the vital processes are disrupted.

If freezing is sufficiently rapid, the formation of ice crystals in an aqueous fluid may be delayed, a phenomenon known as *supercooling*. Supercooled solutions either contain minute crystals or are amorphous and highly viscous, but in either form they seem to be less injurious than ice crystals. Supercooling enables some frost-hardy insects to survive sub-zero temperatures, as in the case of larvae of the sawfly *Cephus cinctus* and its parasite *Bracon cephi*. The blood of *Cephus* can be supercooled to a temperature of $-33°C$ before ice forms, and in *Bracon* this does not occur until a temperature of $-47°C$ is reached. *Bracon* and some other insects are aided in their resistance to cold by the presence of quite large concentrations of glycerol within their haemolymph, which acts like an anti-freeze by lowering the freezing point. Some deep-sea fishes are also able to supercool to a much more limited extent, and some of them also contain an unidentified solute that helps to lower the freezing point. Supercooling in land animals is presumably a mechanism for resistance to rapid falls in temperature, and it is not observed in species that are subjected to sub-zero temperatures for long periods.

Aquatic molluscs such as *Mytilus* and *Littorina saxatilis* do not super-cool, yet they can survive freezing at or below $-22°C$ in Arctic regions for many months. It is not known how this resistance to extreme cold has been achieved, but it is assumed that these animals must be able to tolerate to an unusual degree osmotic dehydration of the cells and the development of a high salinity within them, and the mechanical effects of ice.

The complex and probably varied causes of cold death are reflected in the fact that death may occur in some animals at temperatures above zero. This is perhaps not unexpected in animals like birds and mammals, which maintain their body temperature at a constant level which is considerable in relation to their freezing point, but in fact the same situation can occur in some arthropods and fishes, and may well be common to all poikilotherms adapted for life in exclusively warm environments. Little is known about the causes of cold death at temperatures above zero, except that in birds and mammals there is a breakdown in the mechanism for maintaining the body temperature, which is to be described later. Attempts to produce universally applicable theories of cold death have not met with any general agreement.

8.8 Heat death

The causes of heat death are often as difficult to evaluate as those of cold death, and again it may be that the emphasis should be on different factors in different cases. Some Arctic animals die at temperatures as low as 16–18°C, indeed the Arctic fish *Trematomus* dies at about 10°C; and the upper temperature limit for many marine animals in temperate areas does not exceed 25°C. These heat death temperatures may be contrasted with those that operate in the case of desert insects and hot-spring proto-zoans, which can withstand temperatures up to 50°C.

It is evident that although enzyme denaturation and coagulation might be operative factors in the heat death of animals in very hot habitats, they could not be the causes of heat death in animals from temperate or cold habitats. As in the case of cold death, attempts have been made to arrive at a general theory that would cover all cases of heat death. It has been observed that excess heat appears to cause an increase in the viscosity of protoplasm, and that this is frequently accompanied by the appearance of vacuoles and the release of calcium ions within the cells. It has therefore been suggested that since protoplasm is thought to contain complexes of lipids and proteins bound in some way to calcium, the effect of excess heat is to break down these complexes. The calcium thus released would then exert a disruptive influence on the cell, for example by affecting the permeability of the plasma membrane and the intracellular membranous systems. There is no doubt that profound permeability changes do precede heat death in many animals. There is evidence that heat death in *Astacus* and *Calliphora* is due to changes in the lipoprotein complexes of the plasma membranes of their body cells, or to changes in the enzymes that maintain the structural integrity of these membranes, and this could be an alternative explanation of the changes in permeability that seem to precede heat death.

In some cases heat death may be associated with an inability to deal with waste products. It has been shown that the heat death of desert scorpions and beetles is accompanied by a change in the pH of the haemolymph, which becomes acid in contrast to its normal strongly alkaline pH. The details underlying this observation, and the extent to which it may be general, are not known.

In mammals, heat death can be bound up with an inability to osmoregulate. Apart from animals like the camel, in which water is lost from the tissues rather than from the blood (§6.6), water loss in mammals causes the blood to become more viscous. This both imposes a strain on the heart and impedes the easy flow of the blood, with the result that the heat produced in metabolism cannot be carried round the body rapidly enough for it to be dissipated at the skin surface. The internal temperature therefore rises at increasing speed, and ultimately *explosive heat death* ensues, the cellular basis of which is no doubt similar to that of other forms of heat death, although the viscous blood will also give rise to an oxygen lack.

8.9 Hibernation

Most homeotherms pass into a state of torpor if the temperature is lowered sufficiently. In general, a rectal temperature of 15°C or less for a reasonably long period in mammals results in their death. Shorter periods of *hypothermia* do not necessarily kill, but in recovery from extremes of cold much depends on the re-warming process, and this may also be true of poikilotherms. Warming causes an increase in metabolic processes, but the speeding up of respiratory and circulatory movements tends to lag behind and there is a shortage of oxygen. Thus, metabolites may be abnormal, and not removed at sufficient speed.

Instead of passing into a torpid state of this kind, certain mammals such as squirrels, dormice, and some bats and a few birds such as humming birds spend the winter in a dormant state of *hibernation*, in which they appear to be asleep and during which no food is eaten. In this condition they can withstand much lower rectal temperatures than other homeo-therms, and these are generally in the region of 4–5°C for long periods.

Most bats that hibernate have an unusual form of hibernation in which there is a diurnal cycle instead of a seasonal one. Their body temperature rises in the evening, when feeding is to take place, but is close to the environmental temperature for the rest of the day. The period of hiber-nation in other mammals is one of several months, which carries them over the worst part of the winter. During this period they live on food reserves which they have accumulated before the onset of hibernation. These reserves often take the form of a special *brown fat*, which is easily metabolized at low temperatures, although some animals such as the hamster accumulate no fat but build up external food stores—a procedure also followed by some mammals that do build up fat, like the squirrel.

Although fat may be consumed during hibernation, the amount which can be accumulated in the preparatory period is quite insufficient to sustain the hibernating animal at its normal rate of metabolism. For example, the marmot *Arctomys* is probably the largest hibernating mam-mal, with an average body weight of about 3 kg. In the non-hibernating state it uses the equivalent of 10 g of fat in 24 h, so that in 100 days the amount of fat required would be 1 kg, or one-third of its body weight. Hibernating animals are small, which is no doubt one of the reasons for the phenomenon, since small mammals burn far more fuel per unit body weight than large mammals. The proportion of fat would therefore need to be much higher in other hibernating mammals than it would be in *Arctomys*, and it becomes an impracticable proposition. One of the most striking features of hibernation is therefore a marked damping down of metabolism, the animal becoming virtually poikilothermic.

The changes involved are complex. Metabolism is reduced some 20–100 times below the normal level, for example the marmot produces 11.7 kJ (2.8 kcal) heat/kg body weight/h at 10°C, but during hibernation this becomes 0.38 kJ (0.09 kcal)/kg/h. Respiration rate and heart beat both become slow and irregular, and in the Arctic ground-squirrel *Citellus* the heart beat falls from 200–400/min to 7–10/min. The need for oxygen is correspondingly greatly reduced. The thyroid gland exhibits regressive histological changes, and the pituitary and adrenal glands no longer re-spond to cold stress (Chapter 13). The blood magnesium level is higher than normal, and this possibly accounts for some of the torpor, through its effect on excitable cells and the neuromuscular junction (Chapter 9).

Entry into the hibernating state is usually slow, the body temperature being seen to drop at some 2–5°C/h in most cases, although some are faster. The body temperature never falls quite to the environmental level but remains some 1–4°C above it, with some variation between different parts of the body. This indicates that there is some residual ability to thermoregulate, but this may be insufficient to prevent animals from becoming frozen at really low environmental temperatures, and so many of them burrow into the ground before hibernation or take over the abandoned burrow of another animal, and thereby position themselves below the level at which the ground freezes.

The most common factor in the initiation of hibernation seems to be a drop in the environmental temperature, but other seasonal factors may play some part, especially light. This ensures that hibernation begins before the food supply becomes scarce and while food reserves are at a maximum. The marmot appears to be exceptional in delaying hibernation until its food becomes scarce.

Some hibernating animals awaken periodically, usually to eat and drink where this is possible. Hibernating animals become increasingly sensitive to arousal stimuli as the period of hibernation lengthens. The arousal from hibernation at the end of the winter represents a reversal of the changes that initiated it, but the rate of recovery is much faster than the rate of entry into hibernation, being less than 3 h in most hibernators. The metabolic rate and oxygen consumption rise above the normal non-hibernating rate for a short period, suggesting that a brief hypersensitivity of the thermoregulating mechanism occurs immediately after arousal. The stimulus for arousal is usually a rise in the external temperature, but other factors can be important in the laboratory, and this may explain the ability to awaken at intervals during the period of hibernation.

Hibernation is found in those homeotherms which are relatively poor at maintaining their body temperature by physical regulation, and which are small enough to have to maintain a high rate of metabolism per unit body weight under normal circumstances. In cold conditions, their predominantly chemical methods of temperature regulation would add an impossible burden to their already high metabolic rate. Hibernation may therefore be regarded as a metabolic economy on the part of already extravagant metabolizers in conditions in which a higher metabolic rate would be coupled with scarcity of food. It differs from cold narcosis and from sleep in that it is a total body phenomenon. For example, the respiratory organs are adapted to hibernation whereas, in cold narcosis, death is usually due to respiratory failure. Perhaps hibernation is only possible in poor thermoregulators, the others being unable to terminate their regulation in extreme cold. In sleep there is no lowering of metabolism, although there is a reduction in nervous activity and a slowing of heart beat and lung ventilation rate. For this reason, the *winter sleep* of bears and skunks is distinguished from true hibernation because it is not accompanied by the characteristic metabolic changes. Nevertheless, it is interesting to note that any stimulus that will arouse an animal from sleep will also arouse it from hibernation if the environmental temperature is increased at the same time. Sleep is still very little understood, but it is known that it, too, is associated with the hypothalamus, and it is therefore not surprising to find that sleep and hibernation share some common factors.

9

Nervous Communication

Behaviour is defined as 'the overall activities of an animal'. These activities usually arise in response to an environmental influence or *stimulus*, or to a number of stimuli, and in their absence an animal does not normally exhibit a pattern of behaviour. More rarely, behaviour may result from internal activity within the nervous system of an animal by the spontaneous rhythmical generation of impulses by *pacemaker* cells.

Undifferentiated cells, such as cells in tissue culture, are not very responsive to outside influences, although they may respond to light or touch. Free-living organisms, even unicellular ones, must be able to respond to a wider range of stimuli, but obviously the responses of the smaller and simpler animals are not likely to be very specific in relation to their environment. The larger and more complex an animal, and the more mobile it is, the greater will be its need for specialized cells and organs that will pick out with precision one or perhaps a few specific stimuli from a very varied environment. Specialized cells or organs of this sort are known as *receptors* (sense organs). For a metazoan animal to make a meaningful response to its environment, the information from many different receptors must be sorted out and integrated together. This is the task of the *central nervous system* (CNS). The CNS must in turn be coupled in some way to the organs whose task is to carry out the necessary responses to the stimuli. Such *effector* organs are usually either muscles, which contract when activated by the impulses that pass to them from the CNS; or glands, which react to such impulses by releasing one or more secretions.

The ability to respond to stimuli and in turn to conduct impulses to other cells is probably a general phenomenon of living cells, and is presumed to underlie the ability of animals that do not possess nervous

systems, such as protozoans and sponges, to respond to the environment
in a co-ordinated manner. But in all metazoans, except sponges, specialized cells have been evolved for the purpose of communication between
receptors, the central nervous system, and effectors, and these are the
neurones. Receptors are also often highly modified neurones, or a complex
of neurones and other specialized cells. A study of the physiology of co-
ordination must therefore begin with the neurone.

NEURONE STRUCTURE AND FUNCTION

9.1 Morphology of neurones

The neurone consists essentially of a cell body (*soma* or *perikaryon*)
from which one or more processes run to other neurones or to effector
organs. It is often not realized that the morphology of neurones differs
in accordance with the tasks they have to perform (Fig. 9.1). Receptor
and effector organs are usually some distance from the CNS, and in order
to cover this distance effectively they possess a long process, the *axon*.
Similar neurones are found connecting distant parts of the CNS; but the
latter also contains many other neurones, whose function is to link up
with those adjacent to themselves, and these do not possess axons. Instead, their cell bodies bear numerous short branching processes known
as *dendrites*. Such cells are often termed *internuncial* neurones or interneurones, in contrast to the *sensory* neurones which link the receptors with
the CNS, and the *motor* neurones which link the CNS with the effector
organs. Axons and dendrites are not necessarily mutually exclusive, since a
neurone that conducts information over a long distance by means of an
axon may also need to communicate with other central neurones by means
of dendrites, as in the case of the vertebrate motor neurone (sometimes
contracted to *motoneurone*) shown in Fig. 9.1.

The axon is formed by a core of *axoplasm* enclosed by a plasma membrane, and this in turn is surrounded by a cellular layer that represents
the *Schwann cell*. The exact arrangement varies between different neurones.
Frequently, several axons share one Schwann cell, but in other cases, as
the larger axons of crustaceans, there is more than one Schwann cell
around the perimeter of a single axon. An axon is never totally enclosed
by Schwann cells, since there are channels running through the latter
which allow the fluid that surrounds the axon to communicate with the
tissue fluid outside (Fig. 9.2). The primary function of the Schwann cells
is probably the maintenance of the axon in some way; but in many of
the neurones of vertebrates, and in some of those of prawns and a few
other decapod crustaceans, the plasma membrane of the Schwann cell
becomes spirally wound around the axon during its embryonic development to form a thick *myelin sheath*. Such axons are said to be *myelinated*

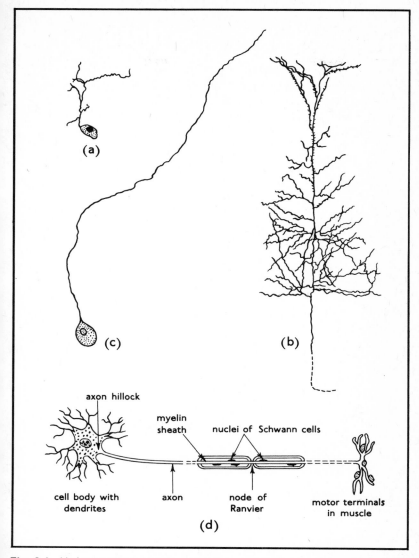

Fig. 9.1   Various neurones to show variations in morphology. (a) and (c) Inter-nuncial and motor neurone respectively of *Nereis*; (b) pyramidal neurone from cerebral cortex of rabbit; (d) vertebrate motor neurone. ((a) and (c) from Smith, 1957, *Phil. Trans. Roy. Soc.*, B, **240**; (b) from Ramon y Cajal, 1909, *Histologie du Systeme Nerveuse*, Vol. 1, Fig. 8. Instituto Cajal, Madrid)

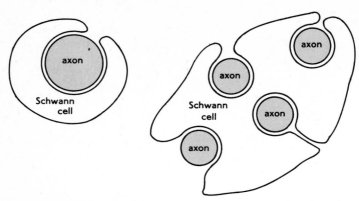

Fig. 9.2 Relationship between unmyelinated axons and the Schwann cell. (After Robertson, 1960, *Progr. Biophys.*, **10**, 343)

(medullated); but, since the Schwann cell is universally associated with axons, it should be understood that the difference between myelinated and non-myelinated axons is one of degree only. Figure 9.3 is intended to make the relationship clearer. The myelin sheath is discontinuous, being interrupted at regular intervals by the *nodes of Ranvier*, at which the plasma membrane is exposed (Fig. 9.1d).

In addition to neurones, the central nervous systems of animals contain a number, often a very large number, of other cells known as *glial* cells. The

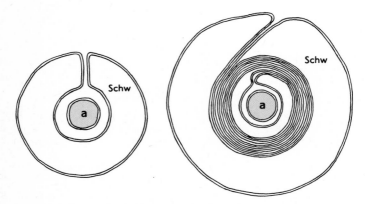

Fig. 9.3 Formation of the myelin sheath by the spiral coiling of the plasma membrane of the Schwann cell. (After Robertson, 1959, *Biochem. Soc. Symp.*, **16**, Cambridge University Press)

function of these cells is still uncertain, although it has excited a good deal of interest. It has been shown that glial cells have a similar electrolyte distribution to neurones (see §9.3) and therefore a similar potential difference across their plasma membranes, showing that they do not, as was once supposed, exchange electrolytes with neurones. They do not appear to conduct impulses. Many workers believe that they must have some role in upholding the neurones, perhaps metabolically, but this is still unproved.

Neurones conduct information along their axons in the form of *impulses*. The passage of an impulse is accompanied by certain physical and chemical changes. These changes have been most extensively studied in the unmyelinated giant axon of the squid, because its large size poses fewer technical problems than that of smaller axons. The available evidence suggests that all axons function in a similar way to the squid giant axon, although myelinated axons exhibit certain specializations which will be dealt with later.

9.2 Electrical model of the plasma membrane

The plasma membrane of neurones is believed to consist, like that of other cells, of a protein/lipid layer some 5–10 nm, thick. Since fat is a poor conductor of electricity, such a structure is bound to affect the passage of an electric current across it. A current of this sort is thought to be the basic phenomenon in the passage of an impulse, and it is therefore important to understand the electrical properties of the plasma membrane. Because of its poor conducting properties, an electric current will encounter *resistance* from the membrane, just as a current flowing in an electrical circuit is impeded by a resistance of wire or carbon particles specially made to conduct less well than the connecting wires of the circuit. In addition, the plasma membrane will have an electrical *capacity*. If we interrupt an electrical circuit by two flat plates made of a conducting material, separated by a fairly small gap and then attempt to pass a current around the circuit, the two plates will act as a kind of store of electricity. If our current is derived from a battery, we shall find that they have become charged like the battery itself and that the charge, which is measured in volts, is the same as that of the two terminals of the battery. One plate will be negative and the other positive, the sign depending on which pole of the battery a given plate is adjacent to in the circuit (Fig. 9.4). Although the plates are separated by a gap, the process of charging the plates requires a flow of current from the battery, which will continue until the charge on the plates has become equal to the charge at the terminals of the battery.

The resistance of a plasma membrane is known as an *ohmic resistance*, because it conforms to Ohm's Law. In an electrical circuit such as the one considered above, the charge on the battery can be thought of as a kind of electrical pressure, because if it is connected up to a suitable circuit it will 'drive' a current around it. The charge is measured in volts, denoted by the letter V; and the current, which is the quantity of electricity actually

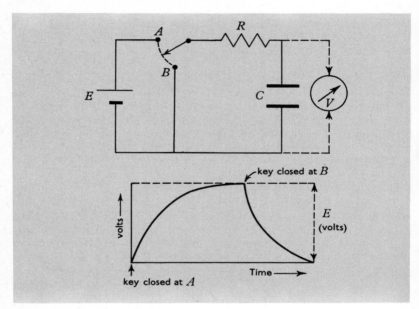

Fig. 9.4 Charging a capacitor by means of a simple electrical circuit. The capicator is denoted by C, the battery by E, and the resistance by R. V is a voltmeter connected across the terminals of the capacitor. When the key is at A, the capacitor charges up through the resistor (the steady rise is shown in the bottom figure), until its charge equals the battery voltage. When the key is at B, the capacitor discharges through the resistor. If the key is between A and B the charge on the capacitor remains steady.

moving in the circuit, is denoted by the letter I. If we denote the resistance in such a circuit as R, Ohm's Law states that

$$R = \frac{E}{I}$$

Thus, by applying a known voltage (E) across the membrane of a cell, and recording the current (I) that passes across it, the resistance (R) of the membrane can be determined. Strictly speaking, we should refer to the *impedance* of a membrane rather than its resistance, because this term describes the total resistance of the membrane to the passage of current across it. The distinction is made because the total resistance contains an element due to the resistance imposed by the membrane capacity, as well as to the ohmic resistance. However, although the capacity of the membrane does affect the electrical properties of the membrane, its value remains constant during the activity of the cell, whereas the value of the resistance alters. For this reason, we tend to concentrate upon the latter when discussing the mechanism responsible for the transmission of the nerve impulse.

The membrane acts like a *capacitor* (a body having an electrical capacity) because the poorly conducting material of which it is made is bounded on either side by an aqueous solution of electrolytes that is freely conducting, the axoplasm on the one hand and the tissue fluid (or experi-

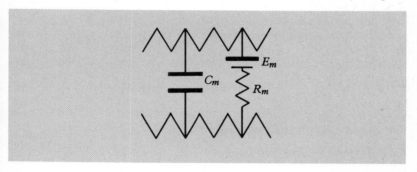

Fig. 9.5 Simple electrical model of the membrane. C_m, R_m and E_m denote the capacitance, resistance and potential difference across the membrane respectively. The longitudinal resistances represent those of the internal and external fluids.

mental saline) on the other. A structure of this sort is called a *core conductor*. In its electrical behaviour, which is considered more fully below, the axon can be likened to a rather leaky electric cable, incorporating a series of capacitor–resistance elements (Fig. 9.5). The electrical model in Fig. 9.5 incorporates a battery in addition to these elements. The reason is that the two surfaces of the plasma membrane of a living cell are like the two poles of a battery, the outside of the membrane being positive, the inside negative. If we were to connect the two sides of the membrane to a voltage-recording instrument such as a galvanometer, we should be able to measure the size of the charge or *potential difference* across it. The instrument used would need to be a sensitive one, because the potential difference across living membranes is small compared to that of, say, a torch battery, and is measured not in volts but in *millivolts* (mV), or thousandths of a volt. Neurones—and, as we shall see, muscle cells—

have developed this characteristic to a greater extent than other cells, because they have utilized it in carrying out their special functions. Thus, most cells at 'rest' exhibit a *resting potential* across their membranes of less than 10 mV; but excitable cells have resting potentials ranging from about 30 mV up to about 100 mV.

We can use this electrical model to help us to understand what happens when a voltage is applied across the membrane that is different in size from its resting voltage. The application of such a voltage results in an alteration of the charge on the membrane capacity, which builds up as the current flows to become equal in size to the applied charge. The flow of current will then cease, except for a small residual flow which 'leaks' through the membrane resistance. The change in potential that occurs under these conditions is known as an *electrotonic* potential. Since two electrodes are used to apply the charge, one positive-going (anode) and the other negative-going (cathode), the sign of the electrotonic potential is equal and opposite at each electrode. The potential change is greatest immediately under each electrode, and falls off exponentially away from the electrode (Fig. 9.6). It must be emphasized that electrotonic potentials are purely passive in nature, and depend upon the physical properties of the membrane, not upon any activity on the part of the membrane.

9.3 Origin of the resting potential

In order for current to flow in an electrical circuit, a flow of charged particles in the form of electrons must take place. Since the fluid which bathes both sides of the membrane is rich in mineral ions, it is not surprising to find that in living cells it is the movement of these ions that is responsible for the flow of current across the membrane in experiments like the one referred to above. In addition, it is known that the resting potential is connected with a differential distribution of ions between the axoplasm and the external medium.

Three explanations have been advanced to account for the origin of the resting potential. The first relies on the fact that some cells are able to secrete ions into the external medium or to absorb them from it against a concentration gradient. We have already encountered this kind of active transport in connection with excretion and osmoregulation, and we shall do so again later in this chapter when discussing the position of sodium ions in relation to the resting potential. It is a simple matter to visualize how this process might be harnessed to the production of a resting potential. For example, if the membrane was to secrete positive ions out of the cell, with the result that the axoplasm contained fewer positive ions than the external medium, the inside of the cell would then be negative to the outside. This result could be achieved by the active transport of a single ion, or by the differential secretion of a number of ions, and might involve cations or anions, or both (secreted in opposite directions).

The second explanation also requires that energy be used. It was pointed out in Chapter 3 that oxidation and reduction are processes which can involve the transfer of electrons, and the familiar example of the oxidation of ferrous iron to ferric iron was cited:

$$Fe^{2+} \rightleftharpoons Fe^{3+} + e \text{ (an electron)}$$

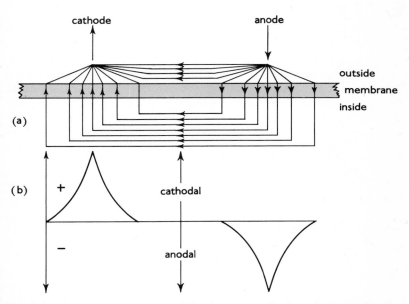

Fig. 9.6 Electrotonus in an axon following the application of a weak stimulus, assuming that current has flowed sufficiently long for a steady-state to be established. In (a) the arrows indicate the flow of current between the stimulating electrodes; in (b) the distribution of membrane potential that corresponds to this current flow. (After Katz, 1939, *Electric Excitation in Nerve*, Oxford University Press)

The ferric iron has a greater positive charge than the ferrous iron, and if a suitable system could be set up in which different proportions of ferrous and ferric irons were contained in two separate solutions, a potential difference would be set up at the interface between the two. In fact, it is theoretically possible to postulate that such a *redox potential* is responsible for the resting potential, utilizing a variety of substances other than iron.

The third possible mechanism requires no source of energy. It depends on a physico-chemical equilibrium condition known as the *Donnan equilibrium*. If we were to take two solutions containing different concentrations of potassium and chloride ions and place them on either side of a porous membrane in a suitable glass vessel, we should find that, provided the pores in the membrane were sufficiently large, the ions would diffuse across the membrane until their concentrations became equal on either side. The same result would not be obtained if one side of the system contained an anion so large in size as to be unable to pass through the pores of the membrane. Let us denote the large anion P^- and assume that it will form a salt KP with potassium ions. If a solution of the salt KP is placed on one side of the membrane and a solution of potassium chloride on the other, the K^+ ions from both sides and the Cl^- ions from the side containing potassium chloride will be free to diffuse across the membrane, but the P^- ions will not. We have said that the normal tendency of the potassium and chloride ions would be to diffuse across the membrane until their concentrations became equal on either side. But in this situation the P^- ions would produce an excess of negative charge on their side of the membrane. In such a situation there will be a tendency for positive ions, in this case potassium ions, to be drawn across the membrane from the other side in order to restore electrical neutrality. In this system there will thus be two opposing forces: the electrical force exerted by the P^- ions, which will tend to result in an excess of K^+ ions on the P^- side; and a chemical force due to the tendency of the movable ions to become equal in concentration on each side of the membrane. The net result is that an equilibrium condition becomes established between the two opposing tendencies, in which there remains some imbalance of ions on either side of the membrane, and the P^- side of the system is more negative than the other, i.e. there is a potential difference across the membrane (Fig. 9.7). It will be appreciated that the electrical effect of the P^- ions will be not only to draw in positive ions, but to repel the Cl^- ions to an equal extent. At equilibrium, therefore, the ratio of potassium ions on the P^- side to those on the other side will be the reciprocal of the corresponding ratio for chloride ions.

Since the system is an equilibrium between two opposing forces, there must be a definite relationship between the relative concentrations of ions on either side of the membrane, and the potential difference across it. This relationship is expressed mathematically by the *Nernst equation*

$$E_m = \frac{RT}{F} \log_e \frac{K_i}{K_o} = \frac{RT}{F} \log_e \frac{Cl_o}{Cl_i}$$

where E_m is the potential across the membrane in volts, R the gas constant (8.3 joules per degree per mole), T is the absolute temperature, F is the electrical constant the Farad (96 500 coulombs per gram equivalent) and K_i and K_o are the concentrations of potassium ions on the P^- side and the other side respectively, Cl_i and Cl_o being the corresponding chloride concentrations. Theoretically, the ions should be represented in the Nernst equation by their activities, but since it is evident that potassium and chloride ions are able to diffuse freely within cells, as in the general model considered here, it is possible to regard their activities as equivalent to their concentrations with little loss of accuracy. Since R and F are known

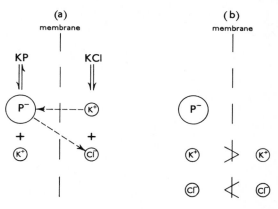

Fig. 9.7   Diagram to illustrate the Donnan Equilibrium. In (a) the ionization of KP and KCl, placed initially on opposite sides of a porous membrane, are shown. The electrical forces exerted by the P^- ions are indicated by broken lines. The final equilibrium situation is shown in (b).

constants and Napierian logarithms can be converted arithmetically into ordinary logarithms, the Nernst equation can be rewritten

$$E_m = 58 \log_{10} \frac{K_i}{K_o} = 58 \log_{10} \frac{Cl_o}{Cl_i} \text{ at } 20°C$$

It follows from the Nernst equation that we should need to know only the ionic ratio of one of the ions involved, since each should be the reciprocal of the other, in order to be able to calculate what should be the theoretical potential difference across the membrane. If the theory is correct, the calculated resting potential should equal the one actually recorded. However, such agreement does not by itself prove that a Donnan equilibrium is responsible for the resting potential. The Nernst equation is derived from the Second Law of Thermodynamics, and it would be surprising indeed if, *at equilibrium*, the two did not agree. But if, for example, the potential difference across a membrane was due to a redox

potential system in the membrane, the ions would still distribute themselves in accordance with the Nernst equation. The distinction between
the two cases is that in one it is the concentrations of ions that determines
the resting potential, due to the P^- ions; and in the other, it is the potential difference that determines the concentrations of ions. It is necessary
to carry out further experiments to settle the point. Such experiments
involve the alteration of the ionic content of the fluid bathing the cell
under observation and the recording of the resting potential at each new
experimental concentration; or to use radioactive tracers to follow the
passage of ions across the membrane. In the first type of experiment, if
the concentration of potassium ions is determining the resting potential
the latter will vary when the external potassium concentration is altered,
and this variation will be predictable from the Nernst equation. There
should be a linear relationship between the logarithm of the external
potassium concentration and the size of the resting potential. We can also
predict the slope of the curve: for a tenfold difference in concentration
between the inside and the outside of the cell, the logarithm of the ratio
of the two concentrations will be unity. Hence, for such a difference in
concentration the membrane potential should equal 58 mV; and for any
further tenfold change in the concentration ratio, there will be a further
change in membrane potential of 58 mV.

The results of such experiments appear to show that, in the majority
of excitable cells, the resting potential can be attributed to a Donnan
equilibrium involving potassium ions. The exceptions are usually muscle
cells, which will be discussed in Chapter 11. However, even in those cells
which show good agreement with theory, the linear nature of the curve
relating resting potential to external potassium concentration is lost at
low external potassium concentrations (Fig. 9.8). The discrepancy is
thought to be due to changes in the permeability of the membrane to
potassium ions at low potassium concentrations, i.e. the permeability may
depend on the resting potential to some extent.

Chloride ions exhibit a less satisfying fit with theory. The equilibrium
potential calculated for chloride ions is frequently found to be lower in size
than the recorded resting potential; and there is increasing evidence that
it is not uncommon for some active transport of this ion to be carried
out by the membrane. Thus, the resting potential or the differential concentrations of ions may often be due to the partial operation of both a
Donnan equilibrium and active transport. So far, however, no convincing
evidence has been obtained to indicate the participation of a redox potential mechanism in the production and maintenance of the resting potential.

The situation outlined above is further complicated by an abundance
of sodium ions in many tissue fluids. If the sodium ions were free to

diffuse across the membrane like potassium ions, they would enter into the equilibrium system with them, and both ions would ultimately attain the same differential distribution. In fact, natural membranes are found to be relatively impermeable to sodium ions. For example, the use of isotopes in studying the exchange of ions across the membrane of the squid giant axon has revealed that the permeability of this membrane to sodium ions is only about one-thirtieth of its permeability to potassium ions. The same work has shown that sodium ions are actually passing into

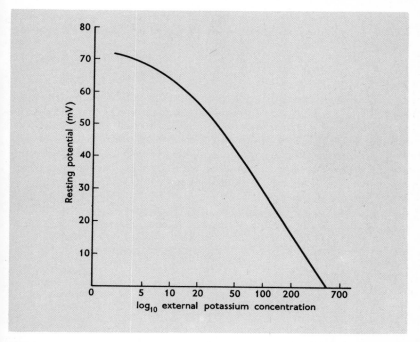

Fig. 9.8 Relationship between the resting potential of squid giant axon and the external potassium concentration. (After Hodgkin and Keynes, 1955, *J. Physiol.*, **128**, 61)

the membrane quite rapidly, but that the membrane sends them back into the external medium before they reach the interior of the nerve. In order to exclude sodium ions in this way, the membrane must expend energy, and the concept has arisen of a *sodium pump* within the membrane. This is pictured as working by means of a *carrier* molecule which can transport the sodium ions back across the membrane once they have entered.

To sum up, the resting potential of neurones seems to depend on a partial Donnan equilibrium system in which potassium ions can pass more or less freely across the membrane, but in which some chloride ions may be actively transported, accompanied by an outwardly directed sodium pump. It will now be appreciated that the Nernst equation as given is too simple a description of the relationship between the resting potential and the ionic concentrations inside and outside the cell. It is necessary to modify the equation to take into account the different permeabilities of the membrane to the major ions. The modified version is known as the *constant field equation*, and it may take a variety of forms. The version which can be most readily compared with the Nernst equation is

$$E_m = \frac{RT}{F} \log_e \frac{P_K \cdot K_i + P_{Na} \cdot Na_i + P_{Cl} \cdot Cl_o}{P_K \cdot K_o + P_{Na} \cdot Na_o + P_{Cl} \cdot Cl_i}$$

in which the terms are the same as those used previously, but the constants P_K, P_{Na} and P_{Cl} have been added to include a component due to the permeability of the membrane to each individual ion. Although the constant field equation is of theoretical importance, for practical purposes the permeability of the membrane to sodium is so low that it is reasonable to ignore this ion, and to consider the resting potential as being determined almost entirely by potassium and chloride ions.

Table 9.1 Ionic ratios between axoplasm and blood of the squid *Loligo*.

	Axoplasm	Blood	Ratio	Calculated E_m
Potassium	400	20	20:1	E_K=75 mV (inside negative)
Sodium	50	440	Na_o/Na_i= 8·8	E_{Na}=55 mV (inside positive)
Chloride	108	560	Cl_o/Cl_i= 5·2	E_{Cl}=41.5 mV (inside negative)

The recorded resting potential is 60–70 mV in size.

The resting potential of the giant axon of the squid *Loligo* is about 62 mV. In Table 9.1 the concentrations of the three main ions, sodium, potassium and chloride in the axoplasm and blood of this animal are given, together with the membrane potentials calculated from the various ionic ratios, using a simple version of the Nernst equation. It will be seen that the nearest approach to what is observed is given by the potassium equilibrium potential, that for chloride being well removed from the actual value. There is good evidence that the discrepancy in the case of chloride is due to an inwardly directed chloride pump that raises the internal concentration of this ion above the level expected theoretically. The figures relating to sodium will be discussed below.

The resting potentials of other neurones are comparable in size to those of the squid giant axon, being generally in the region of 60–70 mV. The resistance of the resting membrane is always high, by comparison with that of the axoplasm or the blood, although the actual figure varies among different species of animals. In the giant axon of the squid the resistance of the membrane taken as a 1 cm cube is $2 \times 10^9 \Omega$; which may be contrasted with a figure of 30Ω for the axoplasm and 20Ω for the external medium. It should be noted that the resistance of a membrane can be expressed in several ways, for example either in terms of a given area or of a given volume of the membrane, and care should be taken to compare only like expressions. A common way of stating membrane resistance is to take a figure based on cubic volume like the one given earlier in the paragraph, and to multiply this figure by the thickness of the membrane in centimetres. This results in a figure of about $1000\Omega/cm^2$ for the squid axon.

9.4 The nerve impulse

During the passage of a nerve impulse, the resistance of the membrane suddenly falls to about $25 \ \Omega/cm^2$ for a very brief period, and then returns to the resting level. The fall in membrane resistance coincides with a flow of current inwards across the membrane, and there is abundant evidence that the current is caused by the inward movement of sodium ions. Thus, for a brief moment of time the membrane becomes freely permeable to sodium ions and then relatively impermeable to them once more. The extent to which sodium ions cross the membrane during this brief period of permeability will depend on the relative concentrations of the ion inside and outside the cell. If the membrane was infinitely permeable to sodium ions and completely impermeable to all other ions at this point, the change in membrane potential would be that caused by the sodium ions moving across the membrane according to the laws of diffusion, in an attempt to reach equal concentrations of the ion on either side of the membrane. The relationship between potential difference and relative concentration for a freely diffusible ion is given by the Nernst equation, and we can therefore predict the change in membrane potential during activity from the sodium version of this equation:

$$E_{Na} = \frac{RT}{F} \log_e \frac{Na_o}{Na_i}$$

where Na_i and Na_o are the resting concentrations of sodium inside and outside the cell respectively. Since the movement of sodium ions is inward, it will tend to make the inside of the fibre less negative. From the figures

given for the squid giant axon in Table 9.1, it can be seen that the resting concentrations of sodium are such that the membrane potential will
actually reverse in sign, the inside of the fibre becoming positive to the
outside. In practice, although the membrane potential always does 'overshoot' the zero potential level during the passage of the nerve impulse
the theoretical degree of change is never observed. The theoretical value
for the squid giant axon is $+54.8$ mV (Table 9.1), but the observed
figure is about $+46$ mV, making a total swing from the resting potential
level of about 108 mV.

There are two reasons why the theoretical sodium potential is not
reached by the membrane. Although, for the brief period of the sodium
influx, the membrane becomes largely impermeable to potassium ions, it
is not completely so; and furthermore, before the peak of the sodium
influx is attained, an outflow of potassium ions begins as the membrane
again becomes more permeable to them. The potassium outflow tends to
oppose the change in membrane potential due to the sodium inflow, and
therefore prevents the latter from having its full effect. The major changes
in membrane potential during the passage of a nerve impulse are thus due
to the movements of the two ions, sodium in and potassium out of the
fibre. The time course for the movements of the two ions is shown in
Fig. 9.9 with the time course of the accompanying change in membrane
potential superimposed on them. The latter is known as the *action
potential* in contrast to the resting potential.

The movements of sodium and potassium ions that produce the action
potential have been measured, using radioactive isotopes. It has been
found that some 3.7 pmol (1 pmol $= 1/1$ 000 000 of a gram-molecule)
of sodium ions enter per cm^2 of surface of the squid giant axon, and some
4.3 pmol of potassium ions leave the same area during the passage of
one impulse. In other words, the movements of the two ions are roughly
equal in opposite directions. In the squid giant axon, the influx of sodium
ions reaches its peak level in 0.5–1.0 ms and is finished by 3.0 ms. The
outflux of potassium ions begins about 0.5 ms after the start of the sodium
influx, just before the latter reaches its peak. Its time course is also slower
than that of the sodium influx, making the restoration of the membrane
potential to its resting level slower than the rising phase of the action
potential (Fig. 9.9).

Because the total area of the pores in the membrane available for the
transport of ions is very small in relation to the total volume of the nerve,
the number of ions which actually moves across the membrane during
the passage of an impulse is only a very small part of the total ions inside
and outside the fibre. It has been calculated that an axon 500 μm in diameter (the size of a giant axon in *Loligo*) could conduct some 5×10^5
impulses before it would need to call on metabolic sources for the further

maintenance of its resting potential. Smaller fibres will be less efficient than this, but on the other hand they tend to recover their resting ionic state faster than large axons. A giant axon may require several hours for the final rearrangement of ions that is necessary for the return of the membrane to its original state after a burst of activity.

Most of the critical work on the nerve impulse has been carried out on the squid giant axon, and modification of some details may be needed

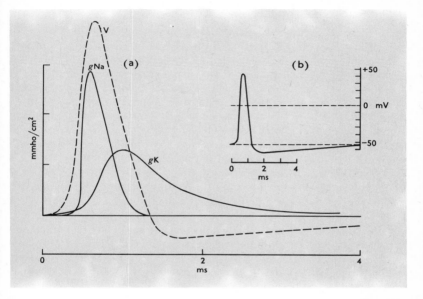

Fig. 9.9   (a) the time-courses of the sodium (gNa) and potassium (gK) conductances of the membrane of the squid giant axon during an action potential, superimposed on the theoretical action potential (V) derived from the known ionic gradients across the resting membrane; (b) the observed action potential. ((a) After Hodgkin and Huxley (1952) *J. Physiol.*, **117**, 500; (b) after Hodgkin and Keynes, from Hodgkin (1964) *The Conduction of the Nervous Impulse*, Liverpool University Press)

to describe the action potentials in the nerves of some other animals. For example, it seems that some of the inward current flow which takes place during the activity of insect nerves may be carried by divalent ions such as calcium and magnesium. Even the outflow of anions is a theoretical possibility. Nevertheless, whatever ions are actually reponsible for the transport of charge across the membrane during activity, we are justified on present evidence in believing that all neurons function in a basically similar manner.

9.5 Recording membrane potentials

The recording of resting and action potentials is usually accomplished nowadays by means of intracellular microelectrodes. The microelectrodes are made from glass tubing that is pulled out under the influence of heat to a fine, tapering open tip, of the order of 0.5 μm in diameter, and filled with a suitable electrolyte, which is usually 3M potassium chloride. Provided that the cell to be studied has a reasonably large diameter in relation to the size of the tip of the electrode, the latter can be inserted into the cell by pushing it through the membrane. The membrane appears to be undamaged by this procedure and being elastic in nature it forms an effective seal around the electrode. A second electrode, of silver or platinum, is placed in the fluid that surrounds the cell, and the two electrodes are connected to the inputs of an amplifying system, made necessary by the small size of the potentials to be recorded. Since one electrode is inside the cell and the other is outside, the potential difference across the membrane is picked up by them, its amplitude is increased by the amplifier, and the signal is then passed to a suitable display system, usually a cathode-ray oscilloscope. The oscilloscope is virtually a small television set, with a tube that produces a small beam of electrons which is focused onto a fluorescent screen. By a suitable arrangement of charged plates inside the tube, the spot thus formed on the screen can be made to move from one side of the screen to the other, forming a continuous line of light. If the amplified voltage from the excitable cell is applied to plates in the tube at right angles to those responsible for the movement of the beam of light on the screen, the beam will be deflected in the vertical plane.

The arrangement is shown in outline in Fig. 9.10. When both electrodes are in the bathing fluid there is no potential difference between them, and the line on the screen represents 'zero potential'. As soon as the microelectrode penetrates the cell, the line changes in position, and the difference in position between the two lines represents the resting potential (Fig. 9.10). The screen can be photographed, recording the two lines; and by means of a suitable calibration system that injects a known voltage across the amplifier inputs, the size of the resting potential can be determined. During the passage of a nerve impulse, the microelectrode will record the changes in membrane potential at the point of its insertion into the cell. The amplified signal from the cell will cause a vertical deflection to be superimposed on the spot moving horizontally across the screen of the oscilloscope, and the shape of the line that results will indicate the time course of the changes in membrane potential. The time course can be accurately measured by superimposing a known time signal like a sine wave on the moving spot (in practice, a second spot, moving at the same speed as the first, is often used).

The technique has been widely used, and indeed it has revolutionized the study of excitable cells. It cannot be used for very small cells because, the smaller the cell, the greater is the relative damage done to it by the penetrating microelectrode. For this reason, the technique cannot be used for the majority of nerve fibres, but the somas of most neurones can be investigated by its use.

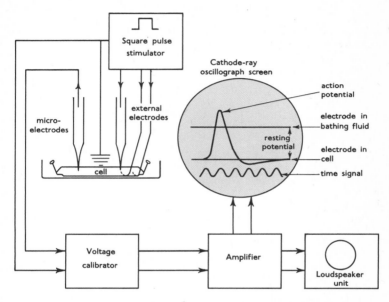

Fig. 9.10   Diagram of apparatus used in recording membrane potentials from single cells. For explanation see text.

9.6 Electrical stimulation of nerve

Because the passage of a nerve impulse is an electrical phenomenon, it is possible to set up an impulse in a nerve by applying a voltage to its surface through stimulating electrodes. A variety of stimulators may be used. The traditional induction coil used by students, which provides a high-voltage shock at low-current intensity, is gradually being replaced by electronic stimulators. Whatever instrument is used, the main require- ment of a biological stimulator is that the voltage should be applied to the nerve over as short a time as possible, for reasons which will become clearer shortly. The stimulus is applied through two electrodes placed on the surface of the nerve, one of which will be positive in sign (the anode) and the other negative (the cathode). At the cathode the pulse of current from the stimulator will make the outside of the nerve negative, and tend to break down the existing polarity of the membrane. If the applied stimulus is sufficiently large, it will bring about changes in membrane permeability similar to those which occur during the passage of a natural impulse, and an action potential will be initiated that will travel along the nerve.

The applied stimulus must reach a certain minimum strength or *threshold* before it can elicit an action potential from the nerve fibre. Once the threshold level has been attained or exceeded, the impulse is found to

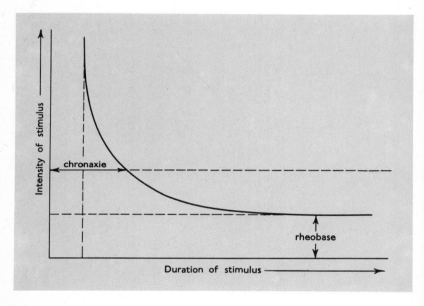

Fig. 9.11   Strength/duration curve of frog gastrocnemius muscle.

reach its maximum height independently of any further increase in the intensity of the applied stimulus. A nerve impulse is therefore referred to as an *all-or-nothing* phenomenon: it requires a minimum intensity of stimulus to elicit it, but once elicited it will proceed to its maximum size. It will be realized that one consequence of this characteristic is that all impulses from a given nerve fibre under constant physiological conditions will be equal in magnitude.

The threshold level of stimulation varies with the time over which a stimulus is applied. Duration and intensity of stimulus are linked by what is known as the *strength/duration curve* (Fig. 9.11). The curve demonstrates that within limits the stronger the stimulus applied, the less the time needed to evoke a response. However, there is an intensity of stimulus which is so low that no response is evoked however long its period of application. This lower limit is known as the *rheobase*. Similarly, there is an intensity of stimulus so great that it always results in the generation of an action potential. This is a stimulus of *maximal* strength. The reader may also encounter another term in the literature, *chronaxie*, which is the duration of an intensity of stimulus twice that of the rheobase. Chronaxie is sometimes used in comparing the properties of different cells or of the same cells under different conditions, but it has no theoretical significance.

Despite the all-or-nothing nature of the nerve impulse, local changes in membrane potential are recorded in the region of the stimulating electrodes when the intensity of the stimulus is below threshold level. The change is greatest immediately beneath the stimulating electrodes, and falls off exponentially away from them. These local changes are the electrotonic potentials which result from the charging of the membrane capacity, and to which we have already referred. We have seen that the electrotonic potential is purely passive in nature, in contrast to the action potential which is an active response on the part of the membrane that involves rapid switches in its permeability to sodium and potassium ions. The transition between electrotonic potential and action potential at threshold stimulus intensity is not a sharp one. A stimulus which is less than half threshold strength produces only an electrotonic response. However, a sub-threshold stimulus of more than half the threshold strength but less than threshold may elicit a further local response which is additional to, and superimposed on, the electrotonic response (Fig. 9.12).

These changes in membrane potential, with different stimulus intensities, can be explained on the basis of movements of sodium and potassium ions. When the membrane is stimulated by a weak stimulus, there is a small flow of current from surrounding areas which are thus made slightly less positive, or *depolarized*. In order to restore the equilibrium in these areas, positive ions must move outwards across the membrane, and a small outward movement of potassium ions takes place for this purpose. A stronger but still sub-threshold stimulus is presumed to cause a slight change in the permeability of the membrane to sodium ions, and hence a slight influx of sodium takes place. This will be opposed by a further outflow of potassium ions, but it is sufficient to delay the repolarization of the membrane by the potassium for a brief additional space of time, and results in a small further depolarization which sums with that due to the electrotonic potential. The active nature of this change, i.e. its dependence on a small change in membrane permeability, is emphasized by the fact that it is recorded only at the cathode. At the anode there is simply a positive electrotonic response (Fig. 9.12). The difference between this *local active response* and the action potential is that in the former the inward sodium current does not exceed the outward potassium current. As soon as the intensity of the stimulus is sufficient to cause more sodium ions to enter than potassium ions to leave the fibre, the permeability of the membrane suddenly switches to allow sodium to enter in a rapidly

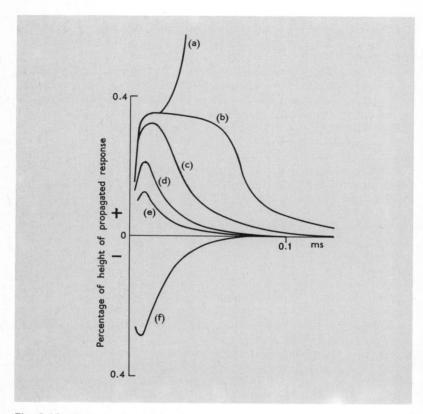

Fig. 9.12 Electrotonic and local active responses in a *Carcinus* limb nerve. (a), (b) and (c) represent successive responses to the same stimulus intensity; (a) is a propagated response, (b) and (c) are local active responses superimposed on the electrotonic response, which cause the repolarization phase of the latter to be delayed. (d) and (e) are responses produced by a lower stimulus intensity and are simply electrotonic responses. All these are responses at the cathode and represent an inward flow of current across the membrane. (f) is the response recorded at the anode, with the stimulus intensity at the same level as that which produced (a), (b) and (c), but the response here is purely electrotonic. (Redrawn from Hodgkin, 1938, *Proc. Roy. Soc.* B, **126**, 87)

increasing amount, and the action potential results. As we have seen, the size of the potential is primarily dependent on the resting concentrations of sodium inside and outside the fibre. These are fixed quantities under given conditions, and it is for this reason that the action potential is an all-or-nothing phenomenon and does not vary in size.

It should be noted that although there are longitudinal current flows inside and outside the nerve during an impulse (Fig. 9.13), it is the *transverse* flow of current across the membrane that initiates the action potential. Once threshold has been reached at the point of stimulation,

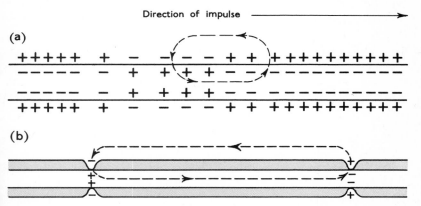

Fig. 9.13 Local current flow during the passage of an impulse along (a), a non-myelinated nerve and (b), a myelinated nerve fibre. (From Hodgkin, 1964, *The Conduction of the Nervous Impulse*, Liverpool University Press)

the action potential will travel along the fibre, and hence it is often termed a *propagated response*. It is propagated because the action potential causes a flow of current from adjacent areas that results in the depolarization of these areas. When the depolarization reaches threshold level in these areas they produce an action potential in their turn. The change is, of course, continuous along the length of the fibre, but pictorially the propagation of the impulses can be visualized as a series of local action potentials arising in turn along the length of the fibre to produce an impulse. Because the action potential is a transient change in the membrane potential, the impulse manifests itself as a wave of depolarization passing down the membrane. It follows from this description of the action potential that a nerve impulse set up at a certain point on the membrane will propagate in both directions from the point of stimulation; and the observed directional flow of impulses through the nervous system is found to depend on other factors that will be discussed later in this chapter.

9.7 Conduction in myelinated nerves

So far, we have been considering the passage of impulses along non-myelinated nerves only. The myelin sheath of nerves, being derived from a spirally wound plasma membrane, is very fatty and a highly effective insulator, and at first sight this would seem to rule out the possibility of an impulse passing along a myelinated nerve. However, refined experimental techniques have shown that the action potential arises in exactly the same way as in non-myelinated nerves, but only at the nodes of Ranvier, where the myelin sheath is absent. Despite the restriction of the transverse current flow to the nodes of Ranvier, the action potential can be recorded at any point along a myelinated nerve, and this and other evidence has led to the conclusion that conduction in such a nerve is *saltatory* ('jumping'). The action potential virtually jumps from one node to the next, and it is aided in doing so by the electrical characteristics of myelin. Myelin possesses high electrical resistance but low electrical capacity, and so it does not take a long time to charge up as the current spreads outwards from the node. The local currents that will set up the action potential at the next node therefore reach the latter faster than an impulse would travel along the same distance of a non-myelinated nerve. Saltatory conduction is compared with conduction in non-myelinated nerve in Fig. 9.13.

One consequence of saltatory conduction is that impulses travel faster along nerves if they are myelinated. For example, non-myelinated axons of the polychaet *Neanthes* with a diameter of about 35 μm conduct impulses at a speed of about 5 m/s, whereas myelinated axons of the frog, which are only about half this diameter, conduct impulses at about 30 m/s. Homeothermy, which demands a higher metabolic turnover and hence a quickening of many body processes compared with those of poikilotherms, is accompanied by a further increase in conduction velocity. For example, myelinated nerves of the cat similar in diameter to those of the frog referred to above conduct impulses at a speed of about 90 m/s. Homeothermy apart, comparisons of the speeds of conduction of nerves in different animals, or even of different nerves in the same animal, shows that myelination and diameter are not the only factors which govern the speed of nervous conduction. The electrical resistance and capacity of the membrane influence speed of conduction because of their effect on the speed of depolarization of the membrane and the spread of local currents along it. The rate of rise of the action potential, which is affected by the sodium concentration of the external medium, and the threshold level for the initiation of the action potential, will also exercise some influence. In myelinated nerves, the thickness of the myelin sheath is a contributory factor to variations in the speed of conduction. Thus, speed of conduction in axons can be said to be proportional to their diameter multiplied by a constant; but the value of the constant varies, and will be affected by the factors mentioned. Since, however, it remains true that considerable increases in fibre diameter are always accompanied by an increase in speed of conduction in spite of these other factors, it is found that in circumstances in which it has been necessary to conduct impulses faster, the necessary increase in speed has been obtained by an increase in fibre diameter. This will be referred to again in the next chapter.

9.8 Refractoriness of nerve

The frequency with which impulses can pass along a nerve fibre is governed by the *refractory period* of the fibre. If an action potential is

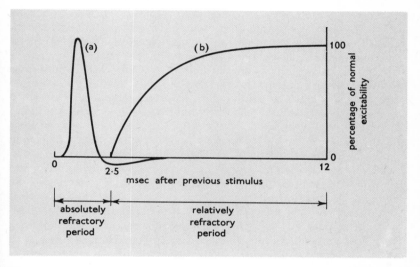

Fig. 9.14 Refractory period of nerve. The excitability of the axon falls to zero during the action potential (a), and gradually returns to normal afterwards, as shown in (b). The degree of excitability determines the refractory periods.

initiated by an applied electric shock, a second stimulus applied during the period when the action potential is reaching its peak will not produce a second action potential, however intense the stimulus. This period of complete inexcitability is known as the *absolutely refractory period*. It is followed by a further period of time during which a second stimulus will evoke a second action potential only if its intensity is much greater than the threshold intensity of the first stimulus (Fig. 9.14). This is the *relatively refractory period*.

The refractory periods are an inevitable consequence of the changes in membrane permeability that are responsible for the production of the action potential. Once the propagated response begins, sodium ions are moving in at the maximum possible rate, and a further stimulus cannot increase the speed or the extent of their movement. Once the sodium influx has ceased, and some restoration of the membrane potential has taken place, a stimulus will again elicit a response; but the outflow of potassium ions which is helping to restore the membrane to its resting state will effectively oppose the depolarizing action of the second stimulus. Therefore, in order to overcome the effect of the potassium outflow, a stimulus of higher intensity than the original is required.

The refractory periods impose limitations on the frequency with which a nerve will conduct impulses. The absolutely refractory period determines the maximum frequency of conduction; but the relatively refractory period also imposes a further limitation, even when the stimulus intensity is raised to a level sufficient to overcome it. The reason is that the membrane conducts at a slower rate than normal during the relatively refractory period, making the minimum interval between each impulse in a continuous train of impulses greater than the absolutely refractory period, whatever the intensity of the stimulus. In practice, most nerves work at well below their maximum possible frequency under natural conditions. In most cases, the highest working frequency observed rarely exceeds 200 impulses per second. This may be contrasted with a maximum theoretical limit of about 500/s in *Carcinus* and about 1000/s in mammalian nerves. In special cases, the functional conduction frequency probably does approach the theoretical maximum. For example, the gymnotid electrical fish of South America navigate by emitting electrical pulses at frequencies as high as 1600/s and then detecting disturbances produced by their surroundings in the resultant electrical field. The pulses are produced by a modified kind of muscular structure known as an *electroplaque*, and the nerves that supply this tissue must be capable of conducting at these high frequencies. Another exception is the mammalian auditory nerve, which will sometimes transmit impulses at a frequency of 1000/s.

9.9 Effects of divalent ions on nerve function

Apart from the three monovalent ions already mentioned, body fluids contain calcium and magnesium ions, and these have been found to affect the activity of neurones. Removal of calcium from the bathing medium leads to spontaneous outbursts of impulses. Calcium ions tend to depress the permeability of living membranes and it may be that their removal brings down the threshold for the action potential. In addition, they alter the permeability of the membrane to sodium and potassium ions. In these effects, as in others, magnesium ions exert the opposite effect to calcium ions; and for the proper functioning of the nervous system a definite balance between them must be maintained in the external medium.

With the possible exception of the method of storage of information, which is discussed in the following chapter, all nervous function has to be related to the membrane changes and responses described above. Transmission of information can take place by means of electrotonic or local active responses only if the distance to be covered is short and the electrical conditions are favourable. For transmission over longer distances, a propagated response is necessary. As we have seen, the propagated response does not vary in size in a given nerve, and it follows that discrimination of afferent impulses within the nervous system can depend only upon their frequency, and on the external spatial arrangements of the receptors and those between their afferent nerve endings and the central neurones with which they form junctions.

SYNAPTIC TRANSMISSION

9.10 Synapses

The junction between two neurones is known as a *synapse*. Muscles are excitable tissues which, like neurones, depend for their functioning on the depolarization of their membrane by the motor nerve that innervates them. For this reason, the junction between a motor nerve and its muscle is also regarded as a form of synapse, and the transmission of impulses from the nerve to the muscle (*neuromuscular transmission*) as a form of synaptic transmission.

Synapses may be formed between any part of one neurone and any part of another neurone. In the vast majority of cases, the presynaptic neurone forms terminal swellings, or *synaptic knobs* on the post-synaptic neurone. The size of the synaptic knobs varies from 0.5–2 μm in diameter, and they are also somewhat variable in shape (Fig. 9.15). Any neurone in the central nervous system is closely covered with synaptic knobs, and in some cases as much as 80% of the soma has been seen to be covered by knobs from other neurones. The knobs are produced by the dendrites of these neurones.

The synaptic knob contains numerous *synaptic vesicles*, which are believed to represent preformed 'packets' of a chemical transmitter substance (Fig. 9.16 b). The vesicles appear to be permanent structures, and are probably the sites of manufacture and storage of the transmitter. They are believed to produce the transmitter up to a maximum concentration, and then to cease producing it until the nerve impulse reaching the presynaptic terminals causes the release of some of the transmitter. Those vesicles whose store of transmitter has been depleted by this process then manufacture more until the maximum level is again reached inside the vesicle. The material from which the transmitter is made seems to be transported from the cell body down the axon to the presynaptic terminals. Little or nothing is known about how the transmitter is formed, nor how it is liberated when the impulse arrives at the nerve terminals.

In the majority of synapses, the pre- and postsynaptic membranes are

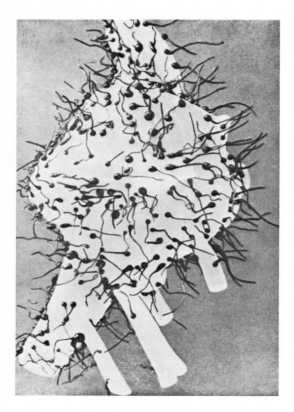

Fig. 9.15 Model of soma and large dendritic stumps of a mammalian motor neurone, showing synaptic knobs, constructed from serial sections. (Haggar and Barr, 1950, taken from Eccles, 1953, *The Neurophysiological Basis of Mind*, Clarendon Press, Oxford)

separated by a gap, the *synaptic cleft*. The neuromuscular junction is similar in its essential details to the neuro-neuronal synapse described above, but in many vertebrates its synaptic cleft is thrown into folds, which form secondary synaptic clefts (Fig. 9.16). This is presumably a device to increase the surface area of the postsynaptic membrane, and hence to speed up the action of the transmitter upon it.

In the central nervous system, the membranes on either side of the synaptic cleft are thickened and more dense in appearance than those of the surrounding non-synaptic areas. These dense patches are believed in most cases to be concerned in some way with the release and subsequent

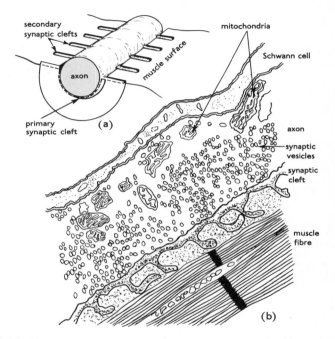

Fig. 9.16 Neuromuscular junction of frog. (a) Diagram of fine motor axon terminal lying in the primary synaptic cleft in the surface of the muscle fibre, and showing the semi-circular secondary synaptic clefts. (b) Tracing of an electron micrograph of a section through the neuromuscular junction. (After Birks, Huxley and Katz, 1960, *J. Physiol.*, **150**, 134)

action of the transmitter substance. They can be divided into two types: in Type I the synaptic cleft is about 30 nm deep, with a plaque of extra-cellular material near the postsynaptic membrane and dense patches which are thicker and more extensive than those of Type II; and the latter, apart from possessing thinner and less extensive patches, has a synaptic cleft about 20 nm deep. There is good reason for believing that Type I synapses are excitatory and Type II synapses inhibitory. The synaptic cleft of the vertebrate neuromuscular junction is about 15 nm deep.

In the 'resting' state, when no impulse is arriving at the presynaptic terminals, small depolarizations of the postsynaptic membrane are still observed. These *spontaneous miniature potentials* (*miniature end-plate potentials* at neuromuscular junctions) vary in size, but rarely exceed 3 mV in height and are usually less. They are produced by the random release of small 'quanta' of the transmitter substance from the presynaptic terminals. The quantitative relationship between these quanta and the transmitter in the synaptic vesicles is not known. When the presynaptic terminals are activated by a nerve impulse, the release of the quanta is greatly increased, and the consequent depolarization is much larger. About 150–300 quanta of the transmitter are synchronously released at the vertebrate motor end-plate under these circumstances.

The depolarization of the postsynaptic membrane through the liberation of transmitter by an impulse in the presynaptic nerve is known as an *end-plate potential* (epp) at the motor end-plate, and as an *excitatory postsynaptic potential* (epsp) at central excitatory synapses.

9.11 Transmitter substances

The transmitter substance liberated at the neuromuscular junction of vertebrate skeletal muscle, and at some central synapses in a variety of animal groups is *acetycholine* (ACh). This is a quaternary ammonium compound, i.e. it can be thought of as an ammonium ion NH_4^+, in which the hydrogen atoms have been replaced by alkyl groups (from the series methyl, ethyl, propyl, etc.). Acetylcholine itself is formed from the quaternary ammonium compound *choline* in combination with the acetyl radical which we have already encountered in connection with cellular metabolism. In the presynaptic terminals it is formed by a special enzyme, *choline acetylase*.

Quaternary ammonium compounds are strong bases, and therefore ionize freely in solution. At first sight, it might be thought that this property could account for the action of ACh as a transmitter. If the membrane was permeable to acetylcholine ions, a sudden release of additional quantities might depolarize the membrane in the same way as an increase in the external potassium concentration. In fact, the concentration of acetylcholine needed to produce an end-plate potential is too small for this explanation to be feasible. A concentration of as little as one part in ten million will elicit an end-plate potential if the substance is applied very close to the end-plate membrane, and the quantity which is naturally effective will be even smaller because externally applied substances tend to diffuse away before they reach the synaptic cleft. Local application of substances is accomplished nowadays by an elegant technique that uses glass capillary microelectrodes of the kind used for intracellular recording. Instead of potassium chloride, the electrode is filled with the chemical under investigation, and brought into contact with the postsynaptic membrane. An electrode of wire is placed inside the glass microelectrode, and another is placed in the bathing medium. The two are connected to an electrical circuit which can be made to produce a brief pulse of current that passes through the electrode. The flow of current causes the ions inside the electrode to migrate, and if the current flow is adjusted to be in the right direction in relation to whether the ion is positive or negative the result is that each pulse of electricity causes a small jet of the ion to be driven out of the tip of the microelectrode. The use of this technique permits the application of very small amounts of the substance to the postsynaptic membrane; and since there is a known relationship between the amount of current flowing and the quantity of ion ejected, the latter can be calculated with precision.

This method has revealed that if the cell is penetrated by the microelectrode, causing the acetylcholine to be applied to the inside of the membrane, it remains unaffected. An epp is evoked only when the transmitter is applied to the outside of the cell. By testing the reaction of the membrane to acetylcholine at a number of different points along the surface of the muscle fibre, it has been shown that the maximum sensitivity to the substance is in the end-plate region; elsewhere, the quantity of acetylcholine needed to produce a response by the membrane is massive by comparison. These facts have led to the concept of receptor areas built into the outside of the end-plate membrane and other postsynaptic membranes, specific in their reaction to the transmitter substance. These receptor areas are thought of as bearing certain molecules which react with the transmitter in some way that alters the permeability of the membrane that contains them. The concept has been used to explain the action of the drug *curare* and its derivatives on synapses at which the transmitter is acetylcholine (*cholinergic* synapses). When curare is added to the saline bathing a frog muscle, for example, the muscle no longer contracts on stimulation of its motor nerve. By intracellular recording it is found that curare depresses the size of the epp below the level at which it will give rise to a propagated response. The reason is thought to be that, because curare is itself a quaternary ammonium compound, it competes with acetylcholine for the receptor sites on the postsynaptic membrane. Thus, the greater the concentration of curare the smaller the epp, and this fact has been invaluable in permitting the study of the epp in isolation from the propagated response, by reducing the epp to a size at which a propagated response does not arise.

The drug *eserine* and its related compounds cause a considerable prolongation of the epp. The reason is that an enzyme *cholinesterase* is present on the pre- and postsynaptic membranes and is especially concentrated in the synaptic cleft. Its action is to destroy the acetylcholine almost as soon as it is released. By this means the transient nature of the end-plate potential is ensured, so that a single impulse in the motor nerve terminals will give rise to a single impulse in the muscle fibre. Eserine is an *anti-cholinesterase*, and acts as an inhibitor of the cholinesterase, with the result that the acetylcholine liberated by a single motor impulse acts for a much longer time than when the cholinesterase is allowed to function.

The change in the permeability of the end-plate membrane brought about by the transmitter is such that the membrane becomes freely permeable to all the ions in the extracellular fluid. The change in permeability is not selective for sodium ions, as it is in the generation of the action potential; therefore, all the major ions, sodium, potassium and chloride, can move across the postsynaptic membrane in accordance with the electro-chemical gradient present in the resting state. As we have seen, the resting membrane is not completely permeable to any of these ions in nerves, and this is true of muscles to an even greater extent. Sudden freedom for these ions to move will be expected to result in a change in the membrane potential different from that which results from a change in permeability to sodium alone. If the resting permeability of the membrane to different ions is known, the extent of the change can be calculated. At the end-plate of a frog sartorius muscle fibre the equilibrium potential expected on this basis would be about 15 mV negative to zero potential, compared with a resting potential about 95 mV negative. The transmitter should therefore cause a depolarization of about 80 mV in the end-plate membrane. The size of the epp necessary to initiate a propagated response is half this or even less, and since it is swamped by the subsequent propagated response its size is difficult to check, but there is other evidence that this theory of transmitter action is correct. The epsp of central neurones needs to attain a height of only 20 mV to produce a propagated response, but otherwise appears to be similar to the epp.

The epp or epsp is thus a local depolarization which, if it reaches threshold level, will initiate a propagated response. It is probable that the actual membrane receptor areas are electrically inexcitable at both neuromuscular and central synapses, being unable to produce propagated responses. In the end-plate region it is the immediately adjacent areas to the receptor areas that are depolarized by the epp, and which then react by producing a propagated response. The epp therefore does not have to travel any distance before it is effective. At central synapses, however, the propagated response may arise some distance away from the synaptic area in which the epsp is produced. The somas and dendrites of neurones have a much higher threshold for electrical stimulation than the adjacent axon hillock or initial segment of the axon. In the motor neurone of the cat, the threshold of the axon hillock and the unmyelinated part of the axon, which are together known as the *initial segment* (IS), has a mean value of 10.6 mV; whereas the soma and dendrites (the SD area) have a mean threshold value to electrical stimulation of 25 mV. The SD area also conducts impulses much more slowly, as slow as 0 1 m/s in some dendrites. For these reasons, the propagated response is normally set up at the axon hillock in vertebrate motor neurones, or at or just before the first node of Ranvier in others, although the epsp originates largely from synapses on the soma and dendrites. The epsp's arising in the synaptic areas summate together, and as soon as the summated depolarization reaches the threshold of the IS-segment, the latter produces a propagated response. Clearly,

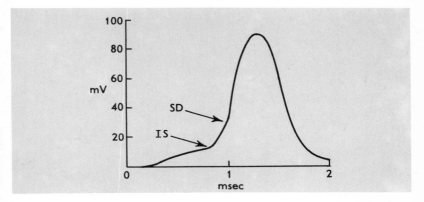

Fig. 9.17 Action potential of a motor neurone, showing its compound nature due to the differences in threshold of excitability of the initial segment (IS) region and the soma-dendrite (SD) region. (After Coombs, Curtis and Eccles, 1957, *J. Physiol.*, **139**, 232)

since the epsp's are local responses which decay in size from the point of origin, the farther away a synapse is from the IS-segment, the less effective it will be in the production of a propagated response. It is noteworthy that the density of synaptic knobs decreases distally from the soma.

Because of these differences in excitability and threshold, the action potential recorded from the postsynaptic neurone is a compound one. The impulse which arises first in the IS-segment by virtue of its lower threshold spreads into the SD area and, being a propagated response, is of sufficient size to exceed the threshold there. The IS action potential is therefore followed by an SD action potential (Fig. 9.17). A system of this sort ensures that the neurone will not generate an action potential because of the action of one or a few presynaptic knobs, and must therefore be taken into account when considering integrative mechanisms. However, it should be understood that the differences in threshold outlined above have been studied largely in motor neurones, and that there may be other neurones in the CNS which do not exhibit them. Indeed, IS-SD differences are not observed in all motor neurones.

The response of a muscle fibre to stimulation of its motor nerve is also a compound one, in many cases. In the majority of vertebrate muscle-fibres it consists of an end-plate potential and a propagated response. Comparable invertebrate muscles, with few exceptions, seem incapable of producing a propagated response; instead, the end-plate potential is succeeded by an active response on the part of the membrane which is

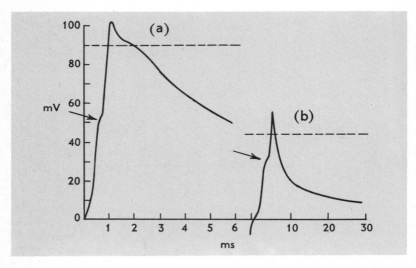

Fig. 9.18 Action potentials of (a) a frog sartorius muscle fibre recorded from the end-plate region; (b) a flexor tibialis muscle fibre of the stick insect *Carausius morosus*. The arrows indicate where the active response arises from the end-plate potential. ((a) from Fatt and Katz (1951) *J. Physiol.*, **115**, 320)

not propagated, and seems to be limited in size (Fig. 9.18). The *active membrane response* in these cases is probably analogous to the local active response which the membrane of a nerve produces in response to a stimulus that is just below threshold. The difference between these two instances is that the invertebrate muscle is incapable of producing a propagated response. This conclusion is borne out by the fact that whereas the vertebrate muscle can be stimulated directly with stimulating electrodes, just like a nerve, the invertebrate muscle cannot. Its membrane seems to have a very limited capacity for the carriage of inward current. However, when all types of muscle responses are considered, it is evident that there is less difference between vertebrates and invertebrates than this account would suggest, and the subject is dealt with more fully in Chapter 11.

The production of acetylcholine by the motor nerve terminals of vertebrates, and the sensitivity of the end-plate membrane towards it, are affected by the quantity and balance of certain ions in the bathing fluid, in addition to any function these ions may have in the transport of charge. Sodium ions appear to influence the quantity of acetylcholine released by the motor terminals, and a reduction in the external sodium concentration leads to a decrease in the quantity of transmitter released. Calcium ions have a similar effect, but whereas the output of transmitter appears to be affected in a continuously graded fashion by sodium, calcium changes the output in discrete 'steps', as though large quantities of transmitter at a time are being affected. One theory has been that acetylcholine ions are released in exchange for the sodium ions transported inwards when the action potential reaches the motor terminals, and that calcium may act by altering the amount of sodium which can enter, one calcium ion being concerned in the transport of a number of sodium ions; but the available information is insufficient to show whether this or other explanations are correct. Calcium ions undoubtedly also reduce the sodium conductance of the end-plate membrane and probably play a small part in the transport of charge involved in the end-plate potential. Magnesium, too, has a complex action that in general is antagonistic to that of calcium. Magnesium ions appear to decrease the amount of transmitter liberated, and to reduce the sensitivity of the end-plate membrane to acetylcholine.

Acetylcholine is not the only transmitter which may promote transmission across excitatory synapses, nor is it the functional transmitter at the majority of synapses studied. Indeed, at some synapses, as in the central nervous system of *Aplysia*, it may be inhibitory in action. *Noradrenaline* acts as a transmitter at the neuromuscular junctions of mammalian smooth muscle, and at certain central synapses. The autonomic nervous system of mammals provides an interesting case of alternating cholinergic and adrenergic synapses: the preganglionic neurones that lead to the sympathetic ganglia are activated centrally by noradrenaline but form cholinergic synapses in the ganglia, from which nerves lead which mostly activate their peripheral effectors by means of noradrenaline.

The substance *5-hydroxytryptamine* is almost certainly an excitatory transmitter in the central nervous system of molluscs. Certain amino acids, notably glutamic and γ-aminobutyric acids mimic the action of excitatory and inhibitory transmitters respectively in arthropods, and it appears that these two at least are naturally produced transmitters. In addition, a variety of other excitatory substances has been extracted from the nervous systems of some vertebrates, although it has not been proved that they are the actual transmitters; and we may expect more to be isolated in the future. At present, there is no reason to suppose that excitatory transmitters in general are different from acetylcholine in their fundamental mode of action at synapses.

9.12 Inhibitory synapses

It has now been shown that inhibitory synapses are a normal feature of the central nervous system, and that there are also some inhibitory neuro-muscular synapses. In their sequence of action they are similar to excita-tory synapses. The impulse arriving at the presynaptic terminals causes them to release an inhibitory transmitter that diffuses across the synaptic cleft to alter the permeability of the postsynaptic membrane to the sur-rounding ions. But whereas the excitatory transmitter increases the per-meability of the postsynaptic membrane the inhibitory transmitter 'sets' the permeability of the membrane so that it tends to become stabilized at an equilibrium potential equal to or near its resting level. Because the precise equilibrium potential at which the membrane is stabilized differs between animals, the *inhibitory postsynaptic potential* (ipsp) may be either positive-going or negative-going in respect of the resting potential. The ipsp of vertebrate motor neurones is more negative than the resting poten-tial, and stimulation of the inhibitory input to them results in a small in-crease in the size of the resting potential. This *hyperpolarization* due to the ipsp contrasts with the depolarization of the same neurones which repre-sents the epsp (Fig. 9.19). In the crustacean stretch receptor, which possesses an inhibitory motor nerve, the equilibrium potential of the ipsp

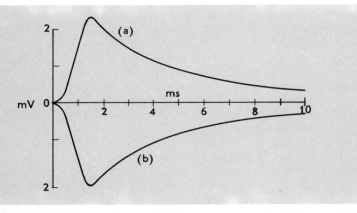

Fig. 9.19 Intracellular recordings of (a) the epsp and (b) the ipsp from the same mammalian motor neurone. (Traced from Curtis and Eccles, 1959, *J. Physiol.*, **145**, 529)

is about 5 mV positive to the resting potential, and is recorded as a small depolarization. Such an ipsp is still inhibitory, because the dendrites of this receptor have thresholds of about 20 mV depolarization for the initiation of a propagated response and the membrane is tending to be stabilized well below this level. In the majority of invertebrates the equilibrium potential of the ipsp is very close to the resting potential level, and little or no electrical change is observed when the inhibitory input is stimulated. The only way in which the nature of such inhibitory synapses can be studied is by artificially displacing the membrane potential from its resting level, when an ipsp can be recorded and changes in the membrane conductance observed. The presence of inhibitory synapses can also be inferred from the absence of a response by the effector organ when the inhibitory nerve is stimulated at the same time as the excitatory nerve.

9.13   Electrical synaptic transmission

The simplest synapses are probably those in the coelenterate nerve net. They appear to lack the complex structure and pharmacology of the synapses described above, and to be similar to the situation in which two unmyelinated axons are laid one across the other. If one of a pair of such axons is stimulated electrically, its action potential sets up an electrotonic potential in the adjacent axon. If the electrotonic potential reaches threshold size, an action potential will be induced in the unstimulated axon. The probability that this will happen is increased if the first axon is stimulated repetitively, when the succession of electrotonic potentials induced in the second axon will summate, much as they do in the soma of a postsynaptic neurone. Direct testing of the method of synaptic transmission in coelenterates has so far proved technically impossible; but there is a strong probability that it takes place in this way. Since nerve nets are found in parts of the central nervous system of higher invertebrates, it would not be surprising if this method of synaptic transmission was found to be more widespread. This is made more likely by the fact that electrically transmitting synapses have been found not only in the central nervous system of invertebrates, but also of vertebrates. Not all of them are as simple in form as the situation described above.

The giant axons that run longitudinally in the ventral nerve cords of some annelids and crustaceans have been shown to be divided by transverse septa. Strictly speaking, these septa should not be described as synapses, because they are intracellular junctions, and it has been proposed that they should be called *ephapses*. However, they are comparable in structure to other electrical synapses, since the septa are continuations of the outer cell membranes of the axon. The effect of the septa is to delay the longitudinal flow of current down the axon and hence to interrupt the local depolarization that precedes the action potential in its passage along the nerve (see Fig. 9.13). The delay amounts to about 0.1–0.2 ms at each septum, but otherwise the septa do not impede the conduction of impulses, which can take place in either direction as in the coelenterate nerve net.

Electrical conduction may take place in either direction at a number of true synapses. In each segmental ganglion of the leech there are two giant neurones that have been shown to conduct between one another by means of electrotonic spread of current, and the same phenomenon has been observed between adjacent neurones in the abdominal ganglia of *Aplysia*. The link in the leech is between axons, and in *Aplysia* between somas, but in both cases a depolarization in the one cell, whether a propagated response or an electrotonic potential, results in the development of an electrotonic potential in the other. If it is sufficiently large, it will initiate a propagated response. The condition in these cases looks primitive, whereas the calciform synapses of the chick exhibit electrical transmission in addition to chemical, and this appears to result from the peculiar morphology of the synapse. The synaptic cleft of a conventional synapse is sufficiently open to the extracellular space for the current in the presynaptic terminals to leak away into the surrounding areas without producing any electrical change in the postsynaptic membrane. In the calciform synapse, the presynaptic membrane also completely envelops the postsynaptic membrane (Fig. 9.20) and such leakage is small. In consequence, the presynaptic membrane sets up an electrotonic potential in the postsynaptic membrane, and this is often large enough to generate a propagated response. The electrical effect is variable and, since chemical transmission always occurs, electrical transmission is not a functional necessity at this synapse.

Conduction both ways across an electrical synapse is presumably a primitive condition, and a more sophisticated type of electrical synapse which conducts in one direction only is found in some of the more complex invertebrates. At the synapses between lateral giant axons and longitudinal giant axons in the crayfish, the postsynaptic membrane acts as an electrical rectifier. This means that it will permit the inflow of positive current required for the initiation of a propagated response, but is much more resistant to current flow in the opposite direction. The physiological mechanism underlying such rectification is unknown. The synaptic cleft is much smaller than at chemical synapses, and ranges from 15 nm down to a negligible amount, with possibly partial or complete fusion of pre- and postsynaptic membranes in places. Because of this reduction in the synaptic cleft, not only electrotonic but also propagated responses can be conducted across the synapse, and the condition is therefore not dissimilar to the ephapse of annelid giant fibres, except for the phenomenon of rectification. A similar synaptic structure has been observed between dendrites of the motor neurones controlling the electrical tissue of certain electric fish, and electrical transmission is known to occur across these junctions.

The occurrence of electrical transmission has been reported in the brain of goldfish, and can be expected to be found in similar circumstances in other vertebrates. The Mauthner cells of the goldfish are two giant neurones situated in its medulla oblongata. The IS-segments of these neurones

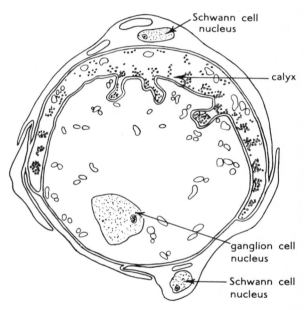

Fig. 9.20   Calciform synapse of chick. The calyx is the pre-synaptic terminal, spread over a large part of the post-synaptic ganglion cell, and contains many synaptic vesicles. (After deLorenzo, 1960, *J. biophys. biochem. Cytol.*, **7**, 31, with permission of the Rockefeller University Press)

are surrounded by a feltwork of fine nerve fibres, which on stimulation produce a hyperpolarization. The position of the fibres is such that the hyperpolarization is at a maximum in the region of the axon hillock, where the propagated response normally arises, and results in a tendency for the latter to be inhibited. The hyperpolarization does not represent a current flow of the kind observed at chemical inhibitory synapses, but is simply an external positive wave which may be likened to the effect produced by the anode lead from a stimulator. The remarkable thing about these inhibitory fibres is that they are joined to others which make chemical inhibitory synapses elsewhere on the soma. The functional significance of having both forms of inhibition active upon the same cell is not known; but there are histological grounds for supposing that this is not an isolated phenomenon.

The central nervous system contains many neurones with processes that ramify among others. The close proximity of such processes raises the question of the extent to which electrical interaction may occur between them, even in those cases in which synaptic contact between their somas is known to be chemical. It is probable that such interaction is far more widespread than was once supposed, for it is certain that the dendrites of adjacent motor neurones can affect the excitability of one another by inducing small electrotonic responses. Similarly, adjacent axons in the same nerve trunk can elicit electrotonic changes in one another of sufficient intensity to increase both their excitability and speed of conduction. Unfortunately, the integrative importance of such changes is still little known.

One further aspect of inhibition in the nervous systems of animals, again about which relatively little is known, remains to be mentioned. So far, we have dealt exclusively with inhibition of the postsynaptic membrane. It is now well established that in some cases inhibitory nerve fibres may terminate on the membrane of the presynaptic terminals, and there bring about *presynaptic* inhibition. They appear to function in the same way as postsynaptic inhibitory fibres, through the release of a transmitter substance that alters the conductance of the presynaptic membrane in such a way that it becomes depolarized to a certain extent. Since the amount of transmitter liberated is related to the degree of depolarization of the presynaptic terminals by the motor impulse, the amount liberated from terminals already partly depolarized by the inhibitory fibres will be less than if the resting potential had been at a normal level. In these circumstances, the epsp of the postsynaptic membrane is less than it would have been. Such presynaptic inhibition is known to take place at the excitatory synapses of the primary afferent fibres of vertebrates; at the neuromuscular junction of the crayfish; and no doubt at other synapses yet to be discovered. In the crayfish it is likely that the transmitter responsible for presynaptic inhibition is the same as for postsynaptic inhibition, γ-aminobutyric acid, which produces an increase in the membrane permeability to chloride ions.

RECEPTORS

9.14 Specificity

Receptors must provide information about the environment for the central nervous system, and since there are many environmental factors to be responded to, the morphology of receptors is very varied. Some receptors may respond to more than one type of stimulus, and these are probably the ones that are least differentiated morphologically, like the free nerve-endings in the mammalian skin which probably respond to heat, tactile stimuli, and pain. The vertebrate eye, on the other hand, is obviously highly specific; and there is a strong teleological argument for supposing that any receptors which are well differentiated histologically will respond to specific stimuli. Most receptors do exhibit such differentiation.

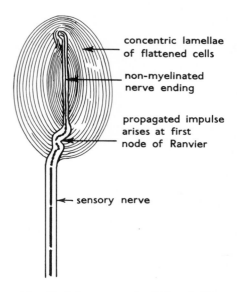

concentric lamellae of flattened cells

non-myelinated nerve ending

propagated impulse arises at first node of Ranvier

sensory nerve

Fig. 9.21 Diagram of a Pacinian corpuscle. (After Quilliam and Sato, 1955, *J. Physiol.*, **129**, 167)

Sometimes a receptor is a single modified neurone, perhaps associated with non-nervous auxiliary cells in some cases, such as the *Pacinian corpuscle* (Fig. 9.21) found in mammalian skin and tendons. This consists of an axon whose non-myelinated tip is surrounded by a fluid-filled capsule composed of concentric flattened cells. More often, a receptor consists of a group of specialized cells, nervous or non-nervous in origin, that synapse with a group of sensory neurones which form the sensory trunk from the receptor leading to the CNS. Examples of this type of receptor are the vertebrate eye and many invertebrate eyes. In either case, the

sensory (afferent) nerves are no different from other nerves in their properties, and relay their information to the CNS in the form of propagated
responses. The task of a receptor is therefore to transform (the technical
term is to *transduce*) some kind of environmental energy, kinetic, chemical
or photic, into electrical energy in the form of an action potential.

9.15 Potential changes in receptors

In the abdomen of decapod crustaceans such as the lobster and the crayfish, each segment is linked to its neighbour by a stretch-receptor whose
function is to signal to the CNS the degree of bending of the joint between
the two segments. The receptor, which is shown in Fig. 9.22, consists of

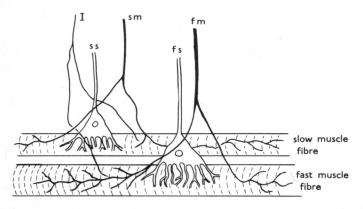

Fig. 9.22   Diagram of stretch receptor in the tail of the lobster. ss and fs, slow
and fast sensory neurones; sm and fm, motor axons to the slow and fast muscle
fibres; I, inhibitory axon. (After Kuffler, 1954, *J. Neurophysiol.*, **17**, 558)

two muscle fibres, to each of which is attached a large sensory neurone
whose dendrites spread over the surface of the muscle fibre. There is
also an inhibitory axon whose endings ramify among the motor nerve-
endings of the muscle fibres. The type of motor innervation is typical of
this group of animals: one muscle fibre is supplied with a fast axon and
contracts rapidly when the latter is stimulated; the other by an axon which
produces a much slower and more gradual contraction. Differential innervation of this type is discussed more fully in Chapter 11. The resting
potential of the sensory neurones has been measured with an intracellular
microelectrode, and found to be about 70–80 mV, and stimulation of the
receptor results in propagated responses with an overshoot of zero potential of 10–20 mV. By stretching the joint, it has been shown that stretch
deformation of the sensory dendrites results in a depolarization of them

of the order of a few millivolts, its exact size being dependent on the degree
of stretch. The greater the stretch, the larger is the depolarization produced. The response is an electrotonic one in the dendrites, which spreads
into the adjoining soma where the responses of the individual dendrites
summate to produce a large depolarization of the membrane. When the
depolarization reaches a critical level in the region of the IS-segment, a
propagated response is produced. The threshold level for this response is
different in the two sensory neurones, being about 20 mV for that associated
with the fast muscle fibre and 10 mV for that associated with the slow
fibre. The propagated response spreads into the SD-segment from the
IS-segment, but the dendrites appear to be electrically inexcitable. The
close parallel between these neurones and vertebrate motor neurones will
be evident: local potentials in the dendrites, summation of a number of
these in the soma, and initiation of the propagated response in the initial
segment of the neurone. The dendritic potential can thus be compared
with an epsp; but there is one important difference, that the dendritic
potential is not produced by the action of a transmitter substance acting
on the sensory cell.

Studies on such different receptors as the Pacinian corpuscle, the
vertebrate muscle-spindle (a postural receptor that signals mechanical
changes) and the eyes of vertebrates and invertebrates, to name only some
of those studied, have confirmed that the basic features of the crustacean
stretch-receptor are common to all receptors, although there are differences of detail related to the morphology of the receptor. In the eye of
Dytiscus the electrotonic potential induced in the light-sensitive retinula
cells by the action of light is not converted by them into a propagated
response. Instead, the electrotonic response is transmitted to the neurones
of the optic ganglion, and it is in these that the propagated responses arise.
In other eyes, the light-sensitive cells are generally too far away from the
optic ganglia for electrotonic spread to be effective. Nevertheless, all
light-sensitive cells seem to be incapable of producing propagated responses and so a second cell is interpolated between the light-sensitive cell
and the ganglion close to the former, which is capable of generating a
propagated response. For example, in the compound eye of *Limulus* each
ommatidium possesses an *eccentric cell* that is close to the retinula cells
and synapses in the optic ganglion. A similar system probably occurs in
the vertebrate eye, although this receptor exhibits one major difference
from all others so far studied in that its light-sensitive cells (rods and
cones) react to the stimulus with a *positive* electrotonic potential.

The terminology applied to the initial electrotonic potential in the sensitive cell is somewhat confused. Some authorities call it the *receptor potential*; and the response set up in the postsynaptic cell with which it synapses
is then called the *generator potential*. Other authorities use these terms in

the exactly opposite sense, and yet others regard them as synonymous. Perhaps the best way out of the dilemma would be to regard the two terms as synonymous and to recognize the electrotonic potentials set up in succeeding cells as the epsp's that they are. There is no doubt that they are produced by a normal process of synaptic transmission, whether chemical or electrical; and if a special term is required for them, 'receptor epsp's' might be satisfactory.

The problem of how receptor cells convert a variety of different stimuli into one kind of electrical response remains to be solved. The reader will by now have come to expect that any change in membrane potential will be brought about by changes in the permeability of the membrane to different ions, but why and how the membrane should alter in permeability in a specific way under specific circumstances are questions still unanswered. If it be supposed that the perception of chemical stimuli (taste and smell) is similar in mechanism to the production of an impulse at chemically-transmitting synapses—and there are difficulties in such a view—it must be admitted that the way in which the latter is brought about is still unknown. The quantity of a substance needed for chemical stimulation of a chemoreceptor is often incredibly small. For example, the feathery antennae of the male emperor moth bear receptors which can detect the scent of females up to a mile away. Again, although we know that certain chemical changes take place in the visual pigments of eyes under the influence of light, we are completely ignorant of the relationship, if any, between these changes and the alterations in membrane permeability of the receptor cells. Nor has any convincing explanation yet been advanced of the way in which mechanical stimuli are transduced into electrical responses.

The sensitivity of some receptors is so high that the threshold at which they will respond to stimuli is at or below their own 'noise' level. *Noise* is a term used to describe the small random movements of molecules in cells due to such factors as thermal agitation, and which, since they involve the release of energy, is manifested in excitable cells by minute random fluctuations in the membrane potential. These fluctuations are measured in terms of microvolts ($1 \ \mu V = 1/1000 \ mV$); and sensitivity of this order must mean that the individual receptor cell is firing irregularly even at rest, and that although perception of stimuli will involve an increase in the rate of firing, it will still be irregular. It is obviously useful for an animal to have those receptors that perceive external stimuli, the *exteroceptors*, as sensitive as possible. If there can be a large number of cells in such a receptor, and a large number of receptor sites on each cell, the picture presented to the sensory input will be of a synchronous discharge due to the stimulus superimposed on the background noise fluctuations. Since the latter will be random, and the former essentially synchronous, the stimulus will be sorted out from the background by the CNS. Most exteroceptors are highly sensitive, especially chemoreceptors; by comparison, *interoceptors*, those receptors that signal the changes in the internal environment to the CNS, are much less sensitive. They will therefore generate impulses at a regular rate in response to a stimulus, and background fluctuations will be absent. The advantage of this state is that only one or a few cells are needed at any one receptor site, and thus a large number can be positioned all over the body, to give the CNS a very detailed picture of the situation of the body as a whole.

9.16 Adaptation of receptors

When a receptor that exhibits regular firing is subjected to a constant intensity of stimulation over a period of time, the frequency of the responses tends to decrease, and in some cases the receptor may cease to respond altogether. This phenomenon is known as *adaptation*, and is presumably to be explained on the basis of changes in the threshold level of the receptor under a constant intensity of stimulation. Initially, the electrotonic response from each dendrite will summate in the soma to produce a steady depolarization of the cell. If this is at or above threshold a propagated response will be generated by the IS-segment. The steady generator potential in the soma will not be able to evoke a further propagated response until the refractoriness of the IS-segment has declined sufficiently for the generator potential to be again at threshold level. Thus, a steady depolarization of the cell will be converted into a series of propagated responses whose frequency will depend upon the ability of the generator potential to overcome the refractoriness of the cell. Since it can be shown that the size of the generator potential varies with the intensity of the stimulus, such that its size increases with an increase in stimulus intensity, the frequency of response will depend on the intensity of stimulus. We should expect, therefore, that a constant stimulus intensity would set up a series of propagated responses in the sensory axon whose frequency would also be constant. To explain the phenomenon of adaptation, we must suppose that the threshold for the propagated response rises with a constant stimulus. There is some evidence that this happens in other excitable cells to a limited extent, although its basis is not understood.

The phenomenon of adaptation enables the response of a receptor to be divided into two phases, a *dynamic* and a *static* phase. The dynamic phase is the immediate discharge following the application of a constant stimulus, and the static phase is the residual impulse frequency that remains after adaptation is complete. Adaptation may be very rapid (*fast-adapting* receptors) or very slow (*slow-adapting* receptors), depending on the nature of the task a given receptor has to perform. For instance, receptors concerned with the maintenance of posture may be required to signal to the CNS the same unchanging position of the body, and hence the same level of stimulation, over long periods. These will be slow-adapting, and their static phase will be little different in frequency of impulses from their dynamic phase. At the other extreme, receptors concerned with touch are fast-adapting, since after the initial contact with the stimulating object the animal may need to carry out some behavioural act in response to it, and this could involve repeated touching. A good example is the Pacinian corpuscle, in which adaptation is so rapid that there is no static phase.

It must not be assumed that a single explanation applies to all cases of adaptation. The vertebrate muscle spindle is stimulated by the stretching of the muscle containing it. Its response consists of two components, one which depends on the velocity of the stretch, the other on the extent of the stretch itself. Such adaptation as there is in this receptor is due to the falling off in the frequency of response to the velocity component, leaving the extension component fully operative. As a result, the dynamic phase can be said to be due to one component, and the static phase to the other. In the Pacinian corpuscle, however, adaptation seems to be a property of the morphology of the receptor. The body of the corpuscle consists of many concentric lamellae (Fig. 9.21), and it is their deformation that sets up the response in the axon contained in the centre of the corpuscle. A constant pressure on the corpuscle does not result in a continuous level of deformation of the lamellae, because they tend to return rapidly to their original state, leaving only a very slight residual deformation. In this instance, therefore, the dynamic phase represents the initial deformation of the lamellae, and the static phase their residual deformation. Practically speaking, the static phase is represented by the absence of impulses.

9.17 Photoreception

The perception of light by animals in the absence of specific photoreceptors is quite common, and occurs also as a property of other parts of the body even in animals which do possess photoreceptors. It is usually a property of the body surface, as in certain protozoans and echinoderms, fishes and amphibians. Parts of the nervous system of some lamellibranch molluscs, fishes and ducks are also sensitive to light. The possession of specific photoreceptors is, however, very common, and there is both a general trend towards a particular type of structure, and an apparent fundamental similarity in the perceptive mechanism. It is not known how apparently non-specific cells are able to perceive light.

9.18 Morphology of photoreceptors

The simplest types of photoreceptors are probably single sensory cells scattered in the epithelium. Thus, earthworms possess epithelial light receptor cells that have a refractile body connected with neurofibrillae on its outer surface, and giving rise at its other end to a nerve fibre that runs

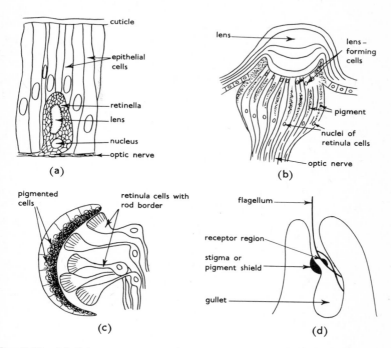

(a)

(b)

(c)

(d)

Fig. 9.23   (a) light-sensitive cell in epithelium of earthworm (Hess, 1925, *J. Morph.*, **41**, 72–7); (b) ocellus of *Aphrophora spumaria* (after Wigglesworth, 1965, *The Principles of Insect Physiology*, Methuen, London); (c) pigment-cup ocellus of *Planaria* (after Hesse, 1897, from Hyman, 1951, *The Invertebrates*: Vol. II, McGraw-Hill, New York and Maidenhead); (d) anterior end of *Euglena viridis*, showing the pigmented stigma and the presumed photosensitive enlargement at the base of the flagellum (from Fraenkel and Gunn, 1940, *The Orientation of Animals*, Oxford University Press).

to the central nervous system (Fig. 9.23 a). Similar receptor cells have been reported in some lamellibranchs. Single cells of this sort, however, are not typical of animal photoreceptors, for there is a general tendency in all groups from the coelenterates upwards for light-sensitive cells to be grouped together, often with non-receptive cells into *eyes*. The simplest eyes are just an aggregation of sensitive cells mingled with pigment cells, as in the coelenterate *Turris* or the Platyhelminth *Planaria* (Fig. 9.23 c). This sort of structure is commonly elaborated in other animals into a slightly more complex structure through the development of a clear cuticular layer over the sensitive cells which is rounded in shape and acts as a primitive lens that concentrates light onto the sensitive cells. Examples of this type of eye are to be found in the coelenterate *Sarsia* and the insect *Aphrophora* (Fig. 9.23 b). In all cases, the function of the pigment is presumed to be to absorb the light and prevent it from being reflected back onto the sensitive cells again, although its value in the very simplest eyes is not apparent. Where, as is often the case, the eye is so shaped and positioned as to have a definite angle of acceptance, the pigment no doubt makes this more precise.

In the most complex eyes a discrete crystalline *lens* is present in addition to the clear cuticular layer or *cornea*. The morphology of lens-containing eyes is very varied, ranging from simple types like the ocelli of some turbellarians and the eyes of *Pecten* to the very elaborate eyes of cephalopod molluscs or vertebrates, which are dealt with in more detail below. The compound eye of arthropods has obvious affinities with an aggregation of ocelli but is much more complex, and it too will be dealt with separately.

It is interesting to note that some of the same trends we have noted above are present at organelle level in the Protozoa. The *stigma* of *Euglena* is simply a mass of granules that shades from the light a swollen photosensitive area near the base of the flagellum (Fig. 9.23 d). There is thus a 'blind' side and a 'seeing' side of the animal, enabling it to orientate to light. In some other flagellates the photoreceptive organelle consists of a clear refractile structure overlying a sensitive area of cytoplasm which takes the form of a cup-shaped mass of pigmented granules.

9.19 The compound eye of arthropods

The surface of the compound eye is divided into numerous square *facets*, each of which represents the outer surface of one of its individual units, which are known as *ommatidia*. In section, the facet is seen to be a curved clear layer or cornea, which overlies another clear structure, the *crystalline cone* (Fig. 9.24). The cornea and crystalline cone together act as the lens. Immediately beneath the crystalline cone is a long refractile structure, the *rhabdom*, which represents the borders of the seven to eight *retinula cells* that surround it. Each retinula cell bears an inner differentiated edge, the *rhabdomere*, which is seen under the electron microscope to be a lamellar structure that interlocks with the lamellae of adjacent retinula cells (Fig. 9.25). The lamellar pattern so formed is believed to enable insects to resolve the plane of polarized light. Each lamella consists

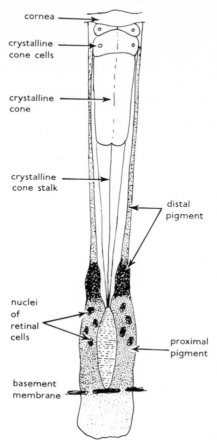

Fig. 9.24   Generalized ommatidium of crustacean compound eye. The eye is in the light-adapted state, with the pigment surrounding the crystalline cone. In the dark-adapted state, the distal pigment collects around the top of the crystalline cone and its cells, and the proximal pigment migrates beneath the basement membrane. In insects the pigment is contained in cells separate from the retinal cells.

of many tightly packed microtubes, and lamellar structures made up of densely packed tubes, rods or granules in association with a light-sensitive pigment have been reported in the sensitive cells of many photosensitive structures from ocelli to vertebrate eyes, and in chloroplasts. It may be that this type of structure, which results in a mass of organized intracellular membranes, is fundamental to photochemical reactions in living organisms.

In insects, each ommatidium is surrounded by pigment cells arranged in two groups. The *proximal* group encircle the retinula cells, and sometimes also the *distal* pigment cells that ensheath the crystalline cone. In the eyes of crustaceans, the pigment is either contained wholly within the retinula cells, or in these together with a distal group of pigment cells. The function of the pigment is to enable the eye to adapt to the intensity of the prevailing illumination. In bright light, the pigment is dispersed and surrounds the entire ommatidium, making it light-tight from its neighbours. In dim light, the pigment moves down the ommatidium,

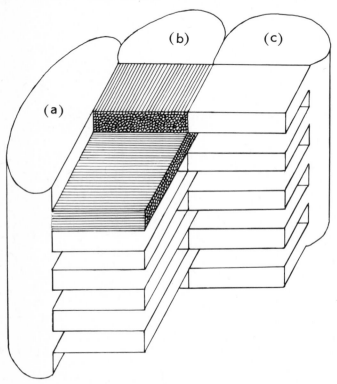

Fig. 9.25 Diagram of the rhabdomeres of three retinula cells of a lobster ommatidium, showing how the plate-like extensions of the edges of the cells, packed with microtubes, interlock to form the rhabdom. There are seven retinula cells instead of the usual eight in this animal, hence the larger size of cell (a) compared with cells (b) and (c). Although the edges of the rhabdomeres where they abut onto the edge of the retinula cell are actually curved, they are here drawn straight. (After Rutherford and Horridge, 1965, *Quart. J. micr. Sci.*, **106**, 119)

leaving the crystalline cone naked, and because of this some of the light entering an ommatidium is reflected sideways to neighbouring ommatidia instead of being absorbed by the pigment cells. In this dark-adapted state, the eye can thus gather the maximum possible illumination, but the spread of images over several adjacent ommatidia means that the image formed in any one of them is not so precise as it is in the light-adapted eye. For this reason, the image formed in the dark-adapted state is known as a *superposition* image, and that formed in the light-adapted state as an *apposition* image.

In *Limulus*, the pigment is contained within the retinula cells, and moves transversely towards or away from the edges of the retinula cells, instead of upwards or downwards.

In many arthropods, both insects and crustaceans, pigment migration seems to proceed according to a fixed diurnal rhythm, for the diurnal pattern of pigment migration persists in the laboratory even when the normal diurnal light changes are abolished or even reversed. The production of this pattern appears to be due, at least in part, to a hormonal mechanism in crustaceans. The sinus gland of the eyestalk (§13.14) contains a substance that will alter the dark-adapted state to the light-adapted state, and another substance that performs the opposite function. Control of the diurnal pattern is presumed to be due to a see-sawing in the balance between these two hormones. There is no evidence for hormonal control of the eye pigment in insects, and the mechanism responsible for their pigment migration is not known.

The principal refractive element of the compound eye is the crystalline cone, the focus of which is fixed and cannot be varied. The main beam of light is directed down the rhabdom and strikes the retinula cells. These are arranged in such a way that one is larger than the others and therefore looks eccentric in cross-section. The *eccentric cell* seems to be the only retinula cell which relays an action potential to the optic ganglion, although all appear to send nerve fibres back to it. Since nobody can believe that these highly-differentiated cells can be non-functional, it is presumed that the retinula cells act as a unit in some undefined manner.

There is no doubt that the compound eye can form an image, to which each ommatidium subscribes a small part. Because the image is essentially a patchwork of small images, it is often termed a *mosaic image*. This does not necessarily mean that this is how the image is realized in the animal's brain, since we do not know what integration of the impulses from the optic ganglion takes place.

9.20 Cephalopod and vertebrate eyes

The primitive cephalopod *Nautilus* has a pinhole eye without a lens which has obvious ocellar affinities. In higher cephalopods, this structure has been replaced by one which approaches the vertebrate eye in its complexity. Much more is known about the functioning of the vertebrate eye than about the cephalopod eye, and so this account will deal primarily with the former, with references to the similarities and differences between this and the cephalopod eye where these are known.

The vertebrate eye (Fig. 9.26) develops as a lateral swelling from the brain, the outer layer of which sinks in to form a cup-shaped structure that is photosensitive, the *retina*. The neuronal elements of the retina are on its inside, and its nerve fibres on its outside, so that the latter come between the light source and the receptive elements. The cephalopod retina is not inverted in this way, but develops from the beginning as a cup-shaped structure. The retinal elements are modified ommatidia, being grouped in four retinular elements at a time, whose inner borders form a typical rhabdom. The vertebrate retina is backed by the *choroid*, a layer of pigment cells, and outside this and covering the whole eyeball is a tough connective tissue layer, the *sclerotic*. At the front of the eye the sclerotic forms the clear *cornea*, through which the light enters the eye. The sclerotic is more than a mere boundary layer, for it acts in conjunction with the turgor pressure of the intraocular fluid to prevent the eyeball from distorting, which would result in an out-of-focus image on the retina. The sclerotic also serves for the attachment of the extra-ocular muscles that move the eye.

The lens is suspended from a muscular *ciliary body* immediately behind the cornea. In terrestrial vertebrates, the cornea and lens together form a compound lens system. The cornea is the main refracting tissue, and being curved it acts as a fixed focus lens. The crystalline lens beneath functions in the fine adjustment of the focus, a process known as *accommodation*. In aquatic vertebrates the cornea cannot function as a lens because the refractive index of water nullifies its curvature, and in these cases the crystalline lens acts by itself.

9.21 Fine structure of the retina

Owing to the inverted retina, light must pass through a layer of nerve fibres before it strikes the vertebrate retina. In practice, the retina becomes elaborated in the development of the embryo, and secondary and tertiary neurones are produced. As a result, light must pass through at least two cell bodies and two synapses, and usually many more, before it reaches the sensitive cells. Many diurnal animals overcome this disadvantage by having a special area of the retina, the *fovea centralis*, in which the layers of secondary and tertiary neurones are reduced or virtually absent, and this area is placed in the optical axis of the lens system. The central vision of such eyes is therefore especially acute.

The resolution of an eye will depend on two factors: the quality of the

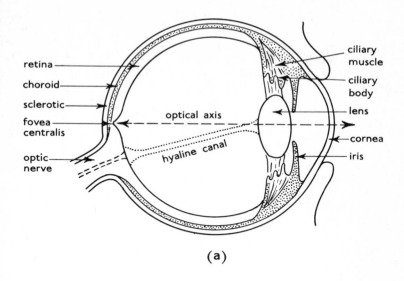

(a)

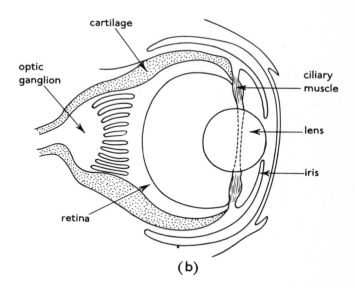

(b)

Fig. 9.26   Diagram of section through (a), a generalized vertebrate eye, and (b), a generalized cephalopod mollusc eye. The two chambers of the vertebrate eye are filled with viscous material (anterior and posterior *vitreous humour*).

lens, and the number of sensitive elements per unit area in the photo-sensitive layer. These factors will be familiar enough to any keen photographer. The closer the retinal elements, the better they will be at separating two points, and the better will be the resolution. There are something like 65 000 sensitive cells per mm^2 in the human fovea, and many vertebrates possess more. Cephalopod eyes show a similar range.

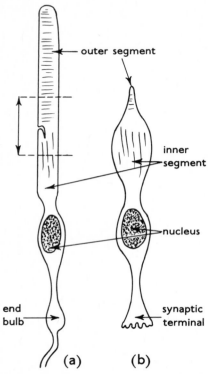

Fig. 9.27 General morphology of (a), rod and (b), cone in a mixed retina. The segment between the two broken lines is shown enlarged in Fig. 9.28.

Two morphological types of photosensitive cell can often be distinguished in the vertebrate retina, the *rods* and the *cones*. They are both built on the same plan, and because of this it can be difficult to determine which is present in a retina that contains only one type. In a mixed retina the distinction is usually fairly plain. Rods and cones consist of an outer segment, an inner segment, a nucleus and a synaptic terminal (Fig. 9.27). As their names imply, cones tend to have a thicker, more conical outer segment, whereas the outer segment of rods is thinner and more stick-like. The outer segment of both types is packed with membranous discs stacked on top of one another (Fig. 9.28). It is not certain whether these discs are separate as shown, or are joined to one another. It is known that if light strikes the receptors at an angle instead of directly in the longitudinal axis, its perception is less effective, suggesting that the orderly arrangement of the discs is necessary for maximum efficiency. The outer segment is connected to the inner segment by a fibrillar structure whose

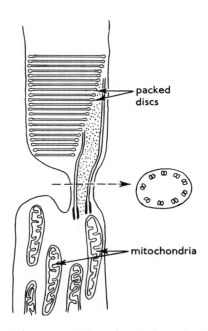

Fig. 9.28 Diagrammatic representation of part of a rod cell, as seen under the electron microscope. The section at the point indicated reveals a structure similar to that of a cilium, but without a central filament. (After De Robertis, 1960, *J. gen. Physiol.*, **43** (suppl.))

fibrils exhibit under the electron microscope an appearance strongly suggestive of cilia (§12.7), leading to a strong presumption that the sensitive cells are modified ciliated cells. The outer segment is known to contain a pigment that breaks down under the influence of light, and the inner segment contains many elongated mitochondria.

The central connections of rods and cones show significant variations. Cones are known to converge on ganglionic cells in the retina, and retinal integration of responses of cones is known to occur, but there are roughly equal numbers of sensory fibres in the optic nerves derived from cones as there are cones; whereas it is usual for many rods—sometimes hundreds—to converge on a single optic nerve fibre. It is therefore not surprising to find that the sensitivity and resolving power of the two types of cells are different. Rods are much more sensitive to light than cones, and their collective sensitivity is increased by their synaptic convergence. The disadvantage of such convergence is a loss in resolving power. In contrast, cones are much less sensitive to light, but their resolving power, or *visual acuity*, is much better.

9.22 Nocturnal and diurnal adaptations

The differences between rods and cones suggest that cones are more suited to daylight vision and rods to night vision, and this is undoubtedly true. In the strongly diurnal squirrel the retina is composed entirely of cones, whereas in the completely nocturnal bush baby it contains only rods. Most vertebrates are not exclusively nocturnal and as a result the retinal pattern in the majority of cases appears to be a compromise in the form of a mixed retina. Rods predominate in such a retina in those animals which are more nocturnal than diurnal, and cones in largely diurnal animals.

Another special development of nocturnal animals is the *tapetum*, a layer that reflects back light which has passed through the retina and would otherwise be absorbed by the choroidal pigment. This increases the amount of light falling upon the photosensitive cells, but it also results in a further loss of visual acuity. The tapetum is the layer responsible for the 'glow' observed from the eyes of nocturnal animals. A similar glow is observed from compound eyes in the dark, due to reflection back from the crystalline cone.

Mainly nocturnal animals which venture out in the daytime, such as the cat, need some protection from the over-bright light. There is commonly a slit-like pupil in such animals, which enables the pupillary aperture to be cut to the absolute minimum when necessary. In addition, the pigmented cells of the choroid extend processes into the retina, and migration of the pigment in these processes can be used to cut down the light passing through the retina. In mixed retinas, the rods and cones may possess some powers of elongation and contraction, the cones elongating in bright light and the rods retracting. The control of these pigment migrations and changes in length of the sensitive cells is not understood. In the higher vertebrates—reptiles, birds and mammals—there is an increasing tendency for pigment migration to be superseded by variation in the pupillary aperture as a method of controlling the amount of light entering the eye.

Diurnal animals have a fovea centralis, which is composed almost or wholly of cones. Their cones contain terminal red or yellow oil droplets in the distal part of the inner segment, and these are thought to act as filters that absorb the blue end of the light spectrum and thereby reduce chromatic aberration. Cones are undoubtedly the older and more primitive type of photoreceptor, since the nocturnal habit is secondary in vertebrates.

9.23 The visual pigments

In all instances in which it has proved possible to extract the visual pigment, whether from vertebrate or invertebrate eyes, it has proved to be a carotenoid pigment conjugated with a protein. The carotenoid, *retinene*, is similar in all animals, and identical in the majority of vertebrates, but the protein is more variable in type. The best-known visual pigments are the visual purples or *rhodopsins*, present both in invertebrates and vertebrates, although they have only been studied to any extent in the latter group.

Retinene (vitamin A aldehyde) may take two forms, designated $retinene_1$ and $retinene_2$. $Retinene_1$ is the one found in rhodopsin, $retinene_2$ being present in visual violet or *porphyropsin*. Porphyropsin seems to be found especially in freshwater fishes, but the idea once held that there was a correlation between habitat and the occurrence of either rhodopsin or porphyropsin is no longer considered valid. As far as is known, porphyropsin reacts to light like rhodopsin.

Rhodopsin and porphyropsin are found in rods. The extraction of the visual pigments contained in cones has so far either proved impossible, or the properties of the extract have been incompatible with the normal functioning of the eye. Some authorities think that the difficulties arise because spectral sensitivity of the pigments of cones may be broad in contrast to that of the visual pigments of rods, but that their breakdown products may be photosensitive to narrow spectral bands.

The action of light on rhodopsin *in vitro* is well known. Light bleaches the pigment, which breaks down into retinene and the protein opsin. This reaction actually proceeds through several steps, and involves two internal rearrangements of the molecule, followed by a hydrolysis. If the exposure to light is prolonged, vitamin A is produced from the retinene, but in any case vitamin A is required for the regeneration of the pigment. These changes in rhodopsin under the influence of light are often presumed to take place *in vivo*, but there is no actual proof of this. Nor is it understood what their significance would be for the production of a receptor potential. Propagated responses have never been recorded from photosensitive cells themselves, and they probably originate in the ganglion cells.

9.24 Colour vision

There is no doubt that colour vision is mediated by cones, rods being monochromatic. This does not mean that animals with cones in their retina are automatically colour-sensitive, for some animals with predominantly cone retinas have unexpectedly poor colour sensitivity, especially at the red end of the spectrum.

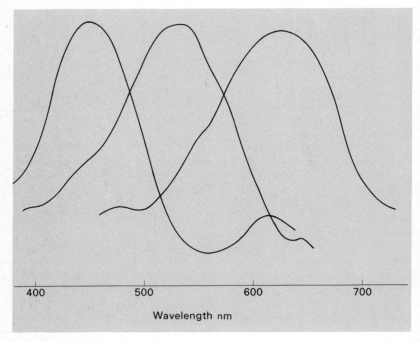

Wavelength nm

Fig. 9.29 Averaged values for the spectral responses of a population of cones in the retina of a goldfish. Three types of cone are present, possessing maximum responses to wavelengths of about 455, 530 and 625 nm respectively. (After MacNichol, 1964, *Vision Res.*, **4**, 119)

When the primary colours red, green and blue are mixed in appropriate proportions they can produce any other colour. At the level of the cones themselves, this appears to be the basis of colour vision. It has been shown that in certain fish and mammal retinas the cones are of three types, having a broad sensitivity to wavelengths in the red, green and blue parts of the spectrum respectively (Fig. 9.29). This *trichromatic* mechanism, however, does not explain all the experimental data satisfactorily, and so the *opponent colour* theory was postulated. This conceives of colour being perceived as one of a pair of opposite colours, black/white, yellow/blue, or red/green. As we have seen, the trichromatic explanation is correct as far as the cones themselves are concerned, but in recent years it has been shown that some of the ganglionic cells of the retina that link with cones exhibit hyperpolarization at one part of the spectrum and depolarization at another, and it may well be that retinal integration of colour perception at post-cone levels does take place in terms of opponent colours.

Colour vision is a recognized fact in a number of invertebrates, but even less is known about its basis than is known about vertebrate colour vision.

9.25 Accommodation

Fishes, amphibians and snakes move their crystalline lens in order to accommodate their eyes to see objects at different distances. Other reptiles, and birds and mammals, do not alter the position of the lens, but instead they change its curvature. Those animals that move the lens may either have eyes that are focused for near objects at rest, as in all fishes except elasmobranchs, or focused for distant objects at rest, as in elasmobranchs, amphibians and snakes. The first group therefore accommodates by moving the lens back the requisite distance, the second by moving it forward. The other vertebrates alter the shape of the lens by adjusting the degree of contraction of the ciliary muscle. The lens is relatively soft and elastic, and is able to be deformed by the contractions of this muscle, instead of being moved by it. This method of accommodation demands a strong eyeball to prevent deformation of the retina, and a highly viscous vitreous humour to prevent backward movement of the lens. In addition to this mechanism, birds have a special *scleral* muscle which can alter the curvature of the cornea, and this combination gives them unique powers of accommodation among the vertebrates.

9.26 Phonoreception

Phonoreception is really a specialized case of mechano-reception, and it is not always easy to distinguish between the two because in both cases the receptor cell is mechanically deformed by the stimulus. Some authors make an arbitrary division on the basis of the frequency of the sound disturbance, low frequencies being regarded as a separate category, *vibration* reception. It is probably simpler to include vibration reception in phonoreception and to define the latter as the perception of any mechanical disturbance external to the animal that involves regular repetition. This definition restricts phonoreception to disturbances with a definite frequency, although this may be a compound disturbance of many mixed frequencies, and certainly need not be a pure wave-form.

9.27 Invertebrate phonoreceptors

Such a definition will not necessarily exclude all tactile organs or other organs which respond to deformation, even though sound perception may not be their primary function. Thus, the primitive invertebrate positional receptor, the statocyst, responds primarily to gravity changes but could also respond to vibration. It usually consists of a sac lined with sensory hair cells and filled with fluid, in which a single large calcareous granule or *statolith* is suspended, or there may be a number of small statoliths or foreign material like sand grains taken in especially for inclusion in the receptor. As the animal moves the statolith alters in position due to the pull of gravity and therefore may press on the hair cells at different places within the statocyst. The force responsible for setting up a receptor

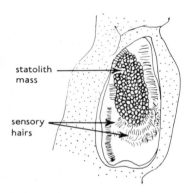

statolith
mass

sensory
hairs

Fig. 9.30 Diagram of the statocyst of the lobster *Homarus*, as seen on removal of the dorsal wall of the basal antennular segment. In this animal a number of small statoliths form a compact mass. (Redrawn from Cohen, 1955, *J. Physiol.*, **130**, 9)

potential is the directional displacement (shearing force) exerted on the cilia, and not pressure. This is why vibration of the statocyst can set up a corresponding vibration in the hair cells. It is certain that statocysts, which are widespread among invertebrates, can contribute to sound reception in some arthropods, and this should be theoretically possible in other invertebrate animals.

Other specific sound receptors in invertebrates have been described and studied almost entirely in the arthropods. In this phylum they seem to fall into three groups, the *hair sensillae*, the *chordotonal* or *scolopophorous* organs, and the *lyriform* organs. Hair sensillae are movable hairs or bristles with articulations on the body surface (Fig. 9.31) which can react to periodic displacement of the air. They seem to be concerned with the perception of low frequency vibrations, or to act as movement proprioceptors. Their sensory axons will respond to sound frequencies between 30 and 800 Hz with synchronous action potentials, and above 800 Hz with asynchronous action potentials up to a maximum of about 3000 Hz, but as the frequency rises the intensity required to elicit a response also increases. Hair sensillae used as tactile and vibration receptors produce a receptor potential only when the hair is actually moved. In other cases, where they are mostly proprioceptors, they discharge the whole time the hair is displaced from its 'resting' position. In the latter case, they often occur in groups (hair beds), that are progressively deformed by the bending of the joint. *Campaniform sensillae*, which act as proprioceptors in insects, are similar to hair sensillae but lack the hair, and instead are attached to areas of the cuticle subject to bending stress.

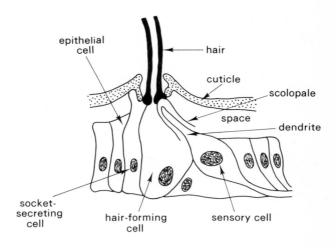

Fig. 9.31 Section of the hair sensilla of an insect (*Gryllus*). Only a very small length of the hair is shown. (After Sihler, 1924, *Zool. Jb.*, **Abt. 2**, *Anat. u. Ontog.*, **45**, 519.)

Chordotonal organs are rather more complex and variable in structure. They are characterized by bipolar sensory neurones attached to apical bodies known as *scolopidia* (Fig. 9.32 a). The most highly differentiated chordotonal organs are the tympanal organs (one is shown in Fig. 9.32 b). Most chordotonal organs are found between the joints of the exoskeleton, but tympanal organs have evolved as special hearing organs, and the two types of chordotonal organ exhibit different abilities in sound reception. Chordotonal organs other than tympanal organs are vibration receptors which are normally sensitive to frequencies in the range 300–2000 Hz. Tympanal organs commonly possess a range of hearing that extends well above that of the human ear, and responses to frequencies of 20 000 to 80 000 Hz are usual. The moth *Prodenia* is said to respond to frequencies up to 240 kHz! The tympanal organs of the Acrididae are exceptional in their low upper limit of 15 000 Hz or even less. However, few insects seem to have even a rudimentary ability to discriminate between different frequencies.

Lyriform organs have been described only in arachnids. They are so named because they consist of parallel slits in the cuticle of varying length,

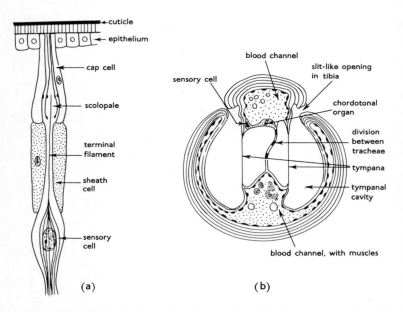

Fig. 9.32 (a) Generalized scolopidium (after Wigglesworth (1965), *The Principles of Insect Physiology*, Methuen, London and Dutton, New York, based on Debauche after Snodgrass); (b) diagram of transverse section across the tibia of *Decticus*, showing the tympanal organ. (After Schwabe, 1906, *Zoologica*, **20**)

like the strings of a lyre, covered by thin cuticle containing nerve endings from a sensory neurone. They are probably mechanoreceptors concerned with signalling the deformation of the cuticle, like the proprioceptive campaniform sensilla of insects. They are sometimes described as phono-receptors but there seems to be no experimental basis for this view.

9.28 Vertebrate phonoreceptors

The most primitive phonoreceptor in vertebrates is the lateral line organ-system of fishes and a few amphibians. The system comprises a line of sensory organs along the side of the fish, which are connected together by a longitudinal groove or canal. The individual sensory organs are groups of hair cells (*neuromast* cells) which project into a gelatinous cavity (Fig. 9.33). The individual cells of each group have widely different thresholds to mechanical disturbances in the water, an arrangement characteristic of complex hearing-organs. A sound source changes in amplitude with distance, and so a longitudinal system with groups of graded receptors should permit an accurate assessment of distance by the central nervous system. It is probable that, besides acting as vibration

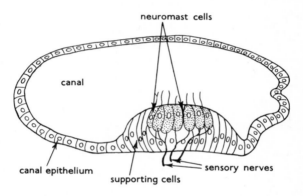

Fig. 9.33   Transverse section of lateral line canal of the elasmobranch *Mustelus canis*. (After Johnson, 1917, *J. comp. Neurol.*, **28**, 13)

receptors, the lateral-line organs signal the speed and direction of the flow of water past the animal, and the localization of moving objects. The maximum frequency of water waves to which a lateral-line system has been found to respond is 180/s, in *Fundulus*, but the upper limit seems generally to be lower than this in fishes.

The vertebrate *labyrinth*, which is contained in the inner ear, is not a sound receptor, but a mechanoreceptor specialized for the perception of position (equilibrium). However, it is generally regarded as a development of the same kind of organs as the ampullae of the lateral-line system, and it is so bound up with the true auditory organ that it is appropriate to accord it a brief description here. The labyrinth consists of two parts, the *utriculus* bearing the semi-circular canals and beneath it the *sacculus*. The sacculus bears a projection, the *lagena*, which becomes progressively larger through the vertebrate series until, in the mammals, it forms the highly complex coiled *cochlea* (see Fig. 9.35). The semicircular canals occupy three separate planes at right angles to one another, two vertical and one horizontal. At the base at one end of each canal is an ampulla containing sensory hair cells which is essentially like the lateral-line ampullae. Acceleration or deceleration in the plane of a canal results in the movement of the fluid within it, and this causes deformation of the cilia of the hair cells by its shearing force. The discharge of impulses from the horizontal ampulla is increased when the ampulla is trailing and decreased when it is leading during rotation in its plane; the vertical ampullae respond to rotation about all three axes. Electrical recording shows that the fluid in the canals, the *endolymph*, is 5–40 mV positive to the perilymph that occurs elsewhere in the labyrinth, and this is additive with the resting potential of the hair cells. When the animal is rotated, the cells of the horizontal ampullae exhibit a depolarization during acceleration towards the ampulla and a hyperpolarization during acceleration away from it, and these are correlated with an increase in the firing of the sensory axons in the first case, and a decrease in the second. Deceleration in either direction results in a return of the membrane to its normal potential. The changes in cellular potential observed are presumed to be the receptor potentials.

9.29 Hearing in vertebrates

Although the lagena gives rise during evolution to the cochlea, and is implicated in hearing in vertebrates of all groups, there is some overlap with the sacculus in this respect in the lower vertebrates, and even in some instances with the utriculus. For example, the skate *Raia* can respond to low frequencies up to 120 Hz, and the corresponding impulses are found in the nerves from the sacculus and utriculus, not those of the lagena. Some teleosts have *Weberian* ossicles, bones that connect the swim bladder with the inner ear. By means of these ossicles the goldfish *Carassius* can hear frequencies up to about 4000 Hz, but if the sacculus and lagena are damaged, the fish can still perceive frequencies up to 500 Hz. Many fish with Weberian ossicles can hear higher frequencies, up to 13 000 Hz in one case. But, in their absence, few fishes can hear above 1000 Hz. A range of 50–10 000 cyles is common in amphibians and reptiles, although there is some variation. Despite the fact that the structure of the lagena in reptiles such as the alligator exhibits some of the features of the cochlea, there is good evidence that no vertebrates except birds and mammals are able to discriminate frequencies. This ability is absent in invertebrates also, and frequency discrimination in the animal kingdom seems therefore to be confined to the cochlea of birds and mammals.

These latter groups possess a good frequency perception range as well as frequency discrimination. They commonly hear up to 20 000 Hz, and some mammals can do better, dogs being able to hear up to 40 kHz, and bats up to 80 or 100 kHz. Bats emit a high-pitched vocal signal, which bounces off obstructions and is picked up again by the animal's ears. This constitutes a highly efficient orientation mechanism.

9.30 The mammalian ear

Although the lagena is large enough to merit the name of cochlea in birds, the structure of the bird ear is otherwise very similar to that of reptiles. In both these groups, as in the amphibians, there is a middle ear in addition to the inner ear, but this contains only one bone or *auditory ossicle* in comparison with the three ossicles of the mammalian ear. In this respect, and in the complexity of the coiled cochlea, as well as in the presence of an external ear bounded by a sound-catching *pinna*, the mammalian ear is unique.

The structure of the mammalian ear is shown diagrammatically in Fig. 9.33. Most readers will already be familiar with its general structure, but one or two details need to be emphasized. The tympanic membrane is more rigidly attached at the top, and thus vibrates to a greater extent at the bottom, where the *malleus* is attached. The malleus rocks with the *incus* about a common axis, and the incus in turn rocks about the *stapes*. The malleus and incus have small muscles attached which act to damp the vibration of these ossicles, and the result of this arrangement is that the movement of all three bones is limited. This means that large vibrations due to loud sounds, which would damage the system, are prevented, but soft sounds are not reduced. Indeed, the ratio of the areas of the tympanic membrane to the malleus, and of the movements of the three ossicles are such that soft sounds are amplified.

The distal end of the stapes fits into the *fenestra ovalis*, which opens into the inner ear, on to a duct that connects with the cochlea. The cochlear spiral is divided into three longitudinal canals by the *basilar membrane* and *Reissner's membrane* (Fig. 9.35). The upper canal, the *scala vestibuli*, contains perilymph and is continuous with the duct that connects with the fenestra ovalis. The scala vestibuli communicates with the lowest of the three canals, the *scala tympani*, through an aperture formed by the stopping short of the basilar membrane before the apex of the cochlea. The scala tympani ends at an elastic membrane, the *fenestra rotunda*, which thus allows the perilymph to move in the inner ear with the movements of the stapes. The middle of the three canals is the *scala media*, which contains endolymph. Because it has a different ionic composition from perilymph, endolymph is about 80 mV positive to the perilymph and surrounding tissues.

The hair cells of the scala media are basically similar to the hair cells of the labyrinth, the lateral line and the auditory organs of other vertebrates. Of the 'hairs' they bear, only one has the filamentous structure typical of a cilium (§12.7) and is known as the *kinocilium*. The others are known as *stereocilia*; they are much stiffer than the kinocilium, and their matrix contains many short fine parallel fibres which at their base fuse to form a granular cuticular plate. In the mammalian cochlea the kinocilium is lost during development, but its basal granule remains. In all cases in the labyrinth and the cochlea the hair cells are orientated in the same direction. This is important, because the direction in which the kinocilium is

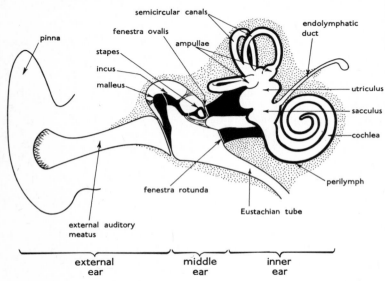

Fig. 9.34 Diagrammatic representation of a generalized mammalian ear. The muscles from the wall of the inner ear to the malleus and incus are shown, but not labelled.

leading is excitatory, displacement in the opposite direction being inhibitory. Hair cells are thus directionally sensitive displacement detectors.

The hair cells of the scala media have a normal resting potential of about 70 mV, so that the inside of these cells is some 150 mV negative to the endolymph. Sound causes the basilar membrane to vibrate, and this movement produces a change in the potential difference of the hair cells that is termed the *cochlear microphonic*, and represents the receptor potential due to the shearing force. The basilar membrane tapers, being much wider at the apical end than at the basal end, and it also varies progressively in stiffness. Any mechanical disturbance therefore sets up a mechanical travelling wave of deformation, which moves at a gradually decreasing velocity from the basal to the apical end of the membrane. Any disturbance produces movement in a particular part of the membrane, but because of its structure low frequency components travel farther than high frequency ones.

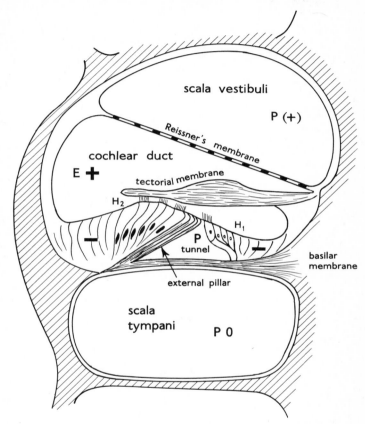

Fig. 9.35 Section through the cochlea of a mammal. The electric charge in each part is indicated (see text).

Asymmetrical sound waves produce an additional response from the hair cells, which is due to the bending force on the hairs and is manifested as a separate potential change from the cochlear microphonic. This is known as the *summating potential*. The hair cells have a threshold of response that varies with the intensity of the sound. The cochlear microphonic, due to the nature of the stimulus, is an alternating potential, whereas the summating potential is not, and this takes the form of a depolarization. The current flow resulting from these two potentials acts as the receptor potential. Clearly, the basis for frequency discrimination lies in the structure of the cochlea, but it is not fully understood, and it is probable that some central integration is necessary in addition.

IO

Nervous Integration

In the previous chapter, we were primarily concerned with the properties of individual neurones, and with their method of communication with one another. We shall now consider how neurones function in the aggregate in producing the pattern of behaviour of an animal. The name *integration* is given to the process of sorting out the impulses derived from the receptors and formulating a set of effector responses appropriate to them, although it may sometimes be applied to the sorting process only. The behavioural response will clearly be dependent on the quality and characteristics of the integrating system—the precision of the receptors in distinguishing different environmental features, the organization of the central nervous system (CNS), which is the hub of the integrating system, and the ability of the effectors to perform a particular response.

10.1 The reflex arc

Except possibly for behaviour which may arise from the spontaneous discharge of neurones independently of any afferent input—a rare event— all behaviour is obviously referable to the sequence receptor→CNS→ effector. This implies that there is always an anatomical pathway involved in any behavioural act. In the case of innate behaviour, the pattern of which is genetically determined, the anatomical pathway is functional as soon as its anatomy is complete. Learned behaviour involves a modification of the behavioural pattern as a result of experience, which means that either a new anatomical pathway must be developed or a pre-existing pathway that has been physiologically non-functional must be brought into use.

In either case, we can think of an established behavioural pattern in terms of a definite nervous pathway, this functional unit being known as a *reflex arc*. Probably the best-known reflex arc is that concerned in the knee-jerk reflex of man, which is figured in most physiology textbooks. Unfortunately, it is an atypical example because its afferent neurone synapses directly with its efferent neurone, whereas the typical reflex arc contains one or more additional neurones interpolated between them (Fig. 10.1). The additional neurones are termed *internuncial* neurones or *interneurones*. The extra synapses introduced by this arrangement provide many more positions on which other neurones in the CNS can converge to modify the passage of impulses through the reflex arc, or from which the processes of other neurones can diverge to run to their cell bodies in other reflex arcs. Such interconnections are essential, because a reflex arc almost always acts in conjunction with others.

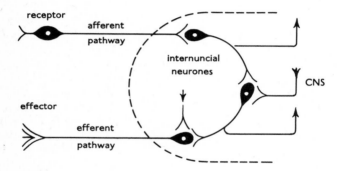

Fig. 10.1   Generalized diagram of the reflex arc. The synapses made by the internuncial neurones enable integration between reflex arcs to take place.

The reflex arc as drawn in Fig. 10.1 represents a *summary* of the pathway through which a behavioural response is achieved. It gives the impression that a single series of neurones is involved, whereas there are actually many parallel pathways, some of which may communicate with different parts of the nervous system from others. Nor does it tell us anything about the grouping, position and relationships of the neurones with the rest of the nervous system. It is important to keep in mind that the reflex arc represents *pools* of neurones and *tracts* containing many nerve axons, and that it expresses the mean of the characteristics of the individual neurones concerned.

10.2 The basis of integration

Integration within the CNS may be considered at several different levels; at the level of individual neurones, and at group neuronal levels of increasing complexity. All mechanisms of integration, at any level, require the conduction of impulses through the nervous system, and as we have already seen this can be described in terms of changes in the electrical charge on the neuronal membrane, both local and propagated. Thus, in describing the properties of neurones in Chapter 9, we have already dealt with some aspects of integration at the neuronal level. The existence of different thresholds of response in soma and initial segment, and the decremental nature of conduction along dendrites will govern the number and position of synaptic knobs required to excite a postsynaptic neurone. Most, if not all, central neurones receive a convergent input from several sources. The differential positioning of these inputs, if at all precise, could obviously be important in determining whether a neurone would develop a propagated response when stimulated by only one input, or whether stimulation by more than one was required.

10.3 Summation and facilitation at synapses

In the situation described in the previous paragraph, other properties of neurones could be important in determining the nature of the post-synaptic response. These include the phenomena of *summation* and *facilitation*. As its name implies, summation refers to the ability of local graded responses such as end-plate or postsynaptic potentials, or generator potentials, to sum in magnitude provided they are elicited within a relatively close time of one another. The combined response is much larger than the individual responses of which it is composed, but it never exceeds the arithmetic sum of the magnitudes of its component responses. The importance of summation lies in the fact that it permits the development of a propagated response from postsynaptic potentials that individually would not reach the threshold size for its production. The only requirement for summation is a repetitive frequency of impulses in the pre-synaptic terminals.

Summation is often confused with facilitation, perhaps because it is not always easy to distinguish between them in some experimental situations. But whereas summation is simply the arithmetic addition of electrical responses, facilitation refers to an increase in the excitability of a synaptic junction without a visible summation of electrical responses (Fig. 10.2). Because of the increase in excitability, succeeding postsynaptic potentials may grow in size. Thus, one impulse may fail to be transmitted across a synapse, but it may raise the excitability of the synapse so that a succeeding impulse is transmitted.

Facilitation is thought to be a general property of most synapses, and its importance is not merely restricted to enhancing the passage of subsequent closely spaced impulses. Its duration can be unexpectedly long. It has been shown that facilitation outlasts the outburst of impulses in the presynaptic fibres to motoneurones by some 10–15 msec, but this may be a very short period compared with that claimed for some synapses. The extreme case is perhaps that of the swimming sea-anemone *Stomphia*, in which facilitation has been stated to persist for up to 7 days. There are a number of other reports of several hours, and it is evident that if long periods of facilitation are normal, it could provide one explanation for the establishment of learning, which is dealt with in a later section.

The cause of facilitation is a matter of dispute, since it has variously been attributed to a slow and therefore incomplete destruction of the transmitter substance, slow conduction in dendrites persisting after the propagated response has been elicited, and changes in the membrane constants of the postsynaptic neurone. Only the latter explanation could

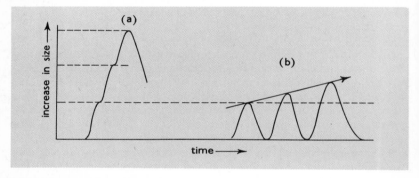

Fig. 10.2   To illustrate the difference between (a) summation and (b) facilitation. In (a) the three successive responses are all of equal size, but are sufficiently close in time to sum together arithmetically. In (b) the three responses are separate in time, but each is larger than the preceding one.

apply to facilitation lasting hours or days. However, there is no reason why facilitation should not be due to multiple causes, and there is certainly evidence that one presynaptic impulse can cause the mobilization of acetylcholine within the presynaptic terminals and hence facilitate the passage of a subsequent impulse. This effect is only short-term.

Despite the operation of facilitation, a repetitive frequency of impulses in the presynaptic terminal does not continue to enhance the passage of impulses indefinitely. At the vertebrate motor end-plate repetitive impulses lead to a drop in the size of the end-plate potential, and if stimulation of the motor nerve is continued for a considerable time this may build up to a level such that the end-plate potential fails to reach the threshold for a propagated response. It has been shown that the number of quanta of acetylcholine released per impulse falls in these circumstances, although the size of the individual quanta remains unchanged. In other words, it looks as if the phenomenon, which is known as *fatigue*, results from some kind of failure in the mechanism that governs the synchronous release of the quanta of acetylcholine when an impulse arrives at the junction. Fatigue is best known at neuromuscular junctions of vertebrates, but it almost certainly occurs at other types of synaptic junctions.

10.4 The afferent input

Integration at the group neuronal level must depend initially on the characteristics of the input. Since nerve impulses are all-or-nothing phenomena, the only variable in the afferent input from a single receptor cell is the frequency of the impulses. We have already seen that the frequency of sensory impulses is directly determined by the intensity of the stimulus and the liability of the receptor to adaptation. We may presume that the afferent nerves are linked to the CNS in a way that takes both these factors into account.

However, the CNS receives continuously inputs from many receptors of different types, and a single receptor cannot really be considered in isolation. There is obviously a need for the CNS to carry out some kind of *comparison* of sensory inputs; comparison between the inputs from a number of like receptors, whether separate or combined into one multi-unit receptor; and once these have been sorted into a pattern, comparison between the sorted inputs from different types of receptors. We may expect the process of sorting to include the simplification of the input by some form of grouping.

These considerations are confirmed by a number of examples. We have already referred to the convergence of rods upon the retinal synapses of the vertebrate eye (§9.20); and although there are obvious reasons for this convergence in relation to the intensity of illumination, it is possible that they may also serve the purpose of grouping the visual input. It is certain that the visual fields of the cat and the octopus are analysed into oval fields, especially in the horizontal and vertical planes, and it is very probable that this analysis is performed by the dendritic trees of the retinal synapses. Many multi-unit receptors, of which the crustacean stretch receptor is an example, send thick and thin nerves to the CNS from units with a different threshold to stimulation and different adaptation times, and this clearly enables the CNS to make a detailed comparison of their inputs.

The related tasks of grouping and comparison are illustrated by two known facts. One is that if the frequency of impulses from a receptor unit is followed through its initial central synapses it is found to decrease showing that some form of grouping has occurred. Secondly, it seems to be a general fact that the number of afferent axons in an animal is always greater than the number of efferent axons, so that some overlap of inputs may be presumed to occur which will permit both grouping and comparison to be made. It is interesting to note that the ratio of afferents to efferents decreases in more highly evolved animals. This is not due to a reduction in grouping or comparison. It arises through an increase in the number of efferent fibres, which is related to the greater complexity and precision of the effector actions of such animals. It is also presumed to enable them to establish many more 'labelled lines' in the nervous system, through the linkage of the already precise afferent inputs with the now more precise efferent output, thus leading to more specific behaviour patterns.

The establishment of such labelled lines may account for some observations made by comparative ethologists. Although an animal's behaviour can only be as good as its receptors, an animal does not necessarily use the whole of its available sensory equipment in the performance of a particular pattern. Thus, the courtship reflexes of sticklebacks can be elicited by crude models which imitate only a few of the characteristics of the adult fish. The beetle *Dytiscus* possesses good eyes, but it never uses them to hunt prey even at close quarters, relying instead on smell and touch for this purpose. There is no question that the unused sense organs are fully functional, and these results can therefore only mean that the input is tested against a central pattern that is specifically related to a particular behavioural act. Such patterning is evident in many known examples of behaviour. For example, certain birds will react to a 'hawk' silhouette in the positions in which they are most likely to encounter them in nature, but not in other positions.

10.5 Central interaction between reflexes

Because of the variety of sensory information arriving at the CNS at any one time, there must be scope within it for interaction between reflexes. One mechanism for interaction is the overlap of sensory inputs, to which we have already referred in connection with the need for grouping and comparison.

The fact that afferent pathways overlap was shown many years ago by Sherrington, who gave the name *convergence* to the phenomenon. He found that stimulation with a maximal electric shock of the afferent input to the group of motoneurones controlling a given muscle resulted in a contraction of the muscle that was much smaller than that obtained by maximal stimulation of its afferent nerve. He deduced from this result that a given afferent input is shared between a number of motoneurone pools. He then proceeded to demonstrate that such inputs overlapped with one another by isolating each afferent input to a motoneurone pool, and stimulating them all simultaneously. The contraction of the muscle was now greater than that resulting from direct stimulation of the efferent nerve trunk. Sherrington illustrated his interpretation by the kind of diagram shown in Fig. 10.3. It should be appreciated that the model represented in Fig. 10.3 is too simple, for it takes into account the overlap of only two sensory inputs. The actual overlap is likely to be multiple and to be achieved by a variety of dendritic connections so complex that it would be impossible to analyse.

One form which interaction between reflexes can take is what is known

afferent
pathways

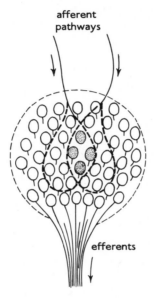

efferents

Fig. 10.3   To show the way in which sensory fields of two afferent pathways to a motor neurone pool might overlap in their effect upon (converge upon) the same motor neurones. (After Creed, Denny-Brown, Eccles, Liddell and Sherrington, 1932, *Reflex Action of the Spinal Cord*, Oxford University Press)

as the *chain reflex*. A chain reflex consists of a series of separate reflexes which occur in a given order, and in which each reflex cannot be elicited before the one that normally precedes it in the sequence has been performed. Many examples are known, including such diverse behaviour patterns as the micturition reflexes of man and the courtship and copulation patterns in many other animals. The completion of any one reflex in such a chain leads to a modification of the sensory situation, which will include interoceptive impulses as well as exteroceptive ones. Thus, the sensory input necessary for any one reflex is produced by the completion of the reflex that immediately precedes it, and there is no need to postulate separate connections within the CNS to ensure the operation of the sequence, although this does not mean that such connections do not exist.

Interesting information about interaction between reflexes is obtained by attempting to elicit more than one reflex action at a time through stimulation of the inputs known to produce them. Where two reflexes are allied in nature, i.e. will lead to the same movement, they will often sum in intensity or area of action or both, but not always. Thus summation of reflexes is not invariable, in contrast to summation at synapses, and this suggests that reflexes can be switched in or out of the operative system. This certainly happens when an attempt is made to elicit antagonistic reflexes simultaneously. Thus, if stimulation adequate for the production of a scratch reflex is given to a dog, and at the same time the sole of the foot of the animal is stimulated, the scratching is inhibited and there is a tendency for the foot to withdraw. The extinction of reflexes in this way is well known to us, for human beings use various tricks which depend on this phenomenon to stifle sneezes and hiccoughs. It is interesting to note that reflex responses to 'dangerous' stimuli always extinguish other reflexes, as might be expected on selective grounds. The mechanism by which extinction is achieved is not known.

10.6 Decision units

Common-sense considerations as well as physiological evidence lead to the conclusion that integration may occur at a number of successive levels, and that at each of these levels a mixed input is taken and integrated, after which a decision must be taken about whether to pass on the integrated material for action or otherwise. Thus, the initial sensory inputs and the output from each integrating centre will always converge upon one or a number of *decision units*, which will determine whether an output shall be passed on for action. The all-or-nothing nature of propagated responses and hence of impulses carried over any distance, and the nature of synaptic transmission imply that in the majority of cases the decision will be of the yes/no variety.

Decision units must be of three types. The simplest type will be the single neurone, and there obviously is a sense in which all neurones act as decision units, just as there is a sense in which all neurones act as integrating units. But in the sense in which the term is used here, for units on which inputs converge and which then determine action, few single neurones will qualify. One class of neurones that does fit the definition comprises the *giant* neurones found at certain sites in the nervous systems of many invertebrate animals and also in some of the lower vertebrates.

The term giant neurone is an arbitrary one, and includes a very wide range of neurones. The criterion is that a giant nerve axon is one whose diameter is much greater than that of the majority of axons in the body of the particular animal, and this, with its cell body or bodies makes up a giant neurone. Thus, in the ventral nerve cord of insects a giant fibre has a maximum diameter of 12 μm, but is still much larger than the other axons present; whereas the giant fibres found in the mantle nerve of the squid may be as large as 800 μm in diameter or even more. Giant axons can originate in two different ways. One type is simply the axon of a single giant cell body and this, which is found in polychaets, nemertines and balanoglossids, rarely exceeds 40 μm in diameter. The other type represents the fused axons of a number of cell bodies of normal size, as in the giant axons of oligochaets, crustaceans and cephalopod molluscs. Functionally, however, they may all be regarded as similar since they both possess two features important for the functioning of the nervous system, namely thick axons and a large surface area.

It was pointed out in §9.5 that the velocity of an axon is proportional to its diameter times a constant. In practice this means that for a given animal, the greater the diameter of the axon the faster it will conduct impulses. Furthermore, the large size of giant axons enables many synapses to be made with them by other neurones in the CNS; and since they are usually long as well as broad they themselves conduct over long distances without the interruption and delay of the impulse through its having to cross a synapse. The result is that they are particularly useful in rapid escape reactions which involve a fairly generalized type of movement by the whole animal. Because the integration they perform is of a crude variety, and is not concerned with small, precise efferent actions, it can be done satisfactorily by a single cell of large size.

Most decision units consist of groups of neurones, and the evidence suggests that they will usually be one of two types. One type is a network of neurones which interact with one another in such a way that the network is *metastable*, or capable of existing in one of two states. In the absence of an afferent input from outside the network, the latter is kept active by the relaying of impulses from its output end to its input, a procedure known as *positive feed-back* (Fig. 10.4). If an input from outside the network was to converge upon it at any of its synapses, each cell would contribute an output to the others that is now related to the input it receives. The normal functioning of the network can be set for a given frequency of operation in the absence of an external input, and its output can be set so that when the external input is applied at a predetermined frequency and intensity, it is sufficient to trigger off the group of effector neurones with which it is anatomically connected.

The second type of decision unit seems to be especially characteristic of what are often called the *higher centres* of nervous systems, those in which the most complex forms of behaviour originate. It consists of a mass of randomly-connected neurones, in which the probability of a connection between any two neurones is a function of the distance between them. Despite this apparent randomness, the mass must still have specific input and output connections. This is the least understood type of decision unit, because of the obvious difficulties of finding out how it works. It probably never functions as a primary pathway in innate reflex behaviour, but seems to have developed as a dominating centre over a wide variety of nervous activity. It is certainly important in learning, and for this reason some of its characteristics are referred to below.

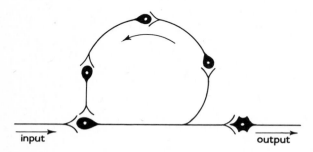

Fig. 10.4   Diagram of a neurone loop producing positive feed-back on a sensory-motor pathway. For explanation see text.

10.7 The physiology of learning

This subject is one of the most interesting that neurophysiologists have to face, and also one of the most puzzling. The problem is usually tackled on the assumption that experience leaves a *memory trace* or *engram* in the CNS. Three possible types of explanation have been advanced to explain the nature of the engram, which are not necessarily exclusive, and none of which can be rejected on present evidence. The first supposes that learning involves structural changes of some kind. It rests on the well-founded view that the differences between instinctive and learned behaviour are ones of degree only, and that learned behaviour is clearly built into the existing pattern of instinctive behaviour. It is therefore assumed that both depend on similar mechanisms, and since instinctive behaviour seems to be related to the existence of well-defined pathways and groups of neurones in the CNS, it is argued that learning must require the development of nervous pathways also.

The relationships between central neurones seem to be too intimate and well ordered for the large-scale growth of processes from neurones, and therefore protagonists of this view presume that the structural changes required must represent the strengthening of pathways or junctions already present but non-functional. The most likely mechanism whereby this might be achieved lies in the change of size of the synaptic knobs. It has been shown that synaptic knobs can increase in size with activity and shrink with disuse. A swollen synaptic knob would decrease the depth of the synaptic cleft and might also liberate more acetylcholine per impulse from its increased surface area, thereby increasing the likelihood that the pathway of which it forms part will be selected. Once established, the pathway would tend to be selected in preference to others, and the engram would reinforce itself. It is quite feasible that new synaptic knobs might be developed in the proximity of other active ones, and the slow growth of short dendrites is not impossible as a long-term explanation of memory. We know surprisingly little about growth of new neurones and degeneration of old ones. It has been suggested that the disappearance of neurones with disuse and the growth of new neurones might even explain learning to some extent.

10.8 Nervous systems as computers

The difficulty about structural explanations for learning is that they are too slow to explain quick or 'instant' learning, such as is seen in higher animals. In the search for a more dynamic process, learning has often been compared to the operation of a computer. Computers are machines into which numerical information is fed and processed according to a predetermined *programme* of additive or subtractive operations (which can be used for multiplication and division, too). The programme may require the storage of information at various stages, which can be brought back into the operation when required. The answer is finally delivered in numerical form. The analogy with the nervous system is obvious, for it also deals with numerical inputs (from the receptors) and outputs (to the effectors), in terms of the number and frequency of nerve impulses; and it also stores information.

Computers may be of two types, *analogue* or *digital*. Analogue computers work by representing the information by some physical variant such as length, voltage or current. Digital computers use absolute units to denote numbers, the difference being illustrated by the difference between a violin and a piano. It has generally been found desirable to make digital computers which use only two digits, o and 1, to represent all numbers from o to 9. In such a *binary* system, different combinations of o and 1 must be used in a code to denote the other numbers. Thus the digital computer tends to be simpler in its physical make-up but requires an elaborate code compared with an analogue computer.

There is, of course, no theoretical reason why a single digit should not be used in a computer, in which case the code used would have to be dependent on the frequency pattern of the digit, which is a kind of analogue system. The similarity with the yes/no decision units of the nervous system and the variable frequencies of nerve impulses fed into it is striking. The frequency of nerve impulses actually observed, however, is much slower than would be possible in a man-made computer, and this suggests that the precise digital method is not the one, or at least not the only one, employed in the nervous system. It has been pointed out that a stable pattern or programme is already built into the CNS in the form of its anatomical relationships, and that this might be organized in such a way that quite broad categories of information are sufficient to switch a decision unit. A mechanism of this sort is an analogue mechanism.

Computers are electronic devices which use electrical pulses to represent the information that is fed into them. They retain information by applying the necessary pulses to the input of an electronic circuit that is capable of keeping the pulses circulating round within itself, requiring only an external source of energy to keep it going. If the analogy with a computer is to be maintained, there ought to be similar circuits in the CNS. In fact, *self-regenerating* or *reverbatory* circuits have often been advanced to explain the engram, without their necessarily being tied to a computer analogy.

There is a network of neurones known as *type B* neurones in the parietal cortex of mammals which has sometimes been considered the antetype of such circuits. A stimulus applied at any point to this network causes a wave of activity to spread over it from the stimulated area. Behind the spreading wave-front the neurons continue to discharge repeatedly for as long as 5 s, and the burst of impulses dies out only because of increasing refractoriness of the neurones. The prolonged discharge by each neurone following a single stimulus is caused by different rates of repolarization in different areas of its surface. This results in a current flow from one area to another, which is sufficient to reach threshold for a propagated response at first, but fails to do so in time because of the increase in refractoriness.

It is doubtful whether a discharge that lasted this length of time could be very important in learning, although not impossible. But a network of this sort does lead to the conception of other networks in which a sustained discharge might be produced by positive feed-back, as in the case of the metastable circuit mentioned earlier (Fig. 10.4). The purpose of such a network would be to provide a level of activity which would sum with the impulses arriving in the afferent input to produce a burst of impulses in the efferent output from the network. If the afferent input was timed to coincide with the circulating impulse(s) this would indeed be the result, but otherwise there would be no summed effect. It should be pointed out that Fig. 10.4 is purely a diagrammatic representation. An actual circuit that conformed to this idea would consist of many more neurones.

It is often supposed that the spontaneous rhythmic discharges observed in the association areas (see §10.17) are due to reverberatory circuits. But these represent the synchronous discharge of many cells, probably thousands and more at a time, and although such discharges are still not understood, they probably do not simply represent such circuits.

10.9 Chemical learning

We have seen that structural changes by themselves are inadequate to explain quick learning; and it may appear that reverbatory circuits would supply a possible explanation for it. Perhaps a combination of the two might be adequate, but it is certain that reverberatory circuits by themselves are not an adequate basis for learning. The reason is that, in some cases, animals have been subjected to anaesthesia or even deep freezing, which has been shown to abolish all electrical activity within the nervous system. Yet their memory of events prior to this treatment are unaffected.

Thus, it looks as though the engram must take the form of a *non-active* information store. Change in size of synaptic knobs is one possibility, but another is storage of information in chemical form. This has always been a theoretical possibility, but it has been brought to the fore recently as the result of some remarkable claims based on experimental work of a similar nature on several widely-different sorts of animals.

It was observed that cannibal planarians appeared to acquire learned behaviour from their victims, and the same effect was noted when extracts of the RNA from planarians which had been trained in a particular conditioned reflex were injected into untrained planarians. The injected animals learnt the conditioned reflex much more quickly than control animals. Subsequent experiments with rats in relation to the operation by these animals of 'reward boxes' also showed that rats injected with RNA extracts from trained rats learned to operate the boxes faster than control rats. It has been argued that RNA is an obvious candidate for a memory-coding chemical, since both heredity and memory involve the storage of information, and it is claimed that ribonuclease, which destroys RNA, abolishes a conditioned response to light when injected into planarians.

Many physiologists remain highly sceptical about such experiments. They find it hard to understand why RNA injected into the body of an animal is not destroyed through normal metabolism. And the same objection would apply to any other substance that might be present in the rather crude 'RNA' extract. It may also be noted that the result of injecting RNA into an untrained animal is that it learns faster, but it does not exhibit the learned behaviour immediately. Although experiments have been performed which seem to demonstrate the specific nature of the learning transmitted through the RNA extract, they do not in fact exclude the possibility that it is a predisposition to learn a particular class of operation that is being transferred. It is impossible to disregard this work completely, but further experiments will be necessary before its significance, if any, can be properly evaluated.

10.10 Characteristics of the engram

In seeking to explain learning, it is necessary to keep in mind the general characteristics of the process as far as they are known. Several features seem to be common to the learning process in animals. One feature is the tendency for a memory to fade with time unless it is reinforced periodically with the training stimulus. It is not possible to conceive a physiological basis for a reverbatory circuit which would have a tendency to run down that could be prevented through reinforcement by further external impulses. On the other hand, both the structural and chemical explanations of learning can encompass fading, in the latter case by depletion of the chemical through normal metabolism. This is an additional reason for supposing that although reverbatory circuits might be important in the initial establishment of a memory, long-term storage of information must depend on some other method.

Perhaps the most puzzling aspect of the engram is that it appears to be both localized and diffuse. If the cerebral cortex of mammals is stimulated at different points on its surface it is found to be divided into motor and sensory areas. Stimulation of motor areas results in well-defined motor actions that are specific to a given area; and stimulation of the sensory areas arouses specific sensations and emotions. The latter observation has been powerfully reinforced by experiments on fully-conscious humans who were able to report their feelings and sensations precisely. As a result of this work, it is possible to map out the cerebral cortex according to the sensory or motor functions with which each of its areas seems to be concerned. An example of such a map is given in Fig. 10.5.

It might be thought that if functions are localized in this way, the engram will be similarly restricted to specific regions. In fact, in some cases, as for example the part of the cortex concerned with visual function, areas not primarily concerned with the sensory-motor aspects of vision, and known as *association areas* can be identified which seem to be concerned with the learning process.

Against this fact must be set many findings that indicate that learning is a function of large areas of the cortex in mammals. For example, rats trained to find their way about mazes or to operate boxes that are opened by latches retain this learned behaviour when a large part of the visual cortex is slashed with criss-cross cuts or removed; and it has been shown that the memory persists when only one-sixtieth of *any part* of the antero-lateral visual cortex remains intact. There are clearly formidable problems

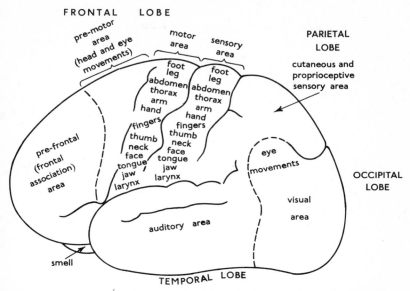

FRONTAL   LOBE

pre-motor area (head and eye movements)

motor area

sensory area

PARIETAL LOBE

cutaneous and proprioceptive sensory area

foot
leg
abdomen
thorax
arm
hand
fingers
thumb
neck
face
tongue
jaw
larynx

foot
leg
abdomen
thorax
arm
hand
fingers
thumb
neck
face
tongue
jaw
larynx

pre-frontal (frontal association) area

eye movements

OCCIPITAL LOBE

visual area

auditory   area

smell

TEMPORAL   LOBE

Fig. 10.5   Localization of function in the cerebral cortex of man. (Adapted from Best and Taylor, 1950, *The Physiological Basis of Medical Practice*, 5th edn., Williams and Wilkins, Baltimore.

relating to the organization of at least the higher centres of learning in animals to which the answer is not at the moment at all clear.

10.11   Gross organization of nervous systems

We now turn from the analysis of function at neuronal and group neuronal level to consider the functioning of nervous systems in relation to the animal as a whole. The most primitive groups, the Protozoa and Porifera, have no nervous system in the metazoan sense, although there is no doubt that some kind of conduction can take place between the basal granules of cilia (§12.8) and these organelles may be grouped in some ciliates through linkage by strands which many workers regard as co-ordinating threads. In sponges, there is some spread of excitation among the collar cells surrounding the osculum, but this is presumably akin to the spread of excitation between smooth visceral muscle (§11.19).

10.12   The coelenterate nerve net

In coelenterates, enteropneusts, echinoderms and ascidians, the major part of the nervous system takes the form of a loosely-organized network of neurones and apparently similar nerve nets are found in parts of the nervous systems of annelids and molluscs, which play a role in the peripheral motor co-ordination of movement. There are in addition networks of fibres, or *plexuses*, in the guts of vertebrates. Little is known about the detailed functioning of many of these networks, and the bulk of the available information relates to coelenterates and echinoderms.

The nerve net of coelenterates comprises a network of neurones not orientated anatomically, except in the cases described below, and interconnected in an apparently random manner (Fig. 10.6). There is evidence

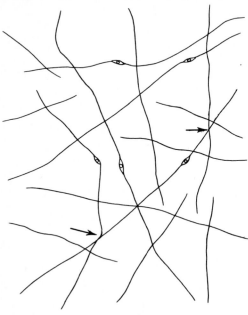

Fig. 10.6 Diagrammatic representation of the nerve net in the mesentery of a sea anemone. Two synapses are denoted by arrows. (Based on Batham, Pantin and Robson, 1960, *Quart. J. Micr. Sci.*, **101**, 487)

that there is syncytial continuity between neurones of *Hydra* and *Velella*, but in general it seems that the neurones of the coelenterate nerve net make synapses with one another. The nature of these synapses, which seem to be akin to the juxtaposition of two unmyelinated axons, has been discussed in §9.12. The junctions are not polarized, and so a nerve net conducts impulses in all directions away from the point of stimulation. The axons are often relatively short, and it may be that propagated responses are not always necessary for conduction between synapses.

Most functional studies of the nerve net have been carried out on the larger coelenterates in the Scyphozoa and Anthozoa. In these groups there are regions of the body in which the nerve net is gathered up into definite tracts in which the axons run parallel to the direction of conduction. These axons are longer and such *through-conducting* tracts therefore possess fewer synapses per unit length than are found elsewhere in the network. Conduction along these axons is much faster than along the other axons, ranging from 0.7–2.0 m/s against a speed of 0.5 *cm*/s in the general network. Whereas conduction by the ordinary neurones may well be non-propagated, that in the fast neurones is undoubtedly propagated. Fast through-conduction systems of this kind are found in those regions of the body in which they confer a special advantage, such as the oral disc (feeding reflexes), the column (defensive and excretory reflexes) of anemones, and the umbrella of medusae (swimming reflexes).

Two characteristics that distinguish the nerve net of the coelenterates from the nervous systems of other metazoans are its marked dependence on facilitation and the phenomenon of *decremental conduction*. The sea anemones *Calliactis* and *Metridium* do not normally respond to a single electric shock applied to their vertical column. But if a second stimulus is applied within 0.2–2.5 s of the first, a marked contraction occurs. Within this frequency range, the closer the stimuli the bigger is the response (Fig. 10.7). Similar effects have been observed in scyphozoans. For example, when stimuli at a frequency of one every 4 s are applied to the subumbrellar muscles of the jellyfish *Rhipolema*, the recorded mechanical responses continue to increase in size up to the sixth or seventh stimulus, as is the case in anemones (Fig. 10.7). Facilitation in these animals is thought to be both neuro-neuronal and neuro-muscular. That it is a labile process is shown by the fact that under abnormal circumstances such as high external potassium concentrations or high temperatures, a single stimulus will elicit a response. However, subsequent responses still increase in size.

On present evidence, decremental conduction is peculiar to coelenterate nerve nets, although it must be admitted that we know so little about nerve nets in other groups that this statement is founded on negative rather than positive evidence. The term is used to describe the fact that the contraction elicited by a local stimulus dies away with increasing distance from the point of stimulation. In other words, the strength of the response in any part of the animal is proportional to its distance from the stimulus. This demonstrates that conduction in the nerve net is not all-or-nothing. It is not clear whether the phenomenon is synaptic, but it is generally believed to be so. The net result of the combination of facilitation and decremental conduction is that more and more of the animal becomes involved as the stimulus becomes stronger or more widespread. This is clearly a satisfactory system for a radially symmetrical animal.

As far as is known, and with the exception of the statocysts of medusae, the sense organs of coelenterates are single cells that are all of one type. It is interesting to note the degree of co-ordinated activity which can be achieved with such an apparently primitive system. At least some sea anemones exhibit complex 'walking' movements over the substrate,

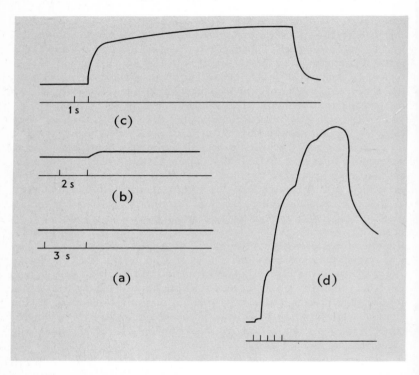

Fig. 10.7 Facilitation in sea anemones. Contractions of the column of *Calliactis* were recorded on a smoked drum. In (a), (b), (c) contractions elicited by pairs of electrical stimuli applied to the column at progressively decreasing stimulus intervals. (d) contractions from a burst of 5 stimuli at $\frac{1}{2}$ s. intervals between stimuli. In (d), the stimulus intensity was twice that used in (a), (b) and (c); note the very small response to the first stimulus.

which involve local concentrations of the column in sequence, and attachment to and disengagement from the substrate. The two together result in a very slow kind of creeping movement which, when captured by slow cinematography and projected at normal speed, present a most complex appearance.

Evidence that reflex arcs occur in the coelenterate nerve net is provided by the fact that contracting muscles appear to be able to re-excite the system to produce further contractions. It may also be presumed that the external sense organs act in reflex arcs with central and effector components.

10.13 Echinoderms and protochordates

The nerve nets found in phyla other than the Coelenterata are mostly peripheral to a more organized central nervous system. Even in coelenterates there is a need for differentiation of the nervous system, which is reflected in the development of through-conducting tracts and differences in neuronal characteristics. Such a system is adequate for small sluggish animals, but larger and more complex animals need more precise receptors and effectors and hence a more complex central integrating system. Because of this, nerve nets have been relegated to local situations in such animals. The nervous system which most nearly approaches that of the coelenterates is probably that of echinoderms, and it is perhaps significant that these are also sluggish animals.

The nervous system of echinoderms consists of a central oral ring, and five radial nerve trunks which link up with a peripheral nerve net that innervates the various appendages—tube feet, pedicellariae, spines, etc. The neurones mingle with the nerve axons in the central system and are not aggregated in any way, a condition that recalls the through-conducting tracts of coelenterates. In echinoids there are some aggregations of neurones which form local reflex centres at the base of the spines, and in this group the net is clearly more differentiated than it is in coelenterates.

In starfish, the peripheral nerve net also contains local reflex centres, which control bending of the spines and aid in the movement of the tube feet in relation to the tension on the arm containing them. In addition to these localized activities, the oral and radial trunks exercise an overall control of movement. The oral ring seems to be the directive agent, and if it is removed each arm moves quite independently towards its base; whereas if a piece of the ring is allowed to remain attached to each arm, it tends to move towards its distal end. When the oral ring is cut into two parts, the two corresponding halves of the starfish pull themselves apart. Experiments of this sort seem to demonstrate that the oral ring controls the direction of movement of the whole animal and that there are probably centres at the junction of the ring and each radial trunk which control the movement for that trunk. Two features demonstrate the relatively primitive nature of the system: a starfish with all radial nerves cut exhibits poor co-ordination, but is nevertheless able to turn itself over; and the cutting of the radial trunk makes comparatively little difference. In both cases, the deficiency is made up by the passage of impulses through alternative pathways in the peripheral net. The other noteworthy point is that impulses will travel in both directions in the oral ring and radial trunks, indicating a lack of polarity in the system. Speed of conduction is actually slower than in the through-conducting tracts of coelenterates, varying between 0.1 and 0.2 m/s in the oral ring and trunks.

It is probable that the nervous system of hemichordates is fairly similar to that of echinoderms. The movements of the proboscis that take place when *Saccoglossus* is burrowing are controlled by the dorsal nerve cord, and those of the body by the main nerve trunks; but in other genera the dorsal cord is lacking, and the co-ordination of proboscis movements is the task of the subepidermal nerve net. In solitary tunicates the large ganglion between the two siphons is necessary for intersiphonal reflexes, but local responses of a siphon continue to be mediated through the nerve net.

10.14 Ganglia

The evolution of the nervous system in metazoans other than coelenterates and echinoderms has followed two main trends; the gathering of the neurones into orientated tracts, with the emergence of a single main longitudinal tract; and the evolution of a brain. The two trends are interconnected, but it will be convenient to consider them separately.

The production of nerve tracts is foreshadowed by the through-conducting tracts of coelenterates, and the oral and radial tracts of echinoderms. But the primary orientation of such tracts in a longitudinal direction is necessarily associated with bilateral symmetry, for in bilaterally symmetrical animals all movement is related to the main axis, and this must necessarily be reflected in the nervous system that controls it. In the more primitive bilaterally symmetrical invertebrates such as the platyhelminthes, and in the more primitive members of higher invertebrate phyla, the longitudinal tracts contain nerve cell bodies throughout their course, and they are joined by many lateral connections that give the nervous system a net-like appearance. Within the platyhelminthes, the number of longitudinal tracts varies between two and eight. In other invertebrates the number is always two, and except in their most primitive members, these have become fused into a single ventral nerve cord.

In addition to the tendency for the ladder-like net to become condensed into a single nerve cord, the evolution of invertebrate nervous systems has resulted in the concentration of the nerve cell bodies within the tracts into discrete masses known as *ganglia*, linked by their axons. Inevitably, the ganglia of metamerically segmented animals are themselves metamerically arranged, since this is the simplest possible functional arrangement, but the production of ganglia was not necessarily dependent on metamerism. The molluscs also possess ganglia, but it is doubtful if the group was ever truly metameric.

The most primitive metameric state is found in the annelids, which consist of numerous segments, each segment containing a single ganglion bearing a number of lateral nerves which contain sensory axons from the receptors of that segment and motor axons to its muscles. These ganglia do not differ in any essential way from the centres present at the junction of oral ring and radial trunk in echinoderms, and are only marginally different from their local reflex centres, because peripheral control of locomotion is still very important in annelids.

The primary cause of locomotion in annelids is an external stimulus of some kind, but the events which then take place are very dependent on events within and between the segments. This was demonstrated in some now classic experiments. These showed that if the weight is taken off the body of an earthworm by suspending it in a liquid of a density that balances it, the peristaltic movements of the body cease, but will reappear if a load is applied to the muscles by stretching the animal. In another experiment two halves of a worm were connected by a piece of thread. If the thread was allowed free play, the peristalsis in the two halves was co-ordinated, but if the thread was not allowed to move they were not. The peristaltic movements had been co-ordinated by the transmission of changes in tension across the thread. If, instead of cutting the worm in half, some segments are removed completely except for the ventral nerve cord, the two parts of the worm exhibit co-ordinated movement even when the body wall on either side of the exposed nerve cord is pinned down (Fig. 10.8). It is evident from such experiments that the co-ordina-tion of movement in an earthworm is primarily due to local reflexes mediated through the ganglia, their sensory input being derived from proprioceptors in the muscles. Obviously, there must be some correlation between antagonist muscles, and some overlap between the fields covered by the segmental ganglia. By varying the number of segments of body wall which were cut out while the ventral nerve cord remained intact, it was found that if more than three segments were so treated, co-ordination ceased. The overlap therefore cannot be more than three segments, a situation which emphasizes the utility of giant axons (§10.6).

Little is known about co-ordination in the lower arthropods, but in the more advanced forms the number of body segments is limited, and different regions of the body often bear appendages which give them a different function from other regions. We should expect these complica-tions to be reflected in the relationship between the body ganglia. Similari-ties with annelid ganglia are, of course observed, as in the case of the severed limb with ganglion intact that executes stepping movements when dragged along a surface, a clear indication of local reflex actions. On the other hand, if the legs of an insect are removed the regular pattern in which the legs are normally placed during walking is upset, but the animal adjusts itself by adopting a suitable pattern with the remaining legs. In similar circumstances, in arachnids, the pedipalps may be enlisted as auxiliary legs. The changes in pattern occur at once, and they must therefore represent existing pathways in the CNS which are not preferred during normal locomotion, but which come into play when the normal pattern is destroyed. These pathways are not merely intrasegmental, but must be intersegmental also. By cementing pegs on to the damaged legs, the usual walking pattern can be restored, suggesting that it depends on the nature of the sensory input. The new pathways are presumably preferred because of the change in the sensory input.

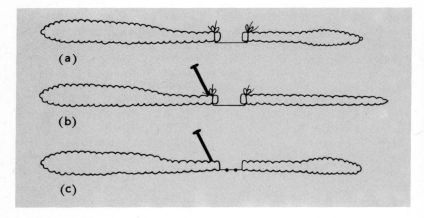

Fig. 10.8 Diagram to illustrate experimental work on peristalsis in the earthworm. (a) The worm is cut in half, but joined by sewing thread; the peristaltic wave passes from the anterior half to the posterior half. (b) The anterior half is now pinned at its front end; no transmission of peristalsis occurs. (c) Several segments are removed from the middle of a worm, but the ventral nerve cord is left intact; although the front half of the worm is pinned as in (c), the peristaltic wave passes to the posterior part of the worm.

In molluscs there is no segmental pattern, and the various body ganglia are largely local in their sphere of influence and autonomous in function, so that overall control is at a minimum.

10.15 Locomotory reflexes in vertebrates

It is interesting to compare the co-ordination of the peristaltic movements of the earthworm with the control of the swimming movements of the dogfish, *Scyliorhinus*. This fish swims by means of a sinuous body movement in which the forward motion of the animal is mainly brought about by the lashing of the tail and the fins are used as lateral and vertical stabilizers. The major swimming movement therefore results from chain reflexes that pass down the spinal cord and in which there must be inter-segmental overlap in both sensory and motor fields.

The influence of the brain can be abolished by transection of the spinal cord just posterior to it, the animal then being known as a *spinal* dogfish. If a spinal dogfish is prevented from coming into contact with the substrate, it exhibits a normal swimming rhythm, together with the appropriate fin movements. If all the afferent inputs to the spinal cord are now severed, a process termed *de-afferentation*, the swimming movement is abolished. If less drastic de-afferentation is performed, involving only some of the dorsal roots, it can be shown that the normal swimming movement is maintained provided that the afferent inputs of a minimum of 25 successive segments anywhere along the spinal cord are left intact. Thus, as in the earthworm, there is an inherent pattern of movement that depends on a minimum degree of afferent stimulation for its appearance. Since the impulses required to maintain the rhythm must pass through the de-afferentated segments, there must also be a considerable degree of overlap between the segments of the spinal cord of the dogfish. However, this overlap does not compare with a simple system such as that found in the earthworm, because the vertebrate spinal cord contains a number of ascending and descending nerve tracts. Similar results have been obtained with amphibians, in which it is found that a minimum of the afferent input from two legs must be present for normal locomotion to take place.

10.16 Cephalic dominance and brains

In bilaterally symmetrical animals the head becomes extremely important, because the anterior part of the animal samples the environment into which the animal is moving before the other parts of the body, and this has resulted in a tendency for the most important receptors to be concentrated at the front end. The nervous system at the head end becomes enlarged to deal with the increased sensory input, and it is not surprising to find a tendency for this anterior part of the nervous system to assume overall directive functions, first of body movements, and then of behaviour as a whole. When the anterior part of the nervous system exercises an overall direction of activity, it is known as a *brain*.

In turbellarians, the anterior mass of the central nervous system is associated with the eyes and with other receptors such as chemoreceptors and current flow receptors, which are concentrated in the head region. At this level of organization it might be thought that the mass functions purely as a co-ordinating reflex centre for these head receptors. However, when the head mass of nervous tissue is removed, in addition to the expected disorganization of the more complex forms of movement the animal becomes hypersensitive to stimuli, which suggests that the head mass normally exerts some kind of overall inhibitory action on movement in this animal.

A similar kind of control is exercised by the supra-oesophageal ganglion in annelids and arthropods. When the supra-oesophageal ganglion is removed from a nereid or an earthworm, the animal becomes hypersensitive and restless, and certain complex movements are either not attempted or are only achieved with difficulty, as in burrowing by the earthworm or feeding by nereids. Although it is difficult to be sure that the disturbance of co-ordination following the operation is not due to the absence of sensory stimuli, the overall restlessness would appear to be the result of the removal of the inhibitory action of the supra-oesophageal ganglion.

The control by the supra-oesophageal ganglion is well illustrated by experiments performed on arthropods in which, instead of severing the ganglion completely from the rest of the nervous system, the circum-oesophageal commissure that connects the supra- and sub-oesophageal ganglia was cut on one side only. This operation resulted in *circus movement* towards the intact side (Fig. 10.9) which may be attributed to the greater motor activity on the cut side of the animal. In crabs this circus movement has been abolished by careful electrical stimulation of the cut connective.

The restlessness of annelids and arthropods that follows the removal of the supra-oesophageal ganglion is eliminated by removal of the sub-oesophageal ganglion. The animals now become abnormally quiet and difficult to stimulate into action. It is concluded that the sub-oesophageal ganglion is responsible for the restlessness of the animals following removal of the supra-oesophageal ganglion. It looks as if the sub-oesophageal ganglion must contain nervous centres that tend to cause an increase in the overall motor activity of the animals, but that these centres are in turn under the overall inhibitory control of the supra-oesophageal ganglion.

The relationship between the supra- and sub-oesophageal ganglia in annelids and arthropods is characteristic of the organization of brains. There is a general tendency in animals for 'primitive' aspects of behaviour, such as the basic locomotor pattern, which are initially the province of local reflex centres, to be directed from motor centres in the brain, and for these 'lower' motor centres in turn to be controlled by 'higher' motor centres. As brains have evolved, the higher centres have come to be associated with special integrating areas that connect them with sensory areas of the CNS. These trends are particularly evident in the evolution of the vertebrate brain, but they are also present in what is perhaps the most complex invertebrate brain, that of the more advanced cephalopod molluscs.

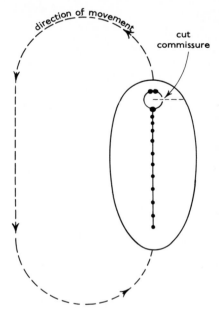

Fig. 10.9 Diagram to illustrate circus movement following the cutting of the circum-oesophageal commissure on one side of a generalized arthropod shown in outline. The movement is towards the intact side, due to increased activity on the operated side.

10.17 The brain of cephalopod molluscs

The brain of squids and octopuses can be divided functionally and anatomically into a supra- and a sub-oesophageal mass (Fig. 10.10). Each mass in turn may be sub-divided into areas that present a lobular appearance externally, and in the case of the supra-oesophageal mass, detailed histological differences. By the classical neurological procedure of extirpation or stimulation of individual lobes or areas, coupled with a study of consequential changes in behaviour, the functions of the different parts of the cephalopod brain have been clearly worked out.

Removal of the whole of the supra-oesophageal mass results in restlessness, as in annelids and arthropods, and the squid on which such an operation has been performed will swim around ceaselessly. If the operation is carried out on one side only, circus movements result. More localized excision or stimulation shows that the higher motor centre for swimming is the *basalis anterior* lobe, which inhibits the action of the lower motor centre in the sub-oesophageal mass, the *palliovisceral* lobe. However, the basalis anterior is not autonomous in controlling the pallio-visceral lobe, for the former is itself subject to inhibitory action by the optic lobe, an important primary sensory centre in cephalopods. This

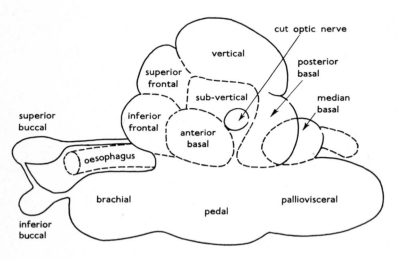

Fig. 10.10   Side view of the brain of *Sepia*. The position of the oesophagus, and of the brain lobes not visible from the outside, are indicated by broken lines. (Based largely on Wells, 1962, *Brain and Behaviour in Cephalopods*, Heinemann, London and Toronto ; Stanford University Press, California)

specific picture is generally true of the actions of the supra- and sub-oesophageal masses. The sub-oesophageal lobes are all 'lower' motor centres that control such structures as the fins, the mantle and the ten-tacles, respiratory movements, and the action of the chromatophores. These are under the inhibitory control of the 'higher' motor centres in the supra-oesophageal mass, which are themselves controlled by the sensory input. The major sensory centres are the optic and olfactory ganglia.

The higher motor centres are all found to be in the posterior part of the supra-oesophageal mass, i.e. in the basal lobes. Stimulation of the anterior part, the verticalis and frontalis lobes, which together are often called the *verticalis complex*, does not produce motor activity, in contrast to the other parts of the brain. If the verticalis complex is re-moved the swimming capacity of a squid is not affected, but there is a reduction in the ability of the animal to perform more complex behaviour. The squid can only attack its prey if it remains within sight, whereas normally it will twist and turn to follow it. Removal of the verticalis complex also abolishes the memory of various pieces of learned behaviour, and these can be relearned only with difficulty, and the memory is retained for a much shorter time. From these and other experiments it is evident that the main centre for the establishment of learning is the verticalis complex, which is the integrative centre for the most complex pieces of behaviour; however the *site* of memory storage is the optic lobe. It will be seen that the verticalis complex is comparable to the *association centres* found in vertebrate brains.

10.18 Vertebrate brains

In the vertebrates, the brain develops as a swelling of the front end of the dorsal nerve tube, the rest of which persists as the spinal cord. The spinal cord is largely an involuntary reflex area, based on a segmental pattern. There is a decrease in its autonomy in the more advanced vertebrates. We have seen that a spinal fish continues to swim normally, but a spinal frog requires some stimulation if it is to swim or walk. A spinal mammal such as a cat is incapable of any motor response for some hours after the operation, and its motor reflexes are depressed for several days. There appears to be a slow adjustment in this case by the spinal cord, presumably through the opening up of existing but unused pathways.

The vertebrate brain is universally based upon the structure known as the *brain stem*, which consists of three parts, the *fore-*, *mid-* and *hind-* brains (*pro-*, *meso-* and *rhomben-cephalons*). Each of these parts has become elaborated during the evolution of the vertebrates. Early in this process, the fore-brain became divided into two parts, an anterior *telencephalon*, which bifurcates to form the olfactory tracts, and the *diencephalon*. The diencephalon differentiated functionally very early on, for its component parts and appendages, the *epithalamus, thalamus, hypothalamus, pineal stalk, infundibulum* and *optic chiasma*—at which the optic tracts from the eyes cross over or decussate before entering the mid-brain—are present in the most primitive vertebrates, as indeed are the olfactory tracts. By contrast, the elaboration of the mid-brain through the development of the roof into *optic lobes*, and of the hind-brain into posterior *medulla oblongata* and anterior *cerebellum* and *pons*, seem to have been more recent developments.

The brain of a lamprey (Fig. 10.11 a and b) exhibits the basic features of the brain stem, with the early fore-brain structures already present, and a very small cerebellum. This brain and that of fishes in general, which usually possess a more highly developed hind-brain than the lamprey, is essentially the major sensory area at the front of the CNS.

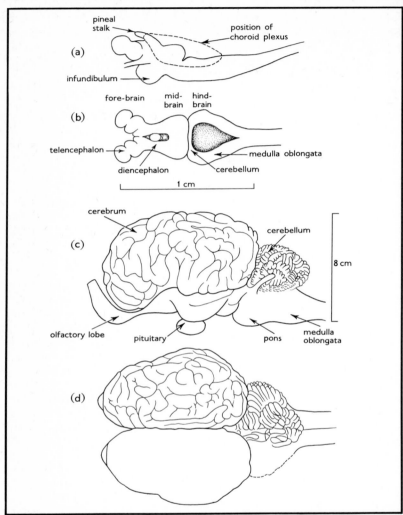

Fig. 10.11 Brains of lamprey (a), (b) and horse (c), (d). The upper drawing in each case represents the lateral view, the lower the dorsal view (horse after Bütschli, Kuenzi and Sisson, from Romer, 1949, *The Vertebrate Body*, 3rd ed., W. B. Saunders, Philadelphia and London).

The fore-brain is largely concerned with taste, the mid-brain with sight, and the hind-brain with balance, and there are not many connecting tracts between the three major areas. One of the most striking features of the evolution of the vertebrate brain is the multiplication and elaboration of such tracts. There is, however, some co-ordination between the actions of the animal resulting from integration in the three primary areas of the brain, the overall direction of which is exercised by the mid-brain in fishes and amphibians, but shifts to the fore-brain in higher vertebrates.

10.19 The hind-brain

The medulla oblongata is a continuation and elaboration of the structure seen in the spinal cord. The only obvious gross difference is that the central cavity has enlarged to become the fourth ventricle, the roof of which is thin and membranous and is the *posterior choroid plexus*. Its major function seems to have been to act as the afferent reception area of the acoustico-lateralis system. Only the labyrinth remains from this system in tetrapods, and the centre persists in them to handle the information from it. In mammals, the centre becomes divided into two parts, one concerned with the labyrinth, the other acting as a specific cochlear centre.

In addition, the medulla contains a number of motor centres which are retained during vertebrate evolution, but become further differentiated. Thus, there is a 'gasping' centre in fishes that produces a rhythmic pattern of inspiration and expiration of water through the appropriate mouth, buccal and gill movements. This persists in tetrapods as a ventilation centre that governs such movements as the swallowing of air in lower tetrapods; and in mammals it is seen to be divided into a number of separate centres, some 'higher', some 'lower', some for inspiration and some for expiration. Another motor centre of importance is one that controls the extent of dilation of the blood vessels.

A striking feature of the medulla of fishes and tailed amphibians is the presence of two enormous giant neurones, the *Mauthner cells*. The axons of these two cells extend the entire length of the spinal cord, and are presumed to provide a generalized control of the movement of the animal, of its extent and speed rather than of its pattern.

The cerebellum is generally believed to be an elaboration of part of the primitive medulla oblongata, and is represented only by a narrow strip in cyclostomes. Its posterior end develops just above the area of the medulla that contains the acoustico-lateralis centre, and its functions are intimately concerned with the information yielded by this centre, among others.

If the cerebellum is removed from fishes the swimming rhythm is not interfered with, but there is a lack of ability to orientate and some loss of general co-ordination. Similarly, frogs and toads can jump and swim after the operation, but there is an increase in spontaneous motor activity and some muscular rigidity. Higher tetrapods are unable to balance properly. Evidently, the functions of the cerebellum are related to the finer control of posture and orientation. It receives impulses from the acoustico-lateralis centre of the medulla, and also from the proprioceptors in muscles and tendons. It integrates this information and directs the contraction of the muscles accordingly.

The area concerned with the acoustico-lateralis centre in the medulla is probably the most primitive. It persists throughout the vertebrates as the *flocculi*, which are laterally placed on the cerebellum. The middle part between the two flocculi exhibits a progressive development in the evolution of the group, for the purpose of receiving and integrating the information arriving from the proprioceptors and other receptors, such as those in the skin, and visual information. The development of the cerebral hemispheres as the dominant area of the brain reaches its zenith in mammals (see below), and here it is necessary for the cerebellum to be connected by both afferent and efferent fibres with the cerebral cortex. This involves the development of many additional tracts entering and leaving the cerebellum, and so the pons beneath the structure becomes especially large and thick in mammals. With the great increase in the activity of the cerebellum in mammals, its size is much increased and it becomes divided into two distinct hemispheres, the surfaces of which are thrown into many convolutions (see Fig. 10.11).

10.20 The mid-brain and diencephalon

The mid-brain is primitively an optic centre. In the lower vertebrates the optic nerves pass straight through the diencephalon after decussating at the optic chiasma and end in the primary visual centre in the roof or *tectum* of the mid-brain. The tectum remains an important association centre throughout the vertebrates, and in the more primitive forms receives sensory information from the cerebellum and from proprioceptors. The motor directions emanating from it after the integration of this data are relayed to the motor centres by a rather diffuse mixture of tracts and neurones, the *reticular formation*. It is these relationships that make the mid-brain the dominant area in the fish and amphibian brain. With the emergence of the cerebral hemispheres as the dominant region of the brain, the tectum becomes subordinated to them, and loses much of its motor-directing function. In mammals, most of the axons in the optic nerves are interrupted by synapses in the thalamus of the diencephalon, and impulses are relayed from there direct to the cerebral cortex. The anterior half of the tectum remains visual in function, and continues to act as an association centre but under the domination of the cerebrum. The posterior part of the tectum changes in function and becomes an auditory centre in association with the cochlear centre in the medulla.

The diencephalon includes a number of important regions. The hypo-thalamus is a relatively small area but is complex in function, even in primitive forms. As we have seen, it forms a link between the pituitary body and the rest of the nervous system in vertebrates generally, influences the blood pressure in reptiles, and in homeotherms develops into a temperature regulating centre, and in some forms a hibernating centre and a sleep centre. The thalamus represents an important relay area throughout the vertebrates, in connecting the fore-brain with the rest of the brain. But in mammals this function is increased, as in the case of the optic nerves cited above; and the epithalamus likewise develops as a relay centre for the cerebral cortex, through which all sensory impulses are relayed except, of course, those of taste and smell.

10.21 The telencephalon

In fishes, the telencephalon seems to be almost, if not entirely, an olfactory area. Its removal has no effect on the general activities of the animal, and may not even lead to the cessation of feeding, since sight and the acoustico-lateralis system can also play a part in feeding activity. Learned behaviour connected with olfactory function may, however, be impaired. The removal of the cerebral hemispheres from amphibians and reptiles does tend to reduce motor activity as well as interfere with feeding. Stimulation of the cerebral hemispheres in lizards results in certain motor responses. There is thus some evidence of additional function in amphibians and reptiles, but this does not amount to a great deal compared with that in birds and mammals.

In the birds, it is the ventral region of the cerebrum that is especially enlarged and added to, and the roof is relatively little altered. This specialized ventral region is the *corpus striatum* (Fig. 10.12), and it seems to be concerned largely with complex patterns of innate behaviour. These may involve some degree of learning, and it is perhaps noteworthy that the most 'intelligent' birds are those in which the new parts of the corpus striatum are particularly well developed. Removal of the cerebral hemispheres of birds results in a general listlessness, suggesting impairment of overall motor function. However, if the corpus striatum is left intact, but the rest of the cerebrum is removed, mating and the rearing of the young and a certain amount of learning can take place.

The development of the cerebral cortex in mammals takes a rather different course from that in birds. It is one that is foreshadowed in many reptiles by the emergence of a new area of cortex in the roof of the cerebrum, the *neopallium* (Fig. 10.12). In even the most primitive mammals the neopallium has expanded over the roof and side walls of the cerebrum, and throughout the evolutionary history of the group it has continued to expand and develop, to become the dominant area of the brain. The corpus striatum is still present and functional as a relay centre for the more automatic reactions, but these and all other bodily functions are

subservient to the cerebral cortex, or can become so through learning.
The 'training' of domestic animals or young children in sanitary habits
is a good example of cortical dominance of what are primitively chain-
reflex actions.

Thus, the cerebral cortex in mammals becomes the area to which all
sensory information is sent, and which dominates and co-ordinates all
voluntary movement and may take over many involuntary movements
as well. The dominance of the neopallium has necessitated the formation
of new connections. The pons has been mentioned already, as has the

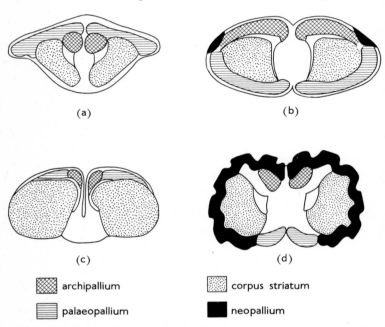

(a)   (b)

(c)   (d)

▨ archipallium   ▦ corpus striatum

▤ palaeopallium   ■ neopallium

Fig. 10.12 Diagrammatic cross-section of the fore-brain of (a) skate; (b) lizard;
(c) sparrow; and (d) a generalized eutherian mammal, to show the major areas.
(Based on Kappers, Huber and Crosby, 1936, *The Comparative Anatomy of the
Nervous System of Vertebrates*, MacMillan, New York)

development of the thalamus as a sensory relay centre. New developments
are especially concerned with the need for intercommunication between
the cortical areas, which is reflected in eutherian mammals by the evolution of a whole new set of tracts that form the *corpus callosum*; and the
need for new motor outlets to the motor centres of brain and spinal cord.
The reticular formation, which in the primitive vertebrate brain is the

main system through which the dominant mid-brain controls motor activity, has been retained as a motor outlet for the cortex, especially in regard to posture and locomotion, for which its connections are to some extent already present. In addition, the cortex transmits motor directions to the brain and spinal cord through a new system, the *pyramidal tract*.

As we have seen, stimulation and excision of the cortex have led to the concept of localized areas of control. These areas form only a small part of the cortical surface, and are interspersed among the so-called *silent areas*, which do not yield primary sensory sensations or motor activity on stimulation. An especially large area of this type occurs at the anterior end of the cortex, which in man is known as the *prefrontal* area. Removal of these areas does not result in a notable depression of normal bodily functioning, but it does lead to an impaired ability to learn in response to environmental situations. We should therefore expect to find, in conjunction with the elaboration of behaviour, and especially of insight learning, that it is these areas that have evolved in the case of primates. The functions of the prefrontal area seem to confirm this expectation. Removal of this area in monkeys results in hyperactivity and a reduction in the ability to solve problems. In man, removal leads to an impairment in the ability to solve more complicated mathematical problems, and to a loss of foresight and insight in relation to problems. There is a corresponding reduction in anxiety in men who are prone to it, and the severance of the prefrontal area from the rest of the brain has sometimes been used in an attempt to calm certain types of mentally sick people. Because of the diffuseness of cerebral and especially prefrontal function, this operation has to be used with caution. It might roughly be likened in its effects to a dose of alcohol. Thus, these association areas, and particularly the prefrontal area, may have contributed to the evolution of man as a social animal, as well as facilitating an improvement in the ability to learn and solve problems.

II

Muscles

Muscles are specialized structures used by animals to move the whole or part of their bodies, or to maintain a particular posture. They consist of elongated cells organized in a variety of ways and bound together by connective tissue. These cells have the power to shorten in length or *contract* when they are activated either by means of a stimulus transmitted down the nerve that innervates them, or in some other way. Sometimes, muscles are unable to change in length when they are stimulated, either because they are attached to a more or less rigid structure, or because of the action of opposing muscles as in the maintenance of posture. Even under conditions of this sort, changes taking place within the muscle and accompanied by the evolution of heat show that the muscle is still expending energy in an attempt to shorten, and it will shorten immediately if the restraint is removed. For this reason, we often refer to the development of *tension* by a muscle rather than contraction, a term that covers all circumstances in which a muscle is expending energy in an effort to contract.

The ability to contract is found in cells from the Protozoa upwards, and is not confined to muscle cells. However, little is known about the functioning of even the muscle cells in most phyla. Isolated pieces of evidence suggest that the mechanism by which tension is developed is fundamentally similar throughout the animal kingdom, but the process of contraction has undoubtedly been most thoroughly studied in a few vertebrates, arthropods and molluscs. Even in these cases, attention has been focused more on one type of muscle, the striated skeletal or voluntary muscle.

The traditional division of vertebrate muscles into *smooth*, *striated* and *cardiac* muscle does reflect certain detailed histological and functional differences; but there are good reasons for preferring a division based on more general functional differences. The vast majority of muscle fibres, invertebrate as well as vertebrate, fall readily into one of two categories, *skeletal* (or voluntary) muscle, and *visceral* (or involuntary) muscle. Voluntary muscle fibres are those that require stimulation through nerves or hormones for the initiation of their activity, each fibre seeming to respond individually to its stimulation. Involuntary muscles include all those that are spontaneously active, although the extent of their spontaneous activity may in some cases be modified by nervous action. Contraction spreads as a wave over a sheet of involuntary muscle, and its conduction does not require extrinsic or hormonal stimulation. Because of these differences, voluntary muscles are often referred to as *neurogenic*, involuntary muscles as *myogenic*.

Such a division on the one hand brings together vertebrate cardiac muscle, which is myogenic but striated, and the smooth muscles of the vertebrate gut and reproductive organs, as well as the partly striated muscle of molluscan hearts; and on the other hand links certain smooth muscles such as those of the nictitating membrane and ciliary and iris muscles of the mammalian eye and the muscle cells of many arteries and arterioles with striated skeletal muscle and certain smooth invertebrate skeletal muscle.

II.I   Vertebrate skeletal muscle fibres

Each individual fibre of a voluntary muscle is usually a syncytium, and is bounded on the outside by a *sarcolemma*, a sheath of connective tissue which has elastic properties and is continuous with the tendinous material that forms the origin and insertion of the muscle on the skeleton. Immediately beneath the sarcolemma is the plasma membrane, and below this is a non-fibrillar layer of cytoplasm, the *sarcoplasm*, in which the nuclei lie (Fig. 11.1). In the centre of the fibre, and taking up most of its area in cross-section, are bundles of longitudinally-running *myofibrils*, that are about 1–2 μm in diameter. Under the light microscope the myofibrils are seen to be transversely striated, and the striations follow a definite pattern which has been further elucidated with the electron microscope. When a muscle fibre is suitably macerated, it will break up into a number of discs or *sarcomeres*, the boundaries of which are formed by the thin *Z-band* seen in the intact muscle. The Z-band lies in the middle of the lightly refracting (*I*sotropic) *I-band*, and running across the middle of the sarcomere is the dark (*A*nisotropic) *A-band*. Sometimes a paler band, the *H-band*, can be seen across the middle of the A-band (Fig. 11.3) and a further dark line sometimes visible in the middle of the H-band is known as the M-band.

Sarcoplasm runs between the myofibrils, as well as being present at the periphery of the muscle fibre. It is largely occupied by an extensively developed endoplasmic reticulum, which in muscle is known as the *sarcoplasmic reticulum*, a system of interconnecting canals and tubules; and by mitochondria. The sarcoplasmic reticulum is organized into a longitudinal system running between the myofibrils, that interconnects with transverse components which in vertebrates are usually found in the regions of the A- and I-bands. In addition, there is a separate transverse or *T-system* of canals at the Z-band or A-I junction, which comes into close contact with the sarcoplasmic reticulum (Fig. 11.1) and is also continuous with the plasma membrane, so that the T-system opens on the fibre surface. The junctions between the tubules of the T-system and the sarcoplasmic reticulum occur at terminal swellings of the latter known as

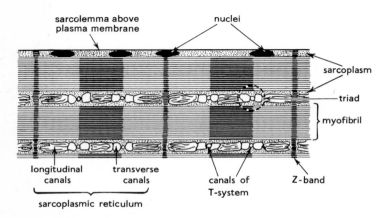

Fig. 11.1   Diagram to illustrate the appearance of part of a muscle fibre in longitudinal section. Individual parts not to scale.

cisternae; and the configuration produced at such junctions has a characteristic appearance in vertebrate muscles that is given the name of *triad*. (Fig. 11.1). The morphology of the T-system is actually rather varied in animals, and may be anything but transverse. To take just two examples, it forms a three-dimensional network in 'slow' frog muscles and is said to be completely absent in the myotomal muscles of amphioxus.

There is no doubt that the myofibrils are responsible for the contraction of the muscle fibres. Much of their protein can be extracted and made to gelate in thin threads by squirting the extract into potassium chloride solution. When ATP is added to these threads, they are observed under the microscope to shorten or contract. This observation has been made the basis of some theories of contraction, but not those currently in fashion. The reader will have noted in any case, that the contraction of a thread of muscle protein in this way may provide us with a model of *rigor mortis*, but not of a living muscle fibre, because it makes no provision for relaxation following contraction.

11.2 Muscle proteins

The bulk of the protein in muscle fibres is of three types, *myosin*, *actin* and *tropomyosin complex*. Myosin is a protein with a basic molecular

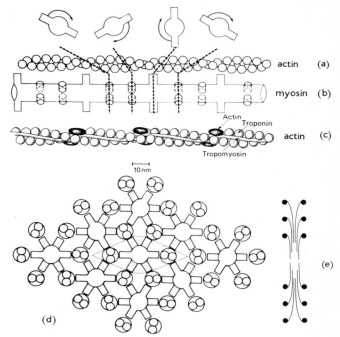

Fig. 11.2 Diagrammatic representation of two actin filaments (a, c) and a myosin filament (b) in a frog sartorius myofibril. The actin filament is a double helix. each strand of which is composed of globular G-actin molecules arranged linearly. The helically-arranged pairs of projections around the periphery of the myosin filament are thought to consist of H-meromyosin and to act as cross-bridges between the myosin and actin filaments. (d) is an end-view of the two types of filaments to illustrate their probable spatial interrelationships. (a–d courtesy R. E. Davies, modified.) A diagrammatic representation of how the individual molecules could be arranged in a myosin filament is shown in (e). Each molecule consists of an H-meromyosin head and an L-meromyosin tail. The molecules are arranged with the heads outwards to form cross-bridges. The arrangement shown would account for the absence of cross-bridges in the centre of the myosin filament.

weight of just over 500 000, but it polymerizes very readily. When it is
digested with the enzyme trypsin, myosin breaks down into two components which, on account of their differing molecular weights, are known
as heavy or *H-meromyosin* and light or *L-meromyosin*. Both myosin and H-
meromyosin can act as an enzyme to split the terminal phosphate group
from ATP, and this is generally regarded as a fundamental reaction in the
contraction of muscle. L-meromyosin does not possess this property, and
H-meromyosin is therefore regarded as the active part of the myosin
molecule. Each unit of H-meromyosin appears to possess a globular head
and a short helical tail, with the head inclined at an angle to the tail. The
heads have been assumed to represent the cross-bridges shown in Fig. 11.2,
which also shows how one authority thinks they might be arranged in the
strand of myosin. Mysoin belongs to the k–e–m–f (keratin–epidermin–
myosin–fibrin) group of proteins, which are capable of existing in two
configurations, the α-form being partially folded, the β-form wholly
folded. Clearly, if the energy released by the enzymic activity of myosin
could be harnessed to change the protein from one configuration to the
other, this might cause a shortening or extension of the myosin.

Actin does not possess contractile properties. Its basic molecular weight
is about 60 000 but it too is easily polymerized, and the monomeric *G-actin*
(globular actin) is readily transformed into the polymeric fibrous or *F-
actin*. ATP must be present for this polymerization to take place, but since
muscle contains abundant ATP, its actin is always in the fibrous condition. It takes the form of a double helix, each strand of which consists of
strings of the globular protein (Fig. 11.2).

'Tropomyosin' is also non-contractile. Crude 'tropomyosin' is actually
a mixture of four proteins, including tropomyosin proper and three others.
This complex of proteins is an important link between excitation and contraction and will be referred to again later in this connexion. A further
protein, α-actinin, can also be extracted from muscle. Its function is uncertain, but it may have something to do with the organization of the
myofilaments.

It is possible to extract myosin and actin from muscle fibres in successive fractions, and to couple this with histological examination of the
fibres at each stage of extraction, and these studies have shown that myosin
is present mainly in the A-band and actin mainly in the I-band. Under
the electron microscope it can be seen that each myofibril consists of a
number of myofilaments, some thick and some thin, arranged in an
orderly fashion. By the process of differential extraction of the proteins,
it can be shown that the thick filaments are made of myosin and the thin
filaments of actin. In vertebrates each myosin filament is between
11 nm and 14 nm in diameter, and in cross-section is found to be surrounded by six actin filaments, each about 4 nm in diameter and arranged

in such a way that they are shared by neighbouring myosin filaments (Fig. 11.3). In longitudinal section (Fig. 11.13) the myosin and actin filaments appear to interleave or *interdigitate* to some extent. The ends of the actin filaments nearest the Z-band adjoin those of the adjacent sarcomere, and the complex method of their joining is responsible for the dark line of the Z-band in vertebrate muscle.

11.3 Energetics of contraction

The ultimate source of the energy required for contraction is, of course, the oxidation of foodstuffs. The immediate energy source is generally

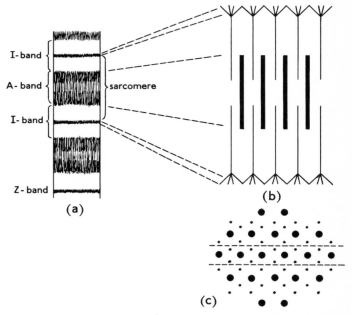

Fig. 11.3 (a) Sketch of a myofibril as it appears under the light microscope. (b) The differences in the refraction of light are due to interdigitating filaments of actin and myosin, revealed by the electron microscope. A lighter H-band is often seen in the middle of the A-band, due to the fact that the actin filaments do not run to the middle of the A-band. (c) The appearance of part of a myofibril of rabbit psoas muscle in cross-section under the electron microscope, showing the arrangement of the actin filaments around the myosin filaments. (Based on Huxley, 1957, *J. biophys. biochem. Cytol.* **3**, 631, by permission of The Rockefeller University Press)

assumed to be ATP, which is broken down to ADP by myosin acting as an ATPase enzyme, or rather by the H-meromyosin part of its molecule.

The contractile threads made from extracted muscle protein which shorten on addition of ATP are known to contain both actin and myosin in some kind of combination. The ATP seems to facilitate the formation of chemical bonds between the two proteins, and the resulting combined form is known as *actomyosin*. Its formation is thought to proceed according to the following scheme:

$$\text{(i)   myosin} \cdot \text{ATP} \rightarrow \text{myosin} \sim \text{P} + \text{ADP}$$
$$\text{(ii)   myosin} \sim \text{P} + \text{actin} \rightarrow \text{actomyosin} + \text{P}_{inorganic}$$
$$\text{(iii) actomyosin} + \text{ATP} \rightarrow \text{myosin} \cdot \text{ATP} + \text{actin}$$

The first two reactions are those that are responsible for contraction, and the third is the reaction which must take place if relaxation is to occur.

The contractile process results in the formation of inorganic phosphate and ADP in cells. The regeneration of ATP from ADP normally proceeds by oxidative phosphorylation (§3.17), but muscle contains a substance for the quick re-phosphorylation of the ATP used in contraction, and its action is so rapid that the level of ATP in muscle does not change detectably during contraction. The substance responsible in vertebrate muscle is the energy-rich compound *creatine phosphate* (CP), sometimes known as phosphorylcreatine. The corresponding substance in invertebrate muscles is usually arginine phosphate. Creatine phosphate can transfer its energy-rich phosphate group to ADP in the presence of a suitable enzyme, with the formation of ATP and creatine. Creatine is regenerated as creatine phosphate by accepting an energy-rich phosphate group from ATP, and at first sight the existence of this cyclic process in muscle seems rather pointless. However, the ATP utilized in contraction forms a complex with the contractile protein, and it is this which is immediately regenerated by CP once relaxation begins. The ATP that is used to regenerate the CP is derived essentially from a source external to the contractile system, presumably the mitochondria of the sarcoplasm. Thus, in a sense the ATP for contraction is a permanent component of the contractile system which must be kept at a constant level, whereas the ATP needed for the regeneration of CP is part of the normal labile energy-rich system of cytoplasm. The need for CP arises from the complex organization of the contractile system, and its importance is demonstrated by the fact that it forms the bulk of the high-energy phosphate of muscle.

The energy for the regeneration of CP is thus derived ultimately from oxidative metabolism. However, a muscle can continue to contract in the absence of oxygen, a situation which is probably not uncommon during periods of rapid activity when the respiratory system cannot supply sufficient oxygen for the needs of the muscle. Under these circumstances the pyruvic acid that would normally be oxidized to carbon dioxide and water during oxidative metabolism (§3.15) is transformed instead into *lactic acid.* Under the influence of the enzyme *lactic dehydrogenase*, hydrogen is removed from reduced NAD and from solution for this purpose:

$$CH_3 \cdot CO \cdot COOH + NADH + H^+ \rightleftharpoons CH_3 \cdot CHOH \cdot COOH + NAD$$

pyruvic acid   lactic acid

The value of ΔF for this reaction is $-25\ 116$ J (-6000 cal), and thus energy is released for the production of high-energy phosphate, but comparison with the yield from oxidative metabolism (§3.17) demonstrates that this is a wasteful method of obtaining energy. However, the lactic acid itself is not necessarily wasted, since when oxygen again becomes available, four-fifths of it can be re-synthesized into glycogen

$$2C_3H_6O_3 \rightarrow H_2O + C_6H_{10}O_5$$

lactic acid   glycogen unit

the remaining one-fifth being oxidized to provide the energy for this process:

$$2C_3H_6O_3 + 6O_2 \rightarrow 6CO_2 + 6H_2O$$

The complete enzyme system for the re-synthesis of glycogen from lactic acid is lacking in muscle, but lactic acid is highly diffusible and hence is able to enter the blood by which it is transported to the liver where the re-synthesis takes place. In general, it will be seen that anaerobic respiration in muscle is more wasteful than aerobic respiration, and involves the depletion of the glycogen reserves of the muscle for a relatively low energy yield, with the consequent need for its glycogen to be replaced from glucose derived from the liver.

The production of lactic acid under anaerobic conditions may be less important in invertebrates than in vertebrates, for under fatigue conditions other pathways may be utilized which are still dependent on oxygen. For example, in the flight muscles of cockroach after about 20 000 contractions, only about 3% of the glycogen utilized has been converted into lactic acid; whereas about 50% has been transformed into α-glycerophosphate, and the rest into pyruvate that is partially converted into acetate. The α-glycerophosphate and remaining pyruvate are then available for oxidation when molecular oxygen becomes available. α-glycerophosphate is undoubtedly important as an alternative to lactic acid in insect muscles, although more lactic acid may be produced in some insects than in the example cited above.

11.4 Ions in the muscle fibre

Certain ions are known to be important in the contractile process. Magnesium ions can be shown to be necessary for both contraction and relaxation, yet injection of magnesium ions into 'resting' muscle fibres has no effect on them. It follows that magnesium ions cannot initiate contraction, but must be interacting with other factors in the functioning of the muscle. On the other hand, the injection of calcium ions into the 'resting' muscle fibre causes it to contract. There is a considerable body of evidence to show that calcium ions are released from the sarcoplasmic reticulum when a muscle fibre is stimulated, and it is also known that the sarcoplasmic reticulum can accumulate calcium ions against a concentration gradient. The pumping in of the calcium ions is an active process, the necessary energy being derived from the splitting of ATP. The calcium pump ceases to operate when the concentration of calcium outside the reticulum reaches a level of between 10^{-8} and 10^{-9} M, the calcium concentration needed to activate the contractile mechanism being about 10^{-7} M or above.

It will be evident that regulation of the calcium content of the sarcoplasm by the sarcoplasmic reticulum could determine the contraction and relaxation of the muscle fibre. Release of calcium from the reticulum on stimulation would cause contraction when it reached a level of 10^{-7} M or more, and relaxation would ensue when, on the cessation of stimulation, the resorption of calcium by the reticulum caused it to fall once more below this level.

There seems to be present in muscle a complex system in which ATP and magnesium ions are bound in some way to the contractile protein. The calcium ions released from the sarcoplasmic reticulum on stimulation of the fibre cause the protein to alter from the dissociated state to the combined state under the influence of ATP. In frogs the stimulus appears to be most effective when it is applied in the region of the Z-band. This is the point in these animals (the position varies in different forms) at which the T-system opens to the surface of the fibre, and it is believed that the stimulus travelling along the membrane of the muscle fibre passes down the T-system to activate the adjacent sarcoplasmic reticulum. The impulse that passes down the T-system is not propagated, and seems to depend on the efflux of chloride ions rather than an influx of sodium ions, thus providing an interesting example of two stretches of membrane continuous with one another but with different permeability properties. The passage of the impulse between T-tubules and reticulum is probably electrical—there are junctional areas at which the two membranes are closely apposed—and causes the cisternae of the reticulum to release calcium ions into the sarcoplasm.

In addition to calcium and magnesium ions, potassium ions must also be present for the proper functioning of the contractile system. They do not appear to be directly concerned in the contractile mechanism, and they are probably needed to keep the contractile proteins in the proper physical state for contraction.

11.5 The sliding filament theory

A theory of muscle contraction which has gained a wide measure of acceptance is the *sliding filament* theory. This takes as its starting-point the fact that myosin and actin filaments are structurally interdigitating, and postulates that the contraction of a muscle fibre does not result from the longitudinal shortening of the myofilaments, but from the sliding between one another of the actin and myosin filaments. The theory has been much elaborated since it was first proposed, and a number of different versions have been advanced in attempts to overcome difficulties that have arisen over the detailed functioning of the system.

It is widely agreed that the projections on the myosin filaments referred to earlier as cross-bridges do in fact link the myosin and actin filaments during contraction, to form the 'actomyosin' mentioned previously. It is also agreed that the cross-bridge shortens and that the $Mg \cdot ATP$ complex near the distal end of the cross-bridge is acted upon by the ATPase sites it bears, to split the ATP. These changes are initiated by the release of calcium ions from the sarcoplasmic reticulum. They act on the tropomyosin complex, which is known to consist of three proteins, besides tropomyosin itself, which were all at one time given the single name *troponin*. One of the three (inhibitory factor) inhibits the splitting of the ATP and the shortening of the cross-bridge; and a second (calcium-sensitizing factor) removes this inhibition under the influence of calcium. The function of the third protein is unknown, but its presence is presumed to be necessary for the proper functioning of the system. Tropomyosin, thought to be bound into the actin filament in some way, is also linked with the inhibitory factor, and its presence is essential for both the latter and the calcium-sensitizing factor to exercise their full activity. These facts are summarized in a pictorial way in Fig. 11.4. It is not clear whether the energy released by the splitting of the ATP acts initially to cause the cross-bridge to adopt its fully-folded, shortened, configuration, or subsequently to cause it to extend once more when the link with the actin filament is broken.

The link formed between a cross-bridge and a myosin filament is thus made and broken rapidly; and its contraction during each step of this process should cause the actin filament to move slightly in relation to the myosin filament (Fig. 11.4). In a system in which each cross-bridge underwent a rapid succession of such micro-contractions, the results would be a smooth movement of the actin filament along the myosin filament. This movement would be maintained as long as the calcium ions remained in the sarcoplasm. As soon as they were withdrawn by the sarcoplasmic reticulum, the inhibitory factor would cause the cessation of sliding and relaxation would take place.

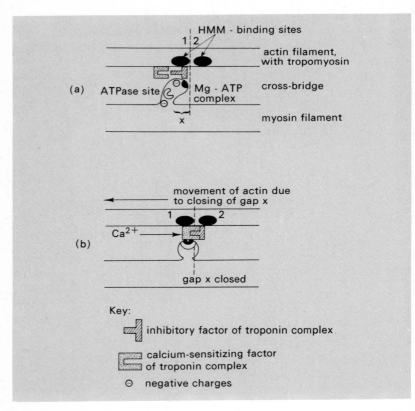

Fig. 11.4 Diagram to illustrate the probable mechanism of sliding filaments during muscle contraction. In (a) the muscle is resting. The heavy meromyosin (HMM) cross-bridge on the myosin filament is extended at an angle due to the electrostatic repulsion between its two ends. The troponin inhibitory factor prevents any interaction between the cross-bridge and the actin filament. In (b) an impulse has travelled down the T-system to cause the release of Ca^{2+} ions from the sarcoplasmic reticulum. The ions both activate the sensitizing factor so that it causes the removal of the inhibitory factor's action, and remove the electrostatic charge from the end of the cross-bridge. The latter therefore contracts, shortening the cross-bridge, and bringing its ATP into contact with its ATPase. The splitting of the ATP causes the extension of the cross-bridge once more, its link with the actin being somehow broken at this point, but the movement produced by its previous shortening means that when it is again extended at an angle, it will be opposite binding site 2. Contraction of the muscle as a whole is due to the combined effect of a whole series of such sequences.

The strongest support for the view that the sliding of filaments does take place during contraction has come from studies with the electron microscope and from other studies that have examined muscles fixed at different degrees of tension. These have demonstrated that the lengths of the myosin and actin filaments remain unaltered at different tensions, whereas the I-band diminishes in length during contraction and the A-band remains more or less constant in size. Such observations are compatible only with the movements of the filaments between one another, even if detailed explanations of the mechanism involved, such as that given above, should prove to be wrong in some details.

There are, however, several observations that do not seem to fit in with the sliding filament theory. In the horseshoe crab the actin filaments seem to run the length of the sarcomere without a break. The giant muscle fibres of the barnacle *Balanus nubilis* contain myosin filaments that pass right through the Z-band into the adjacent sarcomere on contraction, and the appearance of the myofilaments under the electron microscope does not support the presence of regularly interdigitating filaments. It has also proved impossible so far to apply the theory to non-striated visceral muscle, which often does not seem to contain organized myosin filaments. Such difficulties, however, probably arise more from insufficient information than from any defect in the sliding filament theory, the essentials of which are generally accepted.

11.6 Invertebrate skeletal muscles

In general, the transversely striated muscle fibres of invertebrates resemble those of vertebrates, containing thick and thin myofilaments that seem to correspond with the myosin and actin filaments of vertebrates. The number of actin filaments surrounding each myosin filament in vertebrate muscles generally seems to be six, as shown in Fig. 11.3, and when account is taken of the sharing of actin filaments between myosin filaments, the overall ratio of actin:myosin filaments is 2:1. There is much more variation in invertebrate muscle fibres, and the number of actin filaments surrounding each myosin filament commonly exceeds six. The maximum number so far reported is twelve, found in some insect muscles, and this means that the overall ratio of actin:myosin filaments in this case is 6:1. When this ratio exceeds 5:1 the arrangement of the actin filaments round the myosin filaments no longer appears completely regular, and this may be reflected in the stability and efficiency of the contractile system.

Sarcomere length also seems to vary more in invertebrates. Many exhibit sarcomere lengths comparable to those found in vertebrates—up to about 3 μm in length—but in some much larger values have been reported, the largest being found in certain syllid worms, in which the sarcomere length is said to be 33 nm. Filament thickness also varies, and the thick filaments of invertebrate muscles may be as large as 300 μm in diameter, although their actin filaments do not normally exceed about 50 μm.

In some molluscs, and in annelids, echinoderms and ascidians, muscle fibres with helical or spiral striation patterns are not uncommon. The helical pattern is simply the result of the way in which the individually transversely striated myofibrils are orientated with respect to one another, and there is not necessarily any structural difference between these and transversely striated muscles. Other invertebrate skeletal muscles may be non-striated, especially the slower contracting types found in lamellibranch adductors and the retractor and other muscles in annelids and holothurians.

The Z-band is not universally the site at which the T-system opens to the surface in invertebrate muscles, nor even in vertebrate muscles. In lizards and crabs the T-system opens at the junction of the A- and I-bands, and in insects it may be either mid-way between the Z-band and the H-band or, as in asynchronous flight muscles, central in the sarcomere.

Muscles of molluscs and in particular the adductors of lamellibranchs and the anterior byssus retractor muscle of *Mytilus*, have been found to be particularly rich in tropomyosin. Although tropomyosin is normally thought to be associated with actin rather than myosin, electron-microscope studies have revealed that the thick myofilaments of such muscles are very large in diameter and this may be due to the presence of tropomyosin in or on the filaments. These muscles are unusual in a number of properties, exhibiting long filament and cell length, and being able to maintain high tension over long periods, and relax very slowly indeed. Furthermore, a troponin complex does not seem to be present.

Some workers have suggested that the tropomyosin may function in a kind of *catch* mechanism. The idea is that the muscles contract normally, but are then held in the contracted state by the tropomyosin. An analogy is drawn between such a catch system and a mechanical ratchet, but of course the muscle is not held indefinitely, and does eventually relax. The sort of time relationships involved are illustrated by the adductor muscle of the scallop *Pecten*. The scallop adductor contains two kinds of muscle fibres: striated fast-contracting ones that are used in flapping the two halves of the shell for swimming; and non-striated fibres that may be used to keep the shell closed for long periods. The contraction of the slow fibres is five times slower than that of a frog gastrocnemius muscle, and whereas the latter relaxes again in just over 0.1 s, the adductor takes about 90 s.

Evidence for a catch mechanism comes from experiments on the same adductor fibres. Their action in keeping the shell continuously closed can be simulated by delivering a stimulus to the motor-nerve trunk supplying the muscle every few seconds. This would seem to suggest that the need for a continuous background of motor stimulation of the muscle, but if the nerve supply is cut while the muscle is in the contracted state it remains contracted even when forcibly extended. It is, however, possible that the nerve injury potentials will be generating impulses that keep the muscle contracted. A catch mechanism is also a possible explanation for the behaviour of the anterior byssus retractor muscle of *Mytilus* when stimulated directly by either a.c. or d.c. current. The muscle both contracts and relaxes more slowly when stimulated with d.c. than it does when stimulated with a.c. It has been suggested that the a.c. and d.c. are acting on ganglion cells in the muscle, but the presence of such cells has been disputed. Alternative explanations might be that a.c. and d.c. exert different effects on any nerve endings remaining in the muscle, which is isolated from the animal for the experiment; or that there is a catch mechanism that is affected only by d.c. It is clearly impossible to say whether a catch mechanism or ordinary nervous phenomena are responsible for the peculiar properties of molluscan adductors, nor whether tropomyosin is directly involved in them, and a definitive explanation must wait for further information. It may be noted however that mammalian uterine muscle, which is also slow to contract and to relax, contains considerably more tropomyosin than mammalian skeletal muscles.

11.7 Characteristics of muscular contraction

Muscular contraction can be arbitrarily divided into two types, *isotonic* and *isometric*. In isotonic contraction the muscle is not loaded before contraction begins, and so it is able to shorten in response to a stimulus. Isometric contraction results from the muscle being loaded before contraction begins, the weight of the load being such that it cannot shorten when stimulated. In practice, when studying isometric contraction the muscle is allowed to shorten to some extent, otherwise differences in tension cannot be recorded. In the bodies of animals muscular contractions are probably invariably a mixture of isotonic and isometric contraction. The muscles do shorten, except perhaps in some cases during the maintenance of a posture, but they do so against a load afforded not only by the weight of the limb or part of the body they move, but also by the resistance due to antagonist muscles pulling in the opposite direction. This point is further developed below (§11.8).

Many muscles respond to the stimulation of their motor nerve with a short sharp contraction, known as a *twitch* contraction. The tension developed in a twitch contraction is never the maximum tension of which the muscle is capable. The components of a muscle can be thought of as contractile elements in series with elastic elements, the latter being repre-

sented by the connective tissue of the muscle and tendons and any inherent elasticity of the cellular material. The development of tension in a
muscle involves changes in the length of its elastic elements, and the overcoming of the inertia imposed by any external load present at the onset of
contraction. It has been demonstrated that after a single stimulus the
contractile elements develop their maximal activity almost at once, and
this is maintained for only a short time. The overcoming of the load and
elastic elements takes some time, with the result that the activity of the
contractile elements is decreasing before this has been fully accomplished. For this reason, maximum tension cannot be achieved following a
single motor nerve stimulus.

Some drugs, such as caffeine and adrenaline, prolong the activity of the
contractile elements and therefore cause an increase in the twitch tension,
but their effect is still not sufficient for maximum tension to be obtained.
If a second stimulus is applied sufficiently close in time to the first for
the activity of the contractile elements not to have fallen to zero, they are
still exerting a force on the externally resisting elements when the second
stimulus becomes effective. The contractile elements therefore do not
need to overcome the external resistance to the same extent as they did
when the first stimulus became operative, and so the two mechanical
effects *summate* to some extent and the resulting tension is greater than
it was with a single stimulus. The closer in time the two stimuli are applied,
the more effective will be the prolongation of the mechanical effect of the
first stimulus, and the greater will be the tension developed. If a repetitive
chain of stimuli is employed, succeeding mechanical responses will continue to summate and the muscle will develop an increasingly greater
tension, up to the maximum of which it is capable (Fig. 11.5). The summation of successive contractions to produce a continuous increase in
tension, at a frequency sufficiently high to cause their virtual fusion, is
known as a *tetanus*. Only as a result of a tetanus can maximal tension be
developed.

The extent to which the tension developed in a tetanus exceeds that
attained through a twitch contraction varies between muscles in the same
or in different species. For example, in fibres of the frog sartorius muscle,
the tetanus/twitch ratio is about 5 at 20°C, whereas it is commonly much
higher in arthropods, being between 10 and 30 in those insects that have
been examined. These differences are not due merely to differences in the
capacity of these muscles to develop a given tension, but also to the extent
to which that tension can be developed during a single twitch. In terms of
the tension developed per unit cross-section area of the individual muscle
fibres, these and most other muscles seem to give fairly comparable perfor-
mances, the tension usually varying between 2–5 kg/cm², although some
arthropod muscles go as high as 7 kg/cm² and the adductor muscle of

Mytilus, and no doubt other molluscan adductors, can develop a t
of some 8 kg/cm².

The undoubted variations in the speed and efficiency with which animals
move is governed by a variety of factors, and as we have seen, the inherent
contractile ability of the muscle is not usually one of them. Both the
power/weight ratio of the animal and the organization of its skeleton and
muscles are important in this respect. If we look at the tension developed
by the muscles mentioned above in terms of the weight of the muscle,
we find that the maximal tension of the frog muscle is of the order of

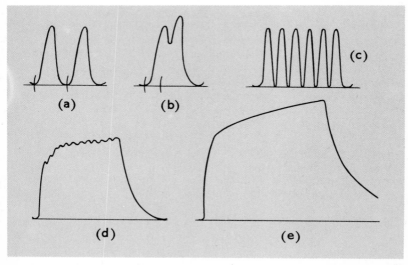

Fig. 11.5   Mechanical summation and tetanus in a frog muscle (gastrocnemius).
(a) Two stimuli to the motor nerve evoke separate muscle responses. (b) The
stimuli are now sufficiently close for a certain amount of summation to take place
between the responses, and the maximum tension of the muscle is greater.
(c)–(e) Repetitive stimulation of the motor nerve to the muscle at 5/s, 20/s and
100/s respectively. Note increase in tension of tetanus with increase in frequency
of stimulation.

1000 g/g muscle weight, compared with a value for the jumping muscle
of the locust of about 20 000 g/g. Even the sluggish stick insect can develop a muscle tension of 13 000 g/g. If we put equally powerful engines
into a light car body and a heavy one we should not be surprised to find
the lighter car moving faster, and the same principle applies to animals.
The analogy is not precise, because the muscles of insects are small in
relation to their bodies and hence are large in proportion to their body
volume, but it still applies because in relation to their power the load they

have to move is less. In addition, the organization of their muscles is
highly efficient (see §11.8).

As might be expected, the velocity of contraction of a muscle decreases

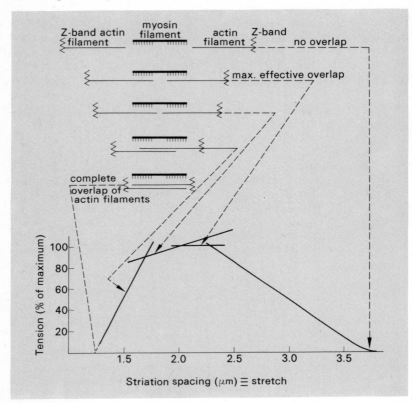

Fig. 11.6 The lower part of the figure illustrates the effect of stretch upon the
development of tension in a single isolated muscle fibre. The shape of the curve is
essentially similar to curves obtained with whole muscles. The upper part of the
figure shows how the different parts of the curve may be equated with the degree
of overlap between the myosin and actin filaments and between the actin filaments themselves. This would suggest that the degree of tension developed
depends on the amount of actin filament available to form attachments with the
myosin filament. This is probably too simple a view, but may represent the major
factor involved. (After Gordon, Huxley & Julian, 1966, *J. Physiol.* **184**, 185)

as the load on it increases. Yet the relationship between the two is not
linear, as it would be with a steel spring, whether held back (damped)
or not. For this reason, models of the contractile system that portray it as

a damped spring are inadequate. Presumably, the series elastic elements can be pictured in this way, but not the contractile elements. The relationship between velocity of contraction and external load is such that the muscle must be adjusting its performance to the size of the load. The energy liberated during contraction varies in a way which demonstrates that the efficiency of the muscle increases as the load increases. This energy may be divided into *activation* heat, *shortening* heat and *relaxation* heat. Activation heat can only be connected with the activation of the contractile mechanism, and since it is only this portion of the liberated energy that alters with changes in the load, it must be the contractile elements which are adjusting to the load in some way that is not understood.

The initial load on a muscle can be varied before an isometric contraction, and since a passive load produces stretching of the muscle, this provides a convenient way of altering its length. By varying its length in this way, or by clamping it at lengths shorter than its normal resting length while it is contracted, the performance of the muscle can be studied over a wide range of initial lengths. It is found that over the range 60–160% of its normal resting length complete reversibility to normal length is attained when the external restraint is removed. Outside these limits, permanent damage to the contractile elements may occur. The tension developed at a given degree of stretch can be correlated with the extent to which the actin filaments overlap with the myosin filaments (Fig. 11.6). It will be observed from Fig. 11.6 that maximal tension is developed by a muscle at its normal resting length, although it will often be the case that, in the body of an animal, a muscle will have to contract at other lengths (§11.8). Its greater efficiency with bigger loads may offset this apparent disadvantage to some extent.

11.8 Skeletons

A skeleton is a rigid structure or a system of rigid structures which functions as a protection for delicate organs, in the maintenance of a distinctive shape, and as an attachment for muscles. It is the latter function that concerns us here. The reader will be most familiar with the endoskeleton of vertebrates, in which the cartilaginous or bony components are often movable in relation to one another at their points of articulation, the *joints*. In the case of a simple joint like the elbow joint of man, movement is possible in only one plane, and is achieved by the action of two muscles, the *biceps* and the *triceps*. The triceps is responsible for the extension of the forearm, the biceps for its flexion (Fig. 11.7 b). In any normal movement of the joint, the two muscles are in opposition to one another, and movement will result from one muscle contracting with greater force than the other. *Both* contract, but to a different extent which determines the direction of the movement. By pitting one muscle against the other, a more precise control of the movement of the joint is effected than would be the case if one muscle was completely relaxed while the other was contracting, one muscle being used to provide additional control of the other. The same principle is commonly employed by engineers in movement control systems. The central inhibition referred to in §9.11 has to be understood in this context.

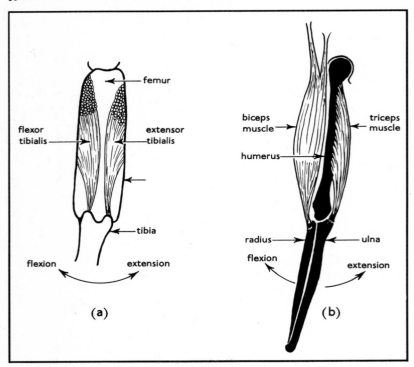

Fig. 11.7 Skeletal systems of arthropods and vertebrates compared. (a) Generalized arthropod leg; (b) human fore-limb.

The principle of opposed sets of *agonist* and *antagonist* muscles applies to invertebrates as well as to vertebrates, although they possess a different kind of skeleton. The exoskeleton of arthropods is also composed of jointed structures, and the system of muscle attachment is thus essentially similar to that in vertebrates, as Fig. 11.7 b demonstrates. An interesting feature of the exoskeleton is that it affords a much greater area for the attachment of muscles than an endoskeleton, and this has led to differences in shape and organization between the muscles attached to them. The bulk of the limb muscle of a vertebrate must be concentrated in the middle, because the area available for its attachment is limited. This area is relatively larger for an arthropod muscle, which is surrounded by its cuticle, and this permits a more splayed-out shape. Sheet-like limb muscles are uncommon in vertebrates, and this is why the frog sartorius muscle, which possesses such a shape, is so popular for experimental work; but in arthropods, muscles are not uncommonly sheet- or fan-shaped. In some cases, these factors have led to the elaboration in arthropods of the *pinnate* muscle that is characteristic of the extensors and some flexors of insect legs. In the pinnate muscle the end of the skeletal element to be moved is extended inside the limb to form an *apodeme*, a rigid chitinous structure to which one end of the fibres of the muscle is attached. The other end is inserted into the cuticle of the exoskeleton at an angle which ensures that when the muscle fibres contract they pull on the apodeme (Fig. 11.8).

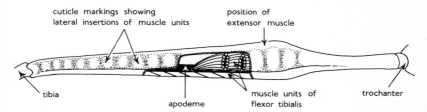

Fig. 11.8 Femur of the stick insect *Carausius,* showing the pinnate arrangement of the flexor tibialis muscle. (Wood, 1958, *J. exp. Biol.,* **35**, 850)

It is relatively inefficient to extend a muscle beyond a certain length, but this situation is inevitable in the longer vertebrate muscles. The arthropod pinnate arrangement enables the fibres to be kept short, and it also permits the maximum number of fibres to be packed into the available space. When this arrangement is coupled to the characteristically long limbs of arthropods and a high lever factor at the fulcrum around which the muscle is acting (Fig. 11.9) it provides a concentration of muscle power unequalled in the vertebrates. It is the pinnate arrangement of the jumping muscle of a locust hopper in conjunction with the power/weight ratio of the animal which enables it to execute a jump some 30 cm high and 70 cm long using two muscles whose combined weight is something like 80 mg. The efficiency of such a system and its practicability is, of course, limited to small animals.

Some animals possess a skeleton which is not made of hard material, but which consists of a fluid-filled space bounded by relatively strong and inelastic walls from which the fluid cannot escape. Such a system is incompressible, and therefore becomes rigid when pressure is applied, and

in this condition it may function as a *hydrostatic skeleton*. The enteron of
coelenterates and the coelom of annelids are examples of a hydrostatic
skeleton which helps to maintain the shape of the animal as well as forming
an incompressible system on which the muscles can act. In the annelids
there are two opposing cylinders of muscle, the circular and longitudinal
muscle, surrounding the coelom. When the circular muscle contracts in
a given region it drives the coelomic fluid away so that the coelom is
elongated and smaller in diameter in the contracting region, and shorter
and larger in diameter in the non-contracting region. In the latter, the
longitudinal muscle aids in the shortening of the coelom by contracting,

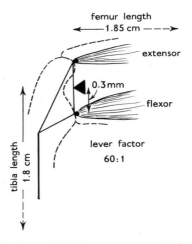

Fig. 11.9 Diagram of the mechanical arrangement at the fulcrum of the tibia
of the stick insect *Carausius*.

and in the region of contraction of the circular muscle the longitudinal
muscle must relax. The use of the two opposing muscle layers in this way
is used to bring about movement in some annelids, and the reader will be
familiar with the way it is used for this purpose by earthworms.

In some cases, the hydrostatic skeleton may be confined to parts of the
body, in which it is employed for local movements. For example, the
extension of the tube feet in echinoderms is achieved by the contraction
of the muscle of the ampulla, which forces the fluid of the water vascular
system into the tube foot, causing it to elongate and swell. Relaxation of
the ampulla muscle and contraction of the longitudinal muscle in the
tube feet produces withdrawal of the tube foot. Spiders and many

molluscs make use of a comparable system, spiders for the extension of their legs, and molluscs for the extension of their creeping foot. In both cases, extension is achieved by forcing blood into the haemocoel of the organ concerned, the muscle of the molluscan foot being spongy in texture due to the presence of many haemocoelic spaces. The opposing actions, the flexion of the spider's leg and the withdrawal of the molluscan foot, take place through normal muscular action, but presumably the blood is removed from the organ at the same time.

11.9 Motor nerve endings

We saw in §9.10 that when an impulse passes down the motor axon to a muscle fibre it causes the liberation of a transmitter substance from the motor nerve terminals which in turn sets up an action potential in the membrane of the muscle fibre. It is the action potential that activates the T-system to produce the contraction of the muscle fibre. It was also pointed out in §9.11 that the transmitter at vertebrate end-plates is acetylcholine, but that invertebrate skeletal muscles do not generally possess cholinergic junctions, several other substances being utilized instead of, or in addition to, acetylcholine.

The morphology of the motor nerve endings in skeletal muscle varies between animals. They are often claw-like, as in the majority of insects and tetrapod vertebrates (Fig. 11.10 b, c), but may also be twig-like as in decapod crustaceans or even circular as in some fish muscles (Fig. 11.10 a). Little is known about the fine structure of many of these endings, although what has been observed with the electron microscope suggests there is no fundamental difference in morphology.

Equally little is known about the detailed working of most neuromuscular transmitters, and again most of the detailed information refers to the vertebrate neuromuscular junction, but it is generally assumed that other junctions with different transmitters work in a similar manner. Even in the case of the vertebrate cholinergic junction, much remains to be discovered. Recent work has shown that the account given in §9.11 of the sensitivity of cholinergically-innervated muscle fibres to acetylcholine requires elaboration. By using the local application technique previously described, it has been found that the sensitivity of the membrane of the fibres of the frog sartorius muscle to acetylcholine rises considerably towards the tendinous ends of the fibres, although it never attains the level of the end-plate region. The functional significance, if any, of this fact remains to be elucidated. Even more interesting is the discovery that the rest of the membrane becomes as sensitive to acetylcholine as the end-plate region when the muscle is denervated. By cutting a sartorius muscle in half, it can be divided into a section containing all the motor nerve endings, and a section devoid of them, and it is found that the section without endings

becomes sensitive to acetylcholine, the section with endings remaining
normal in its pattern of sensitivity. These experiments show that the
whole membrane of the frog sartorius fibre is capable of developing
acetylcholine receptors, but that it is prevented from doing so in a normal

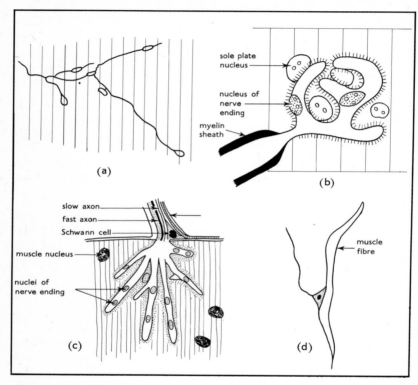

Fig. 11.10 To illustrate the varied morphology of neuromuscular junctions.
(a) From deep lateral muscle of the tench (based on Barets, 1961, *Arch. Anat.
micr. Morph. exp.*, **50**, 92); (b) from a mammal (after Couteaux, 1955, in Thomas
(ed.), *Problèmes de Structure, Ultrastructure et des Fonctions Cellulaires*,
Masson, Paris); (c) from locust muscle (after Hoyle, 1957, in Scheer (ed.),
Recent Advances in Invertebrate Physiology, Oregon University Press); (d) single
nucleated ending on muscle fibre from pharynx of *Dendrocoelum* (from Hoyle,
1957, *The Nervous Control of Muscular Contraction*, Cambridge University Press,
after Monti, 1897).

muscle fibre by the presence of the motor nerve ending. Such an action
by a nerve on a muscle fibre is termed *trophic*, and we shall encounter
another example later.

11.10 Types of muscle responses

In §9.11 it was stated that the response produced by the motor nerve ending in the membrane of the muscle fibre might be one of two types. In one type, the end-plate potential gives rise to a propagated response, which travels from the end-plate along the membrane, and activates the contractile mechanism as it proceeds. Many vertebrate muscle fibres possess responses of this type, a situation which is found to be correlated with the fact that such fibres are innervated by only one, or perhaps two, motor end-plates per fibre. Innervation of this sort requires that a propagated response should travel along the fibre, if the contractile mechanism along its entire length is to be activated.

Other muscles contain fibres each of which is innervated by many end-plates derived from the same motor axon and sited at more or less regular intervals along its length. Each end-plate sets up an action potential when the motor axon is stimulated, and each action potential activates the contractile mechanism in its own region. The result is that many *local* contractions occur virtually simultaneously along the length of the muscle fibre. It will be realized that if the end-plates were sufficiently close together for their spheres of influence to overlap, the local contractions would summate to cause the overall contraction of a fibre. A system of this kind therefore renders propagated responses unnecessary and with some exceptions they are found to be absent in fibres with this kind of *multiterminal* innervation. For example, in many insect muscles the motor axon is found to supply its muscle fibre with end-plates at intervals of from 40–100 μm. The action potential consists of two components, as in the case of a propagated muscle response (see Fig. 9.18), but the active component that follows the end-plate potential is not propagated. The size of this *local active response* is limited, probably because of a fast and early rise in potassium outflux that damps down the effect of the inward sodium current. The membrane is thus electrically excitable, but the ionic conductances limit the extent of any change in membrane potential.

The electrical responses of some muscle fibres lack even an active membrane response, and resemble an end-plate potential in appearance. The membrane in this case is electrically inexcitable. Two of these different types of response—propagated, local active and local non-active—may be found together in the same muscle fibre, or they are confined to separate muscle fibres which are anatomically distinct from one another. At one extreme are the decapod crustaceans, in which some muscle fibres may be innervated by up to five separate axons, one of which will normally be inhibitory and the others excitatory in function. The excitatory responses often provide a spectrum of size and type, at one end of which will be an end-plate potential with a relatively rapid rise time, which is followed by an active membrane response or rarely in this group by a propagated response (although this is functionally unnecessary); and at the other end of the spectrum will be a very small end-plate potential, so small as to be unable to activate the contractile mechanism to an extent sufficient to overcome the inertia of the system and produce any mechanical response with single shock stimulation. This latter type of end-plate will elicit a mechanical response only if its axon is repetitively stimulated, and the extent and speed of the contraction is dependent on the frequency with which the nerve

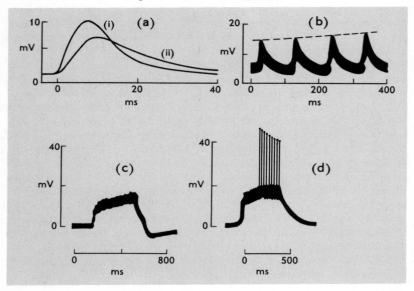

Fig. 11.11 (a) Slow muscle responses of (i) the frog and (ii) the stick insect. (b) Four slow responses from the stick insect in succession, showing facilitation. Facilitation is common in local responses, but does not occur in propagated responses. (c) Fusion plateau of slow responses from stick insect, with repetitive stimulation at 50/sec. (d) The same, but with the fast axon stimulated shortly after the stimulation of the slow axon began. The two responses summate, but note the tendency for the fast response to decline rapidly in size, on repetitive stimulation. (Frog response from Kuffler and Vaughan Williams, 1953, *J. Physiol.*, **121**, 288; stick insect responses from Wood, 1958, *J. exp. Biol.*, **35**, 850)

is stimulated. The electrical response observed with repetitive stimulation at this end-plate is a smooth slow rise in depolarization (*tonic* response) that flattens in time to a plateau, which represents the summation of the many small individual responses. Between these two extremes there may be excitatory axons that exhibit a gradation in type from one to the other. The presence of five separate axons is exceptional, and the majority of crustacean muscles, like those insect muscles which have been studied, are innervated by a smaller number. In these cases the distinction between the types of excitatory response is usually more clear-cut. The basic situation in arthropod muscles appears to be that each muscle is innervated either by an axon that elicits a twitch response from the muscle, and an active membrane response at the end-plate, when stimulated by a single shock; or by an axon that elicits only a tonic response on repetitive stimulation; or by both types of axon (Fig. 11.11). Such axons have often been designated *fast* and *slow* respectively.

In some arthropod muscles, all fibres appear to be innervated by one fast and one slow axon, but in others some of the fibres of a muscle appear to be innervated by only fast or only slow axons. In specialized muscles such as the jumping muscles of locusts or the claw muscles of crabs, the number of excitatory axons is greater than two, which is presumably a reflection of the flexibility required to enable a more delicate control of contraction to be exercised in these cases. This is probably also the reason for the development of inhibitory axons, which are found in many muscles of arthropods. These produce effects similar to those at inhibitory central synapses (§9.12).

Although detailed studies have yet to be made of the muscles of many invertebrate groups, evidence is accumulating that different kinds of muscle responses can be distinguished in many groups of invertebrates, including sea anemones and jellyfish, annelids, molluscs and echinoderms.

11.11   Fast and slow responses in vertebrates

In some arthropod muscles, different types of responses are confined to different muscle fibres. Thus, in the eyestalk levator of the crab *Podophthalmus vigil* there are two populations of muscle fibres contained in separate bundles. One type of fibre is pink, the other white. The pink fibres have a sarcomere length of about 13 μm and possess more mitochondria with more densely-packed cristae than the white fibres. The white fibres have a sarcomere length of about 8–10 μm, and the difference in colour undoubtedly reflects differences in enzyme concentrations. The pink fibres produce 'slow' (tonic) contractions and the white fibres produce 'fast' (sometimes termed *phasic*) contractions. Such morphological and biochemical differences between 'fast' and 'slow' fibres have been described in other invertebrates, including differences in diameter, 'slow' fibres often being smaller in diameter than 'fast' fibres. However, they have not always been observed in muscle fibres exhibiting different kinds of responses.

'Fast' and 'slow' responses are also found in vertebrate muscles. So far these have never been observed in the same muscle fibre, and fibres possessing different kinds of responses always appear to be different in morphology and biochemistry. Many fibres are clearly of the 'fast' (twitch) type, responding to a single stimulus applied to their motor nerve. These usually possess one, or sometimes two, motor end-plates per fibre, and function by means of propagated responses; although multiterminally-innervated twitch fibres also occur in some vertebrates. Other muscle fibres are of the 'slow' type, and these always exhibit multiterminal inner-vation and small non-propagated action potentials similar to those of arthropod 'slow' muscle fibres.

There are other differences, too, similar to those found in inverte-brates. For example, the parietal myotomal muscles of the hagfish contain thick fibres about 100 μ in diameter, with a single motor nerve ending at one end of each fibre. Other fibres are thinner, about 40 μm in diameter, and are innervated by a series of end-plates along their length, whose foci are about 100 μm apart. The membrane of the thin fibres has an electrical resistance over three times greater than that of the thick fibres, and the lower permeability of the thin fibres which this fact seems to indicate is reflected in different values of resting potential. The resting potential of the thin fibres is about 45 mV, in contrast to a figure of 70–75 mV obtained from thick fibres. The action potentials of the two types of muscle fibre are also different, being 5–30 mV for the thin fibres and as much as 120 mV for the thick fibres. As might be expected from these figures, the action potential of the thin fibres resembles an end-plate potential and that of the thick fibres is a propagated response.

Similar differences between muscle fibres have been observed in other fishes and in frogs. The twitch muscle fibres in frogs bear a single end-plate as in the hag-fish. Smaller muscle fibres may be scattered among the others singly or in groups, and are innervated by motor end-plates at intervals of about 300 μm. They respond to a single shock applied to their axon with a small action potential that resembles the end-plate potential of the other type of muscle fibre, being some 7–15 mV in size. The mechanical response is negligible until a stimulation frequency of about 20/s is reached, when the electrical responses summate to produce a depolarization plateau with an upper size limit of about 30 mV, and a slow, steady mechanical response. The slow response is thus very similar to that of many arthropod muscles (Fig. 11.11).

Less is known about the muscle responses of reptiles and birds, but histological studies suggest that fast and 'slow' fibres occur in these also. Apart from other characteristics, 'slow' fibres usually exhibit a more zig-zag transverse striation pattern and their myofibrils have a less distinct appearance in cross-section than is the case in fast muscle fibres. The reader may find the names *Fibrillenstruktur* (fast) and *Felderstruktur* (slow) applied to them in the literature because of their different appear-ance.

Mammals appear to have specialized in a different way from the other vertebrates. Their muscle fibres are, with few exceptions, histologically of the fibrillenstruktur type, bear single motor-nerve endings, and function by means of propagated responses, yet they exhibit variations in mechanical response which enable them to be divided into fast and slow types. For example, the *soleus* muscle of the cat exhibits relatively slow, long contractions when its motor nerve is stimulated by a single shock, whereas the *flexor digitorum* of the same animal gives a normal twitch contraction. If the nerves to each are severed and allowed to re-innervate the other type of muscle, the soleus virtually becomes fast in its response and the flexor digitorum virtually becomes slow. The response of the soleus also changes from slow to fast when its tendons are cut at one end, and some workers believe that this is due to a loss of activity in the muscle. The transition from one type of muscle to the other is not quite complete in the sort of experiments described above, showing that there are some inherent differences between them, but there can be no doubt that the major factor in determining their type is their innervation. These externally influenced features appear to be related to the liberation by the motor terminals of trophic factors, which appear to be substances other than the transmitter.

The existence of fast and slow responses in invertebrates is usually presumed to be due to the need for delicate control. Vertebrate muscles contain many thousands of fibres, and although each motor axon to such a muscle may innervate several hundred muscle fibres, the total number of individual axons running to the muscle is still quite large. Graded control of contraction can be achieved in this case by varying the number of axons along which impulses pass. If the sciatic nerve trunk to the gastrocnemius muscle of the frog is stimulated by electric shocks of progressively increasing intensity, beginning at a level at which no mechanical response is elicited, and if the increase in stimulus intensity is carefully controlled, the muscle is observed to contract correspondingly to an increasing extent up to a maximum level. The rise in the tension of the muscle is not continuously graded, but stepwise. The reason is that the axons to the muscle contained in the sciatic trunk possess different thresholds to electrical stimulation, and are brought successively into play as the stimulus intensity increases. This experiment provides a crude illustration of the way the central nervous system can control the degree of contraction of such a muscle, by the simple expedient of regulating the number of axons active at any given moment.

In most invertebrate muscles, the number of muscle fibres they contain precludes such a mechanism of control, except in the largest muscles. Where there are few muscle fibres, it is an advantage to have them innervated by both fast and slow axons, when control of their contraction can be achieved by utilizing either fast or slow responses, or a combination of the two. In the larger invertebrate muscles, there are sufficient muscle fibres for some to be only fast and others only slow; and some approach the vertebrate situation in that there may be several fast axons to a muscle that produce identical responses in the fibres they innervate, but supply different groups of fibres. This is the situation in the jumping muscle of the locust.

Considerations such as these raise the question of the function of slow muscle fibres in vertebrates. They are presumably not necessary for the delicate gradation of movement. It is often suggested that their function is 'postural', i.e. they develop sufficient tension to be used in maintaining the muscle in a particular position in the absence of movement on the part of the animal. This may well be their function in the land vertebrates, but the need for postural muscles of this type in fishes must be very small because they are buoyed up a good deal by the water in which they live, and teleost fishes possess an additional buoyancy mechanism in the form of a swim bladder. The suggestion has been made that the function of slow muscle responses in fishes is to effect an economy in respiratory activity. The argument is that the use of fast fibres in normal cruising movement would be uneconomical in a medium that is short in oxygen by comparison with air. The slow fibres would therefore be used for cruising, and the fast fibres would only be brought into play when quick darting movements were needed. On this theory, the emergence into air made slow fibres redundant in relation to their original purpose, but they formed a readily adaptable tonic system for the land tetrapods to use.

11.12 Involuntary (visceral) muscles

The majority of voluntary muscle fibres are multicellular and syncytial in structure, and the individual fibres function independently at the muscular level. The fibres of involuntary muscles are unicellular and discrete, but function together like a syncytium. They are smaller than voluntary fibres, rarely attaining a diameter greater than 10 μm and often less, and with a length of 50–100 μm. Most voluntary fibres are well above 10 μm and may be as much as 100 μm in diameter, and they are many times as long as they are broad.

11.13 Vertebrate cardiac muscle

Vertebrate cardiac muscle is made up of branching transversely striated muscle fibres (Fig. 11.12), which at one time were thought to form a structural syncytium. Electron microscope studies have shown, however, that the so-called *intercalated discs*, across which the myofibrils of adjacent sections were thought to be continuous, are simply transverse extensions of the ordinary cell membranes, and form a definite boundary between the muscle cells, each of which contains a single nucleus (Fig. 11.12).

The myofilaments in cardiac muscle are similar to those in skeletal muscle. There are no T-tubules, and the sarcoplasmic reticulum is sparse and only rarely comes close to the plasma membrane. This means that the electrical continuity between plasma membrane and reticulum found in skeletal muscle is lacking in cardiac muscle, and that contraction cannot be due to the release of calcium ions from the reticulum triggered by the action potential. As calcium ions are transported inwards and indeed

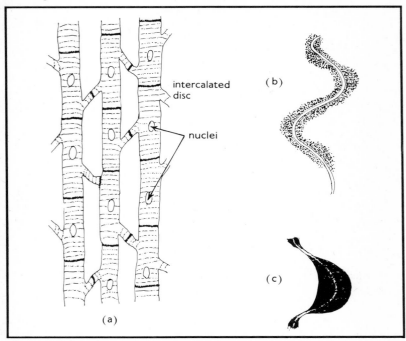

Fig. 11.12 Structure of cardiac muscle of mammal. (a) Diagram of appearance of a group of cells from the sinoatrial node. (b) Appearance of intercalated disc from same area, showing the dense aggregation of material on either side of the cellular membranes. (c) Organization of dense material into discrete bodies or *desmosomes* in intercalated discs of Purkinje fibres; the cleft between the cells is actually a little wider than elsewhere, but tends to be obscured by further dense material. Desmosomes are uncommon elsewhere than in Purkinje fibres. (Based on Rhodin, del Missier and Reid, 1961. *Circulation*, **24**, 350, 365. By permission of the American Heart Association Inc.)

contribute to the inward current responsible for the action potential in cardiac muscle (see below), it is believed that it is those ions that activate the contractile elements, whether directly or by causing the release of other calcium ions bound to the inner surface of the plasma membrane, or both. The tropomyosin of cardiac muscle shows some differences in composition from that of skeletal muscle, and is more akin to the tropomyosin (sometimes called *paramyosin* or tropomyosin-A) of molluscan muscles. Little is known about the details of its action, but it is presumed to function in the same way as the tropomyosin of skeletal muscle.

The contraction of the vertebrate heart has its origin at the sinus venosus in fishes and amphibians, and at the sino-auricular node in higher vertebrates. From here, the wave of contraction spreads out over the auricles and then over the ventricle(s). In birds and mammals a band of connective tissue separates the auricles from the ventricles, and in these animals there are specially modified muscle fibres, the *Purkinje fibres*, which pass from the auricles down between the ventricles and conduct the contraction wave to their apex. This elaboration is no doubt primarily a device to overcome the difficulty of conducting across connective tissue, but it has meant that the ventricles of birds and mammals also empty more efficiently, because they contract from the apex upwards and squeeze the blood out into the aortic arches.⏌

Like skeletal muscle, cardiac muscle exhibits refractoriness following a contraction. The absolutely refractory period lasts for most of the duration of contraction (known as *systole*), and the muscle is relatively refractory for the remainder of systole and the period of filling (*diastole*). If the muscle is stimulated electrically during the relatively refractory period, the contraction is not maximal, and it is followed by a *compensatory pause* that is longer than the diastolic pause of the natural beat. The frequency of the beat can vary under circumstances described below, and when this happens it is found that the refractory period also changes in length, in such a way that it is inversely proportional to the frequency of the beat. These factors mean that it is very difficult for cardiac muscle to enter into a tetanus under normal circumstances. Like fast voluntary muscle, the contraction of cardiac muscle is all-or-nothing, and once stimulated at or above the threshold intensity, it contracts maximally for the state it is in at that time.

Students are commonly required to carry out the *Stannius ligatures* on the frog's heart. The First Stannius ligature is tied between the sinus venosus and the auricles. Immediately after it has been tied only the sinus continues to beat, but after a variable period of time the auricles resume beating, although at a lower rate than previously. The Second Stannius ligature is later tied between the auricles and the ventricles. If already beating, the ventricle ceases to contract as a result, but usually it begins to contract again after an interval, but at a slower rate than the auricles. This experiment demonstrates that the whole heart is myogenic, but that normally the auricles and ventricles are subordinate to the sinus venosus, which is accordingly termed the *pacemaker*. In the mammalian heart, local warming of a part of the auricle or ventricle can convert it temporarily into the pacemaker. It will be appreciated from the account in §11.14 below that such externally-induced alterations must be accompanied by changes in the membrane permeability of the regions affected.

11.14 Membrane potentials of cardiac cells

Despite the small size of cardiac cells, a number of workers have been able to insert microelectrodes into them. The damage inflicted by this procedure cannot be very great, since the resting potential of 85 mV recorded from the frog sinus venosus agrees well with the theoretical potassium equilibrium potential of 92 mV calculated from the internal and external potassium concentrations by means of the Nernst equation

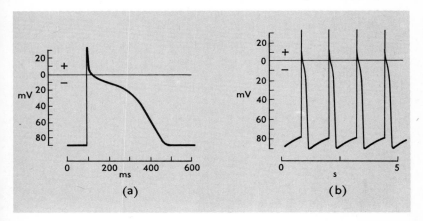

Fig. 11.13 Action potentials of cardiac cells; (a) away from the pacemaker region; (b) at the pacemaker region, showing the initial slow pacemaker potential before each propagated response. (After Draper and Weidmann, 1951, *J. Physiol.* **115**, 74)

(§9.2). The action potential has a characteristic shape (Fig. 11.13), due to the prolongation for 300–500 ms of the recovery phase. The rising phase is a little slower at about 1 ms than that of the fast voluntary muscle fibres of vertebrates, but the size of its overshoot is similar at about 30 mV. The length of the recovery phase is clearly related to the long refractory period.

The rising phase of the action potential recorded from cells of the frog's sinus venosus consists of two components. By moving external electrodes over the surface of the sinus, the point where the beat originates can be detected, and it is found that in this region the first of the two components is at its maximum size of about 15 mV. It is generally believed that this initial depolarization (Fig. 11.13 b) represents the *pacemaker potential*.

In the pacemaker region the sodium conductance appears always to be high, in rest as well as during contraction, and this has suggested a basis for the rhythmic beat of the muscle. Following the action potential, the membrane is assumed to restore its potential to the resting level through the outward movement of potassium. As the outward potassium conductance decreases and the membrane tends towards its resting potential, the permanently high sodium conductance of the membrane causes the depolarization of the membrane to begin again. The depolarization is slow at first, with a local response that is the pacemaker potential, and this in turn generates a propagated response when it reaches the critical threshold. Thus, the rhythmic depolarization in the pacemaker region is essentially due to a permanently high sodium conductance in conjunction with a variable potassium conductance. The pacemaker potential is not seen elsewhere than in pacemaker regions, and in other parts of the heart the main ions involved in the inward current of the action potential seem to be calcium ions. Although sodium ions do also make some contribution, the permanently high sodium conductance seen in the pacemaker region is not found elsewhere.

The initial, rapid, phase of repolarization is largely due to an inward movement of chloride ions. The potassium outflux characteristic of other excitable cells is also present in cardiac muscle, and by the time the plateau phase is reached, this outflux is beginning to be significant. However, it is never very large, and it is because of this, coupled with the slow time course of the decrease in the inward current, that there is a long plateau phase. The final drop to resting potential level is probably due mainly to the cessation of the inward current.

11.15 Control of cardiac muscle

Although cardiac muscle contracts spontaneously, the rate and amplitude of its beat are modified by nervous control in all vertebrates except myxinoids. Presumably, such regulation ensures co-ordination between the myogenic beat of the heart and events elsewhere in the body. The lamprey is unusual among vertebrates in having a heart innervated by a cholinergic excitatory supply. The elasmobranch heart has two inhibitory cholinergic nerves and there are areas of chromaffin tissue in the region of the postcardinal sinus that may be homologous with the adrenal medulla of tetrapods. Teleost hearts also have an inhibitory cholinergic innervation, and probably an adrenergic supply in addition, but there is some dispute about this. In tetrapods, regulation of the heart beat is performed by autonomic nerve fibres in the vagus nerve, the parasympathetic fibres being inhibitory, the sympathetic fibres excitatory. Because the parasympathetic fibres are more numerous, stimulation of the frog's vagus nerve leads to a slowing of the beat, and if the frequency of stimulation is high enough, the heart stops beating altogether (Fig. 11.14). The parasympathetic fibres inhibit the beat by altering the potassium permeability of the heart muscle membrane, which is accompanied by a decrease in membrane resistance and an increase in the size of the resting potential. In the pacemaker region of the frog's heart, hyperpolarizations of up to 33 mV in size have been recorded during vagal stimulation. The parasympathetic nerve endings achieve this result by the release of acetylcholine, and their inhibitory action is therefore comparable with that at inhibitory central synapses (§9.11).

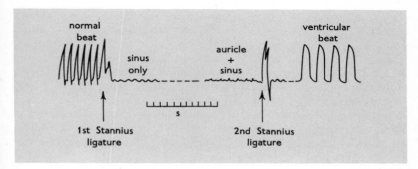

Fig. 11.14 Tracing of recording of frog heart, showing effect of Stannius ligatures. Effect of First Stannius ligature is similar to that of vagal stimulation. Tracing begins on left with normal beat. See text for explanation.

The excitatory sympathetic nerves liberate noradrenaline at their endings when stimulated, and this appears to increase the permeability of the membrance to sodium ions even above that of the normal state. The increased influx of sodium ions causes the membrane to depolarize earlier, and so the rate of beat is stepped up.

11.16 The effect of ions on cardiac muscle

The classic work of Ringer and Locke demonstrated that the presence of sodium, potassium and calcium ions in certain proportions was necessary in the bathing medium if the normal functioning of the vertebrate heart was to be maintained. The effect of reducing or increasing the quantity of one of these ions tends to be complex, since the balance between them is important as well as their absolute concentration.

An increase in potassium ions results in the depolarization of the membrane as in other excitable tissues, and therefore leads to an increase in excitability if the extra amount of potassium is small, and complete cessation of the beat if it is large. A large quantity of potassium causes such a high level of depolarization that repolarization cannot take place, and the heart stops in systole, permanently contracted. However, the depolarizing action of potassium can be lessened to some extent by raising the calcium concentration of the fluid.

Extra sodium in the bathing fluid leads to a faster rate of rise of the action potential and an increase in its size, and therefore to a higher frequency of beat and greater development of tension. A bathing medium of pure sodium chloride increases excitability to such an extent that the heart may again cease to beat and remain in a permanently contracted state.

Calcium is antagonistic to the action of sodium and potassium, and it may act in opposition to them in two ways. It is known to alter membrane permeability, and an increase in the external concentration of calcium would be expected to effect a reduction in the permeability of the membrane to potassium ions, with a corresponding increase in the resting potential, and hence reduce the influence of an increase in external potassium. It would also reduce the permeability of the membrane to sodium ions, thereby reducing excitability and opposing the effect of an increase in the external sodium concentration. Secondly, calcium will exert an influence on the contractile mechanism which is opposite to that of sodium. An increase in the entry of calcium into the cardiac cell on stimulation would increase the development of tension.

Bathing media for tissues are usually made up largely in the form of the chlorides of the necessary cations. As we have seen, ions do not merely serve a passive role in the bathing fluid. Reduction in the external chloride concentration not only results in a change in repolarization time, but also increases the rate of rise of the pace maker potential, showing that it is at least partly involved in the latter. These and other results indicate that chloride ions make a significant contribution only to the slow phases of the action potential, when the *net* cation current is relatively small. This conclusion is in harmony with the finding that cardiac cells are not very permeable to chloride ions.

11.17 Molluscan hearts

Molluscan hearts resemble vertebrate hearts in a number of ways. They consist of a thin-walled auricle and a thick-walled muscular ventricle and, in prosobranchs, at least, the ventricular muscle is striated. The muscle fibres run both along and around the wall of the ventricle, implying that the movement of blood must be peristaltic. It might be expected on grounds of efficiency that such hearts would contract from the auricular end forwards, with a pacemaker in the auricle. This appears to happen in snails, but in many other molluscs the beat seems to start anywhere in the heart and contraction may be local or complete. This state of affairs is typical of the lamellibranchs, and more complete information may indicate a correlation between the level of activity of the animal and the efficiency of its heart beat.

In cephalopods, the branchial hearts at the base of the ctenidia aid in the filling of the auricles, but in the other classes it is unlikely that the venous blood, which has passed through two sets of capillaries, body and respiratory, would be under sufficient pressure to accomplish the filling of the auricle by means of ventricular pressure alone. It has been suggested that the filling of the heart in these cases is analogous to the filling of the lungs with air in mammals. The pericardium is a closed fluid system like the mammalian thoracic cavity, with only a small and controllable opening at the entrance of the renopericardial canal. Thus, contraction of the ventricle would lead to the development of negative pressure in the pericardium, and hence to the expansion of the thin-walled auricle, so drawing in blood.

There is no doubt that the contraction of the molluscan heart is myogenic in origin, and there are similarities with vertebrate hearts in their reaction to drugs and ions. The application of acetylcholine results in the inhibition of the beat, although it does not seem to be the naturally occurring inhibitory transmitter in at least some cases. Adrenaline causes an increase in the frequency and amplitude of the beat, and in some molluscs it may be the natural excitatory transmitter. It is interesting to note that the substance 5-hydroxytryptamine accelerates the beat of some molluscan hearts. In view of the differences between vertebrates and invertebrates in relation to their neuromuscular transmitters (§9.10), it is probably better to treat the results obtained with these and other pharmacological substances with caution until more information is available about the naturally-occurring systems. However, it is quite clear that molluscan hearts are innervated by both excitatory and inhibitory nerves whose actions closely resemble those of the adrenergic and cholinergic nerves to the vertebrate heart.

Variations in the ionic concentrations of the bathing medium often have the same general effect on molluscan hearts as on vertebrate hearts. However, the effect of sodium ions is much weaker, and if this means that the membrane is not very permeable to sodium it follows that any pacemaker system must be different from the vertebrate one.

II.I8 Neurogenic hearts

Hearts in which the beat is initiated by nervous stimulation and not myogenically are known as *neurogenic*. Even in groups which are generally thought of as possessing neurogenic hearts, such as the arthropods, some members have been shown to have myogenic hearts, and it may be that this condition is widespread in the embryos of these groups, nervous initiation of the beat developing later; and therefore a primitive feature. For example, the hearts of *Artemia* and *Daphnia* do not possess an extrinsic nerve supply, and are presumed to be myogenic. The heart of a *Limulus* embryo begins to beat before any neurones have differentiated, and the adult heart will continue to beat after its ganglion cells have been removed, although at a much reduced rate, and this does not appear to be an isolated phenomenon among invertebrates.

11.19 Other involuntary muscle

Arteriolar smooth muscle fibres may be as small as 2 μm in diameter and 15 μm in length, but most smooth muscle fibres are more like those of the vertebrate gut in being about 5–10 μm in diameter, and uterine muscle cells are enlarged during pregnancy to a diameter of 9–14 μm and about 600 μm in length. Thus, although they are small, most such fibres are a little less difficult to penetrate with a microelectrode than cardiac fibres. However, the recording of a genuine resting potential is made difficult by the fact that isolated gut and uterine muscle is continuously active, and this activity is accompanied by fluctuations in the membrane potential which amount to as much as 60 mV in extreme cases. For theoretical purposes, it is probably reasonable to accept the largest recorded membrane potential as being closest to the resting potential that would be observed in the complete absence of activity. This is equivalent to a resting potential of about 80 mV in certain uterine muscle from the rat, and in the majority of smooth muscle fibres is of the order of 40–55 mV. Where the theoretical potassium equilibrium potential has been calculated from the internal and external potassium concentrations it amounts to about 80 mV, which suggests that the recorded figures represent only a partial potassium equilibrium potential, and there is almost certainly some active secretion of chloride ions in addition.

The variation in 'resting' potential inevitably results in similar variations in the size of the action potential. The peak of the action potential is generally up to 10 mV on either side of the zero potential level (Fig. 11.15). Its shape may exhibit some variation, but two components can be distinguished, a slow depolarization lasting 0.5–1.0 s, and an active response lasting about 15 ms. Both components are produced at the same frequency, but the relationship in time between them is rather variable (Fig. 11.15) showing that they do not depend on identical mechanisms. The active response may occur a little before the peak of the slow depolarization or a little after it, or at the peak of the slow depolarization. It is believed that the slow depolarizations represent pacemaker potentials and that the active responses grow from these when they reach a critical threshold level.

Some smooth muscle cells do not fit the distinction between skeletal and visceral muscle made at the beginning of this chapter. The muscle cells of the vas deferens of rat and mouse are innervated by motor nerves, and contract by means of fast co-ordinated responses, although they are multiterminally innervated and their action potentials are not propagated. Other cells, like those of the guinea pig vas deferens and the circular muscle of the vertebrate small intestine, are also multiterminally innervated by motor nerves, but appear to function normally by means of electronic spread of impulses between cells. This is also the method found in the longitudinal muscle of the vertebrate small intestine and in most vascular smooth muscle, which have few or no motor nerve endings (although, curiously enough, they produce propagated responses).

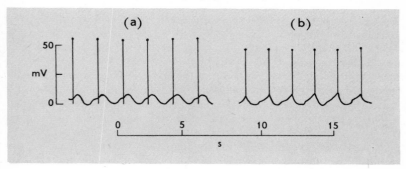

Fig. 11.15 Action potentials of taenia coli muscle of guinea-pig. (a) and (b) represent two different recordings, showing the variation in timing between the slow and fast components, probably due to the circumstances of recording. (After Bülbring, Burnstock and Holman, 1958, *J. Physiol.*, **142**, 420)

II.20 The effect of ions on visceral muscle

The common cations affect vertebrate gut muscle in a comparable way to cardiac muscle, although there are differences of detail. The relationship between the logarithm of the external potassium concentration and the resting potential is linear only above an external concentration of about 30 mmol/l, which is a much higher concentration than the naturally occurring one. Nor does the slope of the linear part of the curve suggest a very high permeability to potassium even in this range. There is some evidence that sodium ions, and possibly some anions, also contribute to the resting potential. Since potassium ions do play a part in determining the resting potential, a sufficiently high level of external potassium will depolarize the muscle sufficiently for the action potentials to die out.

An increase in the external sodium concentration causes a rise in the frequency of contraction and an increase in tension. In very high concentrations of sodium the action potentials disappear after about half an hour, and zero external sodium produces the same result after about 50 min. Only 18 mmol/l or more sodium in the medium is required to maintain the production of action potentials, and they are not completely abolished until the level drops to below 2 mmol/l. The rate of rise of the action potential is not affected by changes in the external sodium concentration. These results suggest that sodium cannot be the only ion involved in the inward current flow of the action potential.

Although there is a longitudinal T-system, it is much sparser than in skeletal muscle, and so is the sarcoplasmic reticulum. It is therefore not surprising that, as in cardiac muscle, it is believed that excitation-contraction coupling takes place largely through the influx of calcium ions, again working through the release of bound calcium from the plasma membrane. An increased calcium concentration will thus enhance contractility. What the calcium does, once inside the cell, is by no means clear. Actin filaments 30–80 μm in diameter are always present in smooth muscle cells, attached to the plasma membrane at various points, but lying predominantly in the longitudinal axis. But although myosin is present, it is not often seen as filaments, and perhaps the extent of its polymerization in these cells is not great. As stated earlier, more tropomyosin is present than in skeletal muscle. The way these features hang together in a co-ordinated system is not at present understood.

Calcium ions also undoubtedly contribute to the inward flow of current during the action potential, thus resembling cardiac muscle and some invertebrate skeletal muscles in this respect. It will be seen that the versatility of the calcium ion can sometimes render difficult an understanding of its effects. It may affect cell permeability, especially permeability to sodium; it may carry inward current; it may activate the contractile mechanism; and it facilitates the release of transmitter at neuromuscular (and possibly, other) junctions.

11.21 The effect of drugs on visceral muscle

The application of acetylcholine to gut muscle leads to a depolarization of the membrane and an increase in the frequency of the action potentials, which are smaller with a steeper prepotential. The changes caused by this drug appear to be complex, and their relationship to normal functioning of the muscle are not clear. The excitability of gut muscle is certainly modified by the action of extrinsic nerves which may liberate either acetylcholine (excitatory) or adrenalin or nor-adrenalin (inhibitory). Atropine will abolish the effect of applied acetylcholine, but it has no effect on spontaneous activity.

Adrenaline causes a decrease in excitability, often accompanied by the abolition of action potentials, and the cessation of contraction. Its mode of action is unknown.

12

Amoeboid and Ciliary Movement

All metazoan animals move by means of specialized muscle cells acting with a skeleton. They may, however, contain within their bodies cells which either cause liquid to flow past them through the movement of their attached cilia, or which move independently through their body fluids by a form of amoeboid movement. Protozoa may possess organelles that appear to resemble certain muscle proteins in their ability to contract, but the unicellular nature of their body precludes the use of these organelles for movement analogous to that of metazoans, and they move either by ciliary or amoeboid movement. These are evidently primitive methods of locomotion, but too little is known about them to show whether the contractile mechanisms of metazoan muscles bear any evolutionary relationship to them.

AMOEBOID MOVEMENT
12.1 Definition
We tend to think of amoeboid movement as being exemplified by the Sarcodina, but the term has been applied to other protozoan groups as well as to the wandering cells or *amoebocytes* of metazoans. Protoplasmic streaming has often been considered a form of amoeboid movement by some authors, and at first sight this is not unreasonable, for at least one theory of amoeboid movement makes use of this phenomenon. However, we shall take the view that amoeboid movement has as one essential feature movement in a particular direction, involving a change in the position of the animal. This definition excludes protoplasmic streaming in stationary cells, and also the filamentous pseudopodia of Heliozoa and Radiolaria, which are virtually permanent structures.

In all instances which fall within this definition of amoeboid movement, progression is by means of *pseudopodia*, protrusions of the wall of the cell into which the protoplasm flows. Some cells, such as the lymphocytes of vertebrate tissue fluid and certain amoebae, exhibit progressive polarized movement at all times. Many of these move by pushing out a single pseudopodium in the direction of movement and in amoebae like *Amoeba limax* and *Vahlkampfia* this imparts to them a movement that appears almost slug-like. Other cells, such as macrophages and the well-known amoeba of laboratory fame *A. proteus*, put out a number of pseudopodia in all directions, and although in a sense they may be said to be moving as they do so, their net position may change hardly at all. Hence the original name of *Chaos chaos* for *A. proteus*! Nevertheless, such cells may exhibit polarized movement when stimulated. *A. proteus*, if stimulated mechanically in the same place a number of times, will move off in a direction opposite to that from which the stimulus is coming.

12.2 Characteristics of amoeboid movement

Because of the relatively uncomplicated nature of their movement, the locomotion of amoebae has been studied largely in those forms that advance by means of a single pseudopodium, and it is generally assumed that the same mechanisms operate in those that produce many pseudopodia. It will be evident to the reader that this assumption is too simple as far as some of the theories which have been postulated to explain amoeboid movement are concerned. The following account therefore refers to monopodial amoebae.

The outer layer, the *plasmalemma*, seems to possess adhesive properties that enable it to stick to the substrate, an essential requirement for this type of movement. Beneath the plasmalemma is a thin clear layer that appears to be fluid in character and is apparently continuous with the much thicker clear layer, or *hyaline cap* at the front end of the animal (Fig. 12.1). The bulk of the cytoplasm is divided into an outer *ectoplasm*, which is partly non-granular, and an inner *endoplasm*, which contains many granules. The ectoplasm is seen to rupture frequently at the front of the advancing pseudopodium, with the result that endoplasmic granules flood into the hyaline cap. This observation suggests that some kind of change takes place in the ectoplasm of a pseudopodium.

Until recently, it had been supposed that the outer hyaline layer of the cytoplasm (not precisely equivalent to the ectoplasm) was in the gel condition and relatively viscous, whereas the rest of the cytoplasm was in the sol state. This is to some extent true, and the gel can be seen to get thicker from the anterior end backwards. But recent studies have shown that the position is not quite so clear-cut as this. It is possible for the cytoplasmic proteins to exist in states intermediate between those of gel and sol, and there is good evidence that the central core of the endoplasm is more viscous than that immediately around it in the ectoplasm. This central core of more viscous material moves forward in the endoplasm to build up at the front end and then bend round to flow backwards (Fig. 12.1). The velocity of flow of this viscous core is faster towards the bend and slower as it moves backwards away from it.

It is generally agreed that sol–gel changes play a part in the production of pseudopodia. For the pseudopodium to flow forward, it must consist

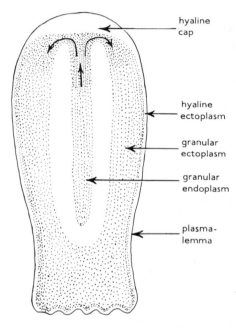

hyaline cap

hyaline ectoplasm

granular ectoplasm

granular endoplasm

plasma-lemma

Fig. 12.1 Structure of an amoeba (see text). (After Allen, 1960, *J. biophys. biochem. Cytol.*, **8**, 379, by permission of the Rockefeller University Press)

principally of the less viscous sol, although the boundary of the pseudo-podium will always be in gel form. Thus, there is a forward flow of sol which gelates at its outer edge where the pseudopodium is extending. To provide the necessary forward flow of sol, the gelated proteins at the posterior end are converted continuously into the sol form. The hyaline cap represents the region in which the sol builds up and is converted to gel. During this process, *syneresis*—the extrusion of water from a gelating protein—occurs, and this water must be passed back to the posterior end to enable more sol to be produced. It seems highly probable that this backward flow of water takes place in the clear layer beneath the plasmalemma.

12.3 Theories of amoeboid movement

Early theories of amoeboid movement, produced in the first part of the nineteenth century, bore a much closer resemblance to modern theories than those which were postulated later in that century during the period when an extreme mechanistic climate of opinion laid too much stress on animals as machines. These early theories recognized that the cytoplasm was differentiated into outer and inner layers, and in the 1860's it was realized that these were not permanent layers but consisted of interconvertible material. Most theories up to this time supposed that the pseudopodium was extended through the exertion of a contractile force, although there were differences as to whether this was a squeezing from the rear or a pulling from the front.

Much the same views exist today, and it is still impossible to be certain how amoeboid movement is achieved. All the likely theories, however, do postulate that there are filamentous proteins which, like the actomyosin threads of muscle, are capable of contracting to a shortened form, or alternatively that contraction results from gelation of the sol through the removal of water and its accompanying shrinkage. The general view nowadays is that the contraction required is greater than could be accounted for by sol–gel changes, however, and so we shall restrict ourselves to a consideration of theories that involve active contraction of 'myofilaments'.

These really fall into two groups, one in which contraction is presumed to take place at the rear of the animal and thus squeeze the pseudopodium forward; and those which presume that contraction takes place at the front of the animal, so pulling the sol forward. The idea that contraction takes place at the rear end derives primarily from the observation that the 'tail' of some amoebae has a wrinkled appearance as though it is contracting. Experiments in which an electric current was applied to an amoeba resulted in wrinkling and appropriate movement, and this was claimed to be similar to what happens during natural amoeboid movement. There is certainly evidence of a standing electric charge on the membrane of an amoeba that changes during movement, but the correlation between this phenomenon and the production of pseudopodia has never been made very clear, if it exists.

The idea that the forward flow of the sol is due to its being pulled forward by contraction at the anterior end springs from the fact that the streaming of cytoplasm in an amoeba continues even when the cell is ruptured, suggesting that the contractile force is not exerted by the outer tube of the animal but is connected with streaming. There is no direct evidence on which to support or reject this view, which firmly equates protoplasmic streaming and contraction. Protagonists of this viewpoint

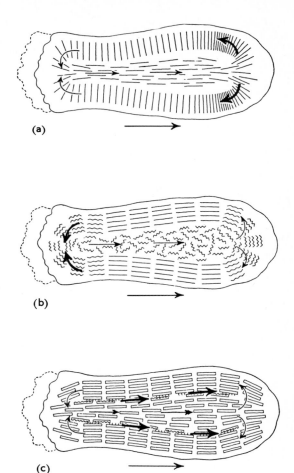

(a)

(b)

(c)

Fig. 12.2 Theories of amoeboid movement. (a) The cortical gel contracts and pulls the central core of endoplasm forward; (b) the tail contracts and squeezes the endoplasm forward. The wavy lines are contracted molecules of endoplasm. (c) Sliding filament ratchets on the inner edge of the gel push molecules of endoplasm forward and produce a forward movement of it. The thick arrows indicate the point at which force is applied. In all three diagrams endoplasm turns to gel at the head and gel to endoplasm at the tail. (From Hayashi, 1961, *Sci. Amer.*, **205** (9). © 1961 by Scientific American, Inc. All rights reserved)

have proposed two different methods of contraction at the anterior end, one through the contraction of 'myofilaments' there, and the other through the sliding of filaments against one another, an idea borrowed from the sliding filament theory of muscular contraction (§11.5). The three variants of the contraction theory are shown in Fig. 12.2. At present, there is insufficient evidence to be able to decide between them.

Finally, it should be mentioned that experiments to determine the relationship between temperature, the sol–gel ratio of the cytoplasm and viscosity do not exhibit the expected correlation. The mean viscosity is higher at higher temperatures, yet the rate of locomotion is also higher. Nor does the viscosity of the cytoplasm seem to be correlated with the sol–gel ratio. These results do not fit very well with any of the proposed theories, but their meaning requires further elucidation.

12.4 Control of amoeboid movement

It is obvious that the production of pseudopodia for polarized movement must be directed in some way by the cell. Response to a stimulus might possibly be due to local depolarization of the surface of the amoeba leading to the setting up in the stimulated area of a temporary organizing centre. A stimulus that affected the whole of the body might achieve the same result if it was nevertheless directional and therefore capable of setting up a stimulus gradient. The same mechanism, the establishment of an organizing centre, could be responsible for directed movement in the absence of a stimulus, although it is not at all obvious why such a centre should exist in these circumstances.

The necessity for an organizing centre to govern the flow of the sol by directing the contraction of the 'myofilaments' has been a feature of a number of theories. Clearly, the simplest supposition is that it exists in the region where contraction occurs. This means that the posterior contraction theory requires a posterior organizing centre, and the pulling and sliding theories an anterior one. There is no direct evidence for the existence of such centres, nor any indication of the form they might take.

CILIARY MOVEMENT

12.5 Cilia and flagella

Cilia are hair-like projections from the surface of a cell which are capable of movement independently of the cell which bears them. They vary from 0.1 to 0.3 μm in diameter, and are usually about 10–20 μm in length. Flagella are similar in structure and diameter, but are much longer up to 2 mm, and whereas cilia usually occur in large numbers on a cell, there are normally only one or a few flagella.

Cilia function by engaging in some sort of lashing motion, which will be considered in more detail below. They therefore achieve their effect by moving against the viscous resistance of water, and this means that they are restricted in their action to a watery medium. If the cell to which they are attached is sufficiently small in relation to the size of the cilia their action will move it. If it is large, or fixed in position so that it cannot move, it is the water that will shift instead. As a locomotory mechanism ciliary movement is therefore restricted to small organisms, and is especially found in the Protozoa. It may be used as an adjunct to other forms of locomotion in larger animals, as in the bigger turbellarians and various creeping molluscs, in which it reinforces the muscular peristalsis of the ventral surface. Its major use in the metazoan animals, however, is to create water currents for feeding or the propulsion of material through various tubular structures.

12.6 Characteristics of ciliary movement

The exact form which the beat of a cilium or flagellum may take shows considerable variation in different circumstances. Cilia vary less than

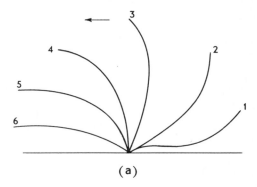

(a)

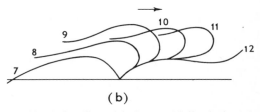

(b)

Fig. 12.3 The normal beat of a cilium. (a) Forward (effective) stroke; (b) return stroke. (After Gray, 1928, *Ciliary Movement*, Cambridge University Press)

flagella, presumably because they are shorter and therefore less flexible. The typical form of the ciliary beat is shown in Fig. 12.3. The effective stroke is made with the cilium held fairly rigid, whereas the recovery stroke is made with the organelle very bent, in which position it offers less resistance to the water than the effective stroke. The effective stroke is also much more rapid than the recovery stroke by some two to ten times. The beat of the cilia on a ciliated surface is co-ordinated to produce metachronal rhythm, with which the reader will already be familiar.

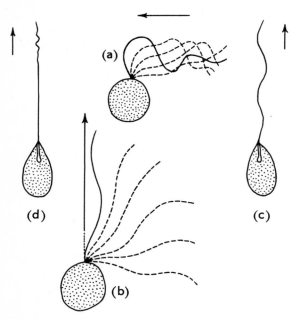

Fig. 12.4 Some examples of flagellar movement. (a) Sideways movement and (b) forward movement in *Monas*; (c) forward movement and (d) forward movement using only the tip of the flagellum, in *Peranema*. ((a) and (b) after Krijgsman, 1925, *Arch. Protist.*, **52**, 478; (c) and (d) after Borradaile *et al.*, rev. Kerkut, 1958, *The Invertebrata*, 3rd edn., Cambridge University Press)

Flagella exhibit a much greater variety of movement than cilia. Rapid forward movement by flagellate protozoans involves movement of the flagellum which is fairly comparable to that of cilia. But flagellates may also indulge in sideways movement, and turning movements by rotation of the tip of the flagellum only (Fig. 12.4), and in some of these more complex movements the beat may alter in amplitude along the length of the flagellum.

12.7 The structure of cilia

The examination of cilia from many sources, both plant and animal under the electron microscope, has revealed that they possess a remarkably uniform structure. There is a central filament that consists of two

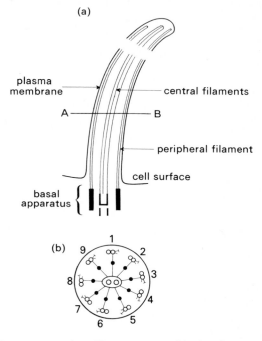

Fig. 12.5   Ultrastructure of a cilium, as seen with the electron microscope. In (a) the cilium is shown in broken longitudinal section. The line AB represents the level at which the cross-section given in (b) was taken. It shows the two central filaments surrounded by nine peripheral double filaments. The latter bear two dynein arms each and appear to be connected to the central filaments by radiating strands that often seem to possess some kind of thickening about mid-way along their length. Below the plasma membrane the dynein arms disappear and the peripheral filaments consist of three members.

tubular elements, and these are invariably all orientated in the same direction in a given piece of ciliated tissue. The central filament is surrounded at the periphery of the cilium by nine filaments, which are often rather thicker than the central filament (Fig. 12.5). These filaments are basically doublets, the two components being designated sub-fibres A and B (Fig. 12.5), but sub-fibre B also bears two projecting arms, composed of a protein to which the name *dynein* has been given and which exhibits ATPase activity. Some authors describe these peripheral doublets as consisting of many spherical sub-units in straight or helical columns. They often appear to be connected to the central filaments by radiating strands of tissue on which thickenings may sometimes be seen, claimed by some to be some sort of sub-filaments. The structure of the central filaments is less well-defined, but they generally seem to be surrounded by dense material that links up with the strands that radiate to the peripheral filaments.

At the base of the cilium just within the cell is the hollow *basal granule*. The top part of the basal granule, the *basal plate* appears to be joined to the base of the filaments, the central filament terminating just short of it. Near their junction with the basal plate, the outer filaments seem to gain a third tube, giving them a triple structure instead of a double one. In some cells the basal granule, which is generally believed to be derived from or to represent the centrosome, sends a striated *rootlet* into the cytoplasm of the cell to end near the nucleus. The basal granules of adjacent cells may be connected by a strand of material. The whole cilium is covered by the plasma membrane of the cell.

12.8 The functioning and control of cilia

Two explanations have been advanced to account for the beating of a cilium. One assumes the outer filaments to be contractile. Bending in a given direction would thus result from the contraction of the filaments of the side pointing towards the direction of bending. In the normal beat of a cilium, the filaments would contract first on one side and then on the other, although the differences in the shape of the effective and recovery strokes suggests local differences in the speed and extent of contraction which are difficult to visualize. This is even more true of the complex movements sometimes performed by flagella. The other explanation suggests that the filaments can slide in relation to one another (Fig. 12.5), but it must be emphasized that precise parallels cannot be drawn with skeletal muscle, since there is no evidence for the presence of thick and thin filaments.

The role of the central filament has excited speculation. It might be the pathway for the excitatory stimulus from the cell, or it might simply be a central elastic core. The fact that the two elements are always orientated in the same direction in a particular tissue could be significant for either view. If the central filament is simply an elastic rod, its orientation might be related to a restriction on the plane of bending. It is even barely possible that the basal granule selects the filaments to contract, and the central filament exerts a local inhibitory effect to produce the complex shapes of the beat. All this is speculation: we have no real evidence on which to accept or reject any of these views. It is noteworthy that the lack of central filaments is usually associated with cilia that have been modified for other purposes, e.g. retinal cells, and have lost their motility.

In the control and co-ordination of cilia, the basal granule is generally agreed to play a fundamental role. If it is removed, the cilia stop beating, and it is noteworthy that modified ciliary cells that are non-motile lack such a structure. Basal granules seem to act as spontaneous pacemaker centres for the ciliary beat, since small pieces of cell with cilia attached continue to exhibit ciliary activity. Yet there must be control of these pacemakers by the cell, because the cilia may beat at different rates or even reverse the direction of their beat, in response to external stimuli. The progressive anaesthetization of the ciliary beat results in a loss first of the power of reversal of beat, then loss of co-ordination and hence of metachronal rhythm, and finally the cessation of activity in the now independently beating cilia, and it has been suggested that these three aspects of ciliary movement are controlled by three separate mechanisms, but the nature of these is unknown. There is, however, no doubt that the beating of cilia is associated in some as yet undefined way with changes in the potential difference across the plasma membrane of the cilium.

Protozoans like *Paramecium*, which exhibit a complex type of movement, often seem to possess strands of material joining up groups of basal granules, or as in *Paramecium* itself, radiating out from a central point, the *motorium*. These strands and motoria have naturally been suspected as having an overall control of the ciliary beat. Unfortunately, direct confirmation is difficult, as cutting them in any way destroys other parts of the cell and makes a definitive interpretation of the results more difficult. Interference with these systems undoubtedly does upset the co-ordination of the ciliary beat. There is no doubt that the ciliary beat is under nervous control in metazoans, in which the cutting of the nerves that control them produces unambiguous results. Experiments involving this technique have demonstrated the existence of such control in a wide variety of animals, from the locomotory cilia of turbellarians and molluscs to the pharyngeal cilia of the frog. But how the nerves exert their effect is still unknown.

13

Hormones

Efferent nerves run not only to muscles, but also to secretory organs known as *glands*. Two types of glands are generally recognized. *Exocrine* glands are those which, like the salivary glands or most of the pancreas of vertebrates, have their secretions conveyed to the site of action by special ducts. *Endocrine* glands lack such ducts, and instead they are richly supplied with blood vessels into which their secretions are passed, to be carried to the site of action by the blood. For this reason, endocrine glands are sometimes called *ductless* glands and their secretions are thought of as chemical messengers. To these secretions the name *hormone* has been given. Although hormones may exert rather diffuse effects on all body cells, these are often secondary effects that result from the carriage of hormones round the body by the blood, and most hormones may be thought of as acting upon a definite *target organ* in the body. Because they must be carried to their target organ by the blood, hormones are usually slow in producing a response by comparison with the nervous system.

Some endocrine glands are not activated by efferent nerves, but by secretions produced for the purpose by neurones. These secretions may either be released into special capillary networks, with which the secretory neurones form discrete *neurohaemal organs*, or released in the close vicinity of their target organs. The production of hormones by neurones is known as *neurosecretion*. Such hormones are distinguished by some workers as *neurohormones*, and placed in a separate category from other hormones. However, there is a whole spectrum of secretions by neurones, from synaptic transmitters (sometimes termed *neurohumours*) to neurohormones; and since some neurohormones are released into the blood circulation like the hormones of non-nervous glands, they overlap in kind with them, and distinctions in nomenclature based on their origin are somewhat arbitrary. In general, more is known about mammalian endocrine glands than about those of any other group. Therefore, any account of hormones must inevitably begin with the mammals and work outwards to those of other groups. We shall begin with the pituitary gland, which liberates into the blood a large number of hormones that affect the secretion of hormones by other endocrine glands. For this reason it is sometimes described as the 'master gland', but it is better to avoid this kind of terminology in describing such a dynamically balanced system.

13.1 The pituitary gland

This gland is a small knob-like body that lies in a depression in the floor of the cranium, just above the roof of the buccal cavity. In all vertebrates, it is derived from two embryological components, the *adenohypophysis*, which arises from the roof of the buccal cavity in the form of a pouch; and the *neurohypophysis*, which develops from the *infundibulum*, a depression of the floor of the diencephalon in the fore-brain. These two rudiments come together during the development of the embryo to form one complex gland. It should be noted that they do not correspond to the 'anterior lobe' and 'posterior lobe' referred to in the older literature on the gland.

A diagrammatic section through the pituitary gland of a mammal is shown in Fig. 13.1, from which it can be seen that the two primary components may be divided into further distinct areas. The adenohypophysis is composed of three such areas, the *pars intermedia*, the *pars distalis* and the *pars tuberalis* (the two latter being the 'anterior lobe' of the older terminology). The pars tuberalis seems to be simply a supporting structure for the gland and its blood vessels, and to be without endocrine function. The neurohypophysis comprises the infundibular process and the median eminence. The pituitary glands of other vertebrates are basically similar in structure, despite differences in the topography and proportions of the individual parts. In those cases in which a part found in the mammalian pituitary is absent in another group of vertebrates, the hormone characteristically secreted by that part is produced elsewhere in the gland. For example, birds and a few mammals lack the pars intermedia, but the hormones secreted by this area in other mammals are formed in these cases within the pars distalis.

The one exception to this substantially uniform picture is the pituitary body of the hagfish *Myxine*. In this animal the adenohypophysis has no recognizable lobes, but consists simply of islets of cells embedded in connective tissue. It is often assumed that the pituitary of the hagfish represents the primitive vertebrate state before the evolutionary elaboration of different areas, but there is no evidence on which to accept or reject this theory.

In the land tetrapods the pars tuberalis is supplied with blood by branches of the internal carotid arteries which form a plexus from which capillary loops extend into the median eminence, where they comprise what is known as the *primary plexus*. From the primary plexus, vessels

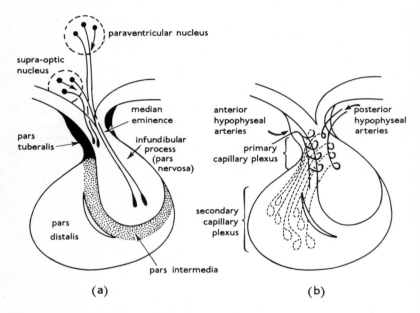

Fig. 13.1 Diagrammatic vertical section through the pituitary body of a mammal; (a) to show the major regions and the link between the neurohypophysis and the hypothalmic nuclei; (b) to illustrate the blood circulation. (Based on Ham, 1965, *Histology*, 5th edn., Lippincott, Philadelphia and Pitman Medical, London)

run down to the pars distalis, where they end in sinusoids that constitute the *secondary plexus*. The median eminence is therefore linked with the adenohypophysis through a common blood supply (Fig. 13.1).

Although the median eminence and infundibular process both receive blood from the internal carotid arteries, the supply of blood to the infundibular process is provided through an independent set of vessels. The infundibular process consists of non-myelinated nerve fibres and non-secretory supporting cells. The nerve fibres originate in the area of the brain dorsal to the pituitary gland, and especially from centres in the hypothalamus. The cell bodies from which they are derived produce hormones which are carried in secretory granules of a colloidal material. These secretions are passed down the axons to their terminals in the infundibular process, and there they are stored. It is believed that these hormones are released from the nerve terminals which contain them when impulses pass down the axon, which means that the hypothalamic neurones concerned are both nervous and secretory in function. Thus, the infundibular process itself is simply a storage depot for hypothalamic neurosecretions, and does not contain its own secretory cells.

By contrast, the adenohypophysis consists of cells which themselves secrete hormones, with the exception of the pars tuberalis whose role is one of support for the median eminence. And whereas the release of hormones from the infundibular process is the result of nervous stimulation, the only nerve supply to the adenohypophysis is the vasomotor innervation of its blood vessels. Unlike the situation in the infundibular process, therefore, any stimulation of the adenohypophysis must be achieved by chemical means, via its blood supply. Such chemical stimulants include not only hormones circulating generally in the blood, but also substances produced by the hypothalamic axons that run through the median eminence, by means of the link between this structure and the pars distalis that is afforded by the two blood plexuses. The hypothalamic neurones therefore provide an important link between the central nervous system and the pituitary body in the integration of behavioural responses.

The median eminence is characteristic of the tetrapods, and is not found in fishes other than Dipnoi (lungfish). Nevertheless, the structure of the gland in fishes is such that the area of contact between the neurohypophysis and the adenohypophysis is large, and to some extent they possess a common blood supply, so that some influence may still be exerted on the adenohypophysis by the hypothalamic neurones.

13.2 The neurohypophysial hormones

Two hypothalamic polypeptide hormones can be extracted from the neurohypophysis of mammals, *oxytocin* and *vasopressin*. In mammals, the two fractions exhibit quite different functions. Oxytocin produces a powerful contraction of the uterine muscle, induces lactation, and facilitates the ejection of milk from the mammary gland. Vasopressin affects the water balance of the animal by reducing its output of urine, and hence is sometimes referred to as the *antidiuretic* hormone (ADH). It exerts a direct effect upon the resorption of water by the descending limb of Henle's loop and the distal convoluted segment of the uriniferous tubule. The controlling factor in the release of vasopressin is the concentration of the blood, but it is not known whether there are special osmoreceptors in the hypothalamus, or whether the hypothalamic cells are directly affected by the blood concentration.

A number of molecular variants of these two hormones is found in other vertebrates. Vasopressin is replaced by *arginine vasotocin* in all groups other than mammals and is probably the only hypothalamic polypeptide hormone in cyclostomes. Elsewhere, at least one other hormone is present; *isotocin*, for example, in teleost fishes and *mesotocin* in certain amphibians. Oxytocin is characteristic of the higher vertebrates, particularly of birds and mammals. These and other variants are all octapeptides, and hence all the neurohypophysial hormones are closely related chemically. It is tempting to suppose that, beginning with one function common to such an octapeptide structure, the functions of the neurohypophysial hormones have been elaborated, for example, and that the reproductive effects of oxytocin have appeared during the evolution of the mammals.

In considering this possibility, it is instructive to note the action of arginine vasotocin on the lower vertebrates. In the Amphibia, water balance is between the uptake of water through the skin, and its loss by evaporation and excretion. Amphibians cannot produce a urine hypertonic to the blood, but they can increase its concentration to some extent. Arginine vasotocin reduces the amount of urine and increases its concentration by facilitating the resorption of water by the kidney tubule, and probably also by constricting the afferent arterioles to the kidney and thereby decreasing the amount of blood available for filtration. In addition, the hormone promotes the active uptake of water by the skin, and is thought to facilitate the resorption of water from the bladder. Thus, in the Amphibia it increases the movement of water across epithelia, presumably by altering their permeability to water as it is believed to do in the mammalian kidney. Its action is more powerful in the most terrestrial amphibians such as toads, and least powerful in permanently aquatic urodeles such as *Necturus* and *Xenopus*.

These facts suggest that the function of arginine vasotocin is to assist in water conservation, and it therefore appears to be related to a terrestrial habitat. Yet it is found in teleost fishes, in which the salt and water balance is probably under the control of the adrenocortical hormones (see below). If, as suggested above, the reproductive functions of oxytocin are a mammalian specialization, and the osmoregulatory function of vasopressin arose with the evolution of tetrapods, we are left with the question of the original function of the neurohypophysial octapeptides. It is possible that the answer may be found in the vascular connection between the neurohypophysis and the adenohypophysis that exists in all vertebrate groups, and has become elaborated into the median eminence of the higher vertebrates. It is known that neurohypophysial hormones can regulate the secretion of hormones by the adenohypophysis through this link, and it may be that this was their original function.

13.3 The pars intermedia and colour change

With two exceptions, all the hormones produced by the adenohypophysis have as their target organ another endocrine gland. Pituitary hormones which affect the secretion and release of other hormones are known as *trophins*, and will be dealt with in conjunction with the endocrine glands they affect.

One exception is the hormone of the pars intermedia, *intermedin* (*melanophore-stimulating hormone*, MSH). This hormone is found in all vertebrate groups, but appears to be functional only in the poikilothermic ones. Unlike the birds and mammals, which depend to a great extent for their

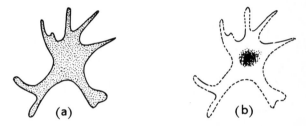

Fig. 13.2 Generalized diagram of vertebrate chromatophore in (a) dispersed condition (skin dark) and (b) concentrated condition (skin light). Individual chromatophores keep the same shape, but vary considerably by comparison with others.

coloration on pigmented fur or feathers that change colour seasonally, if at all, poikilothermic vertebrates are able to adjust their colour in relation to their background, or for the purpose of sexual display.

Colour change of this kind is due to alterations in special pigmented cells contained in the skin. These *chromatophores* are cells with a number of radiating processes that remain in a fixed position within the skin, and which contain within them pigment that can either be in a concentrated condition or dispersed. When the pigment is concentrated in the centre of the cell, the area of the skin which is light is much greater and the skin as a whole appears paler. When the pigment is dispersed throughout the cell, the light area of the skin is smaller, and the skin appears darker (Fig. 13.2). A variety of pigments may be found in different chromatophores, enabling hues of yellow, red, brown and black to be achieved. Some chromatophores contain white crystals or granules of guanine, which impart a white or silvery appearance to the skin. Such *guanophores* are often revealed when the pigment in the coloured chromatophores is concentrated and so enhance the impression of lightness of the skin.

Because the pigments of the red and yellow chromatophores are fat-soluble, they are referred to as *lipophores*. Little is known about the way they function, almost all the information available being related to the brown/black *melanophores*, so named because they contain the dark pigment *melanin*. The mechanism by which concentration or dispersal of melanin within the cell is achieved is unknown, but the agents that control them are well known and are either nervous or hormonal, or both.

Young amphibian larvae become pale in complete darkness and dark in bright light, and the effect can be shown to be independent of the eyes. Later larvae and adults do not exhibit this response if they are blinded, but continue to do so if the eyes remain intact. Evidently, the eyes become progressively more important for the eliciting of the response, and indeed it can be shown that in the adult the response is dependent on the ratio of the light falling directly on the animal (incident light) to the amount of light reflected from the background. If an area of the skin is denervated, it still responds in the same way as the rest of the skin, showing that local control must be hormonal. Such experiments demonstrate that in young amphibian larvae the response of the melanophores to light is under purely hormonal control, but that in later larvae a nervous component is introduced that uses the eye as the receptor organ for the reflex. Although no definite pathway is known between the optical neurones and the pituitary, it must presumably end in the hypothalamus, and be linked with the pars intermedia through the hormones released by the neuro-hypophysis.

Responses such as this are found in many fishes and reptiles, but in some of these the control mechanism seems to be purely nervous, while in some it is probably exclusively hormonal; in others it is both nervous and hormonal.

Intermedin causes the dispersal of melanin within the melanophore. This effect is in contrast to that of two other hormones, both of which produce concentration of the pigment. One of these is adrenaline, and its action on melanophores is less important than its other actions, and may indeed be a by-product. However, as far as is known the other melanin-concentrating hormone exerts no other effect. This hormone is *melatonin*, and it is secreted by the pineal gland on the top of the brain. In some animals, this rather complex and mysterious organ is sensitive to light, and thus the secretion of melatonin in these forms might be a direct reflex.

Intermedin also stimulates the multiplication of melanophores, and an increase in the quantity of pigment inside them, but these reactions are, of course, much slower than the more rapid dispersion effect. It is interesting to note that the lipophores of teleosts are even more responsive to intermedin than are their melanophores, and that the guanophores of amphibians are reduced in number and their pigment concentrated by the hormone. It is possible that the target organs of intermedin have changed during the course of evolution.

The precise chemical nature of intermedin has been determined in only a few mammals, owing to the difficulty of purifying it. Two types of intermedin have been identified in these animals. The α-intermedin type has 13 amino acids in its molecule, and the β-intermedins have 18. Even in the few cases in which the extract has been sufficiently pure for the point to be established, it has been found that the specific amino-acid content of α-intermedin may vary between species. A point of interest is that the α-intermedin molecule is contained within the larger molecule of the trophin ACTH (see below), which acts on the adrenal cortex. It is therefore not surprising that large doses of ACTH have an intermedin-like action, although the high dosage necessary suggests that this hormone has no practical significance in the regulation of colour change. It is clearly possible that intermedin and ACTH had a common origin in the evolutionary past.

13.4 Growth hormone (somatotrophin)

At one time, it was thought that the pituitary gland secreted both a growth hormone and three other hormones that affected carbon metabolism. It is now known that all these effects are exerted by a single hormone. This is a protein secreted by the pars distalis, whose molecular weight is in the region of 22 000.

Throughout the vertebrates, the ablation of the pituitary gland (*hypophysectomy*) leads to retarded growth in the young and the retention of juvenile characters, and in adults to loss of weight, consequences which are abolished by administration of the hormone. Its effects are complex, and its mode of action is not understood. It favours skeletal growth and development, and facilitates the building up of proteins from amino acids. At the same time, it causes a reduction in the quantity of fat stored and inhibits its formation from carbohydrate. The conversion of protein into carbohydrate is also inhibited, and since it promotes the formation of glycogen in liver and muscle, it can cause a transient fall in blood glucose, although this is counteracted by other metabolic hormones. An excess of the hormone will eventually cause the insulin-producing cells of the pancreas to atrophy, and thus produce diabetes (see below). Finally, growth hormone may act as a *synergist* for at least some of the adenohypophysial tropins. This means that although it cannot imitate their effects, it can enhance them.

13.5 The thyroid gland

The thyroid gland originates in the embryo as a median downgrowth from the floor of the pharynx. It is presumed to be homologous with the endostyle of the protochordates, as it can be seen to develop from part of this structure during the metamorphosis of the ammocoete larva of the lamprey. It is very uniform in histology throughout the vertebrates, and consists of vesicles or follicles lined with cubical epithelium which contain within their lumen a colloidal secretion (Fig. 13.3). The base of each

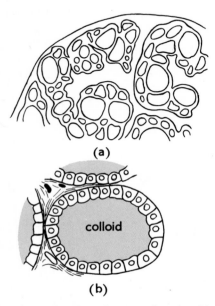

(a)

(b)

Fig. 13.3 Diagram of cross-section of mammalian thyroid gland. (a) Part of gland, showing follicles cut through at different levels, and grouped in lobules; (b) a single follicle.

thyroid cell is folded in a complex way and the thin double-walled folds may run far into its cytoplasm. The apical surface of the cell has small projections or microvilli that increase its surface area. The gland is very compact in mammals, in which it forms a discrete mass above the heart, but may be less so in other vertebrates. In teleosts especially, it forms a loosely organized tissue that extends along the ventral aorta and often into the branchial arches, a situation that makes extirpation experiments very difficult.

The thyroid gland produces hormones which are iodinated derivatives of the amino acid *tyrosine*. The first stage in their production is the active secretion of iodide into the lumen by the follicle cells. Within the lumen, it is probably oxidized enzymatically to iodine

$$2I^- \rightleftharpoons I_2$$

and the iodine is then firmly bound to the protein *thyroglobulin*. The thyroglobulin molecule is very large, with a molecular weight of about 700 000, and must be split into smaller units before its iodine-containing components can become diffusible and pass into the blood. The splitting of the thyroglobulin is brought about by a hydrolytic protease secreted by the

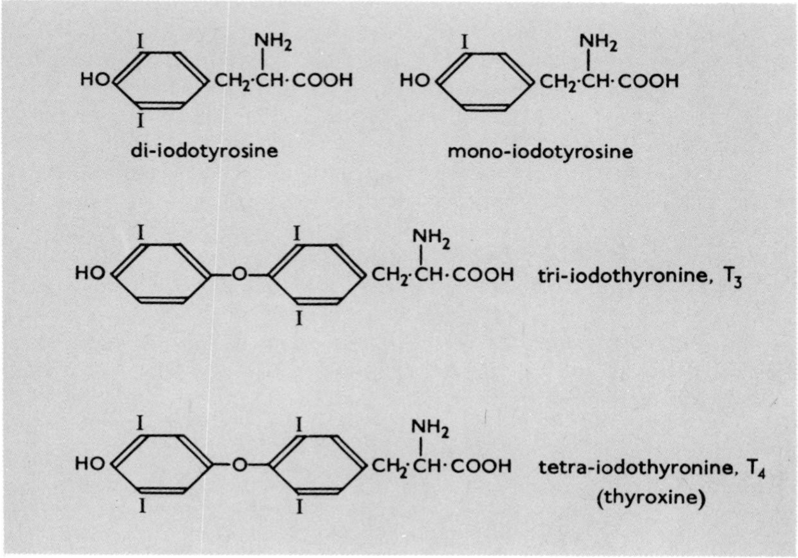

Fig. 13.4   Breakdown products of iodinated thyroglobulin.

follicle cells, and believed by some workers to be derived from their lysosomes, and it results in the production of 4 different iodine-bearing molecules. The chemical structure of these products is shown in Fig. 13.4, from which it can be seen that two, *mono-* and *di-iodotyrosine*, contain 1 molecule of tyrosine and 1 or 2 atoms respectively of iodine; and that the other two are formed by the condensation together of 2 tyrosine molecules, so that the resulting molecule contains 3 or 4 atoms of iodine. These two latter substances are *tri-* and *tetra-iodothyronine*, or they may sometimes be referred to as T_3 and T_4, and the last one also as *thyroxine*.

Mono- and di-iodotyrosine are de-iodinated within the follicles by a specific enzyme, and only T_3 and T_4 pass from the gland into the blood. Both are active hormones, but T_3 is more potent in its effects than T_4. When they are released into the blood they become attached to one of the serum proteins and circulate in this form.

The production of iodinated tyrosine is not exclusive to the thyroid gland. The kidneys and mammary glands of mammals, certain cells of fish gills, the notochord cells of the lamprey, and a variety of tissues from all groups of the invertebrates except protozoans and echinoderms, possess this property, but in most cases, only mono- or di-iodotyrosine are formed. Thyroxine and tri-iodothyronine have, however, been identified in protochordates, and thyroxine in some molluscs. It is not known whether this ability to produce iodinated tyrosine compounds outside the thyroid has any functional significance. Nevertheless, it is evident that the capacity for iodinating protein is widespread in the animal kingdom, and may have been present for a long time before it was utilized in the course of evolution. The derivation of the thyroid gland from a gut rudiment raises the possibility that, as may happen in amphioxus and the ammocoete before metamorphosis, the iodoprotein was primitively hydrolysed in the gut by one of its enzymes.

In mammals and birds, the thyroid hormones produce an increase in the metabolic rate, accompanied by corresponding increases in oxygen consumption and heat output. It was pointed out in §8.5 that when physical methods of keeping warm in a cold environment fail to be sufficiently effective, homeotherms are obliged to resort to chemical heat production through an increase in carbon metabolism. They achieve this increase in metabolic rate by a rise in the output of thyroid hormones, but their action cannot be isolated from that of other hormones. Somatotrophin, ACTH (§13.9) and adrenaline (whose action is potentiated by thyroxine) are also involved. Thyroid hormones and somatotrophin appear to act synergistically in facilitating protein synthesis. One interesting effect of thyroid hormones is their influence on the growth of hair, which can be important in cold regions especially. Heterothermic mammals and those which hibernate do not exhibit increased thyroid activity when they are suddenly cooled and in this, and in the seasonal waxing and waning of thyroid activity in summer and winter, they resemble the poikilothermic vertebrates.

The role of the thyroid gland in heterotherms and poikilotherms is clearly not related to temperature regulation, and it has excited much experiment and even more speculation. It is fair to suppose that it must have some connection with metabolism if we do not restrict that term too far. There does seem to be some link between mineral metabolism and the thyroid in some fishes, but other fishes, in common with some amphibians and reptiles, either make no thyroid hormone at all, or so little that it is difficult to comprehend what effective function it could have. There may also be an association between thyroid activity and the sexual cycles of vertebrates (§13.10), by virtue of the need for an increase in metabolism during certain of their phases.

One of the best-known actions of the thyroid is its effect upon the metamorphosis of amphibian tadpoles. Both T_3 and T_4 are present in the thyroid of the frog tadpole, and as the time for metamorphosis approaches their concentration in the blood rises. The progressive increase in concentration can be shown to be correlated with the sequence of changes preceding metamorphosis—a striking example of different tissues possessing different thresholds to the same hormone. If the thyroid gland is removed from tadpoles they do not metamorphose, but will do so if thyroid extract is administered. If thyroid extract is given to intact tadpoles, their metamorphosis is hastened. Thyroid hormones are also necessary for the proper moulting of the amphibian skin.

There is good evidence that the thyroid is implicated in the metamorphosis of certain teleost fish such as the eel and the salmon, but no evidence that it serves any function in the metamorphosis of the lamprey. It may be that the function of the thyroid in the metamorphosis of some animals is an evolutionary elaboration of a more primitive function in relation to metabolism—metamorphosis and metabolism could obviously be connected—and it is tempting to suppose in view of its homologies that the metabolic function in turn arose from some action of the endostylar secretion upon foodstuffs in the gut.

The pars distalis of the adenohypophysis secretes *thyrotrophin*, sometimes called thyroid-stimulating hormone or TSH, which has the effect of stimulating the thyroid to increase its output of hormones. There is a reciprocal or *feed-back* relationship between the two glands. A fall in the thyroid hormone level in the blood results in an increase in the output of thyrotrophin. The latter stimulates the thyroid to produce more hormones and these in turn inhibit the output of thyrotrophin by the pars distalis. The action of the thyroid hormone on the pars distalis seems to be both direct, by a chemical effect mediated through the blood vessels that pass into it from the median eminence; and through a further chemical influence, this time upon the cells of the hypothalamus, which are presumed to react with secretions that also reach the pars distalis via the median eminence. In homeotherms, there is an additional pathway through which the thyrotrophin-secreting cells are stimulated, resulting from the effect of the environmental temperature on the hypothalamic and skin heat receptors.

As well as the iodinated hormones, the thyroid gland of most mammals produces a polypeptide hormone, *calcitonin*, whose action is to suppress the resorption of calcium and phosphate from bone, which is otherwise a continuous process. The action of calcitonin is thus the opposite of that of the parathyroid hormone (see §13.6 below), and the balance of the two hormones results in a remarkably constant serum phosphate level in vertebrates. The output of calcitonin depends on direct feedback of the calcium level of the blood circulating through the secretory tissue. The presence of this tissue in the thyroid gland of most mammals (it is found in the parathyroid glands of a few species) is peculiar to this group. In other vertebrates it forms the *ultimobranchial bodies*, which are separate endocrine organs although derived from the same pharyngeal rudiment in the embryo as the thyroid and parathyroid glands. The migration of the ultimobranchial tissue into the thyroid gland in mammals is therefore a secondary feature.

13.6 The parathyroid glands

These glands are closely associated physically with the thyroid glands in all vertebrates except fishes and the permanently neotenous amphibians such as *Necturus* and the Mexican axolotl, from which they are absent.

The parathyroid hormone, a straight-chain polypeptide, acts in opposition to calcitonin, tending to raise the serum calcium level through the deposition of bone. It also produces increased reabsorption of calcium by the kidney. Vitamin D assists in the process of bone deposition by increasing the absorption of calcium and phosphate by the small intestine. As in the case of calcitonin, the output of parathyroid hormone is regulated by direct feedback of serum calcium.

13.7 The adrenal medulla (chromaffin tissue)

The adrenal glands of mammals are composed of two structures that are anatomically and functionally separate. These are the outer *cortex*, which is derived from mesoderm, and the inner *medulla* which it surrounds, which is formed in the embryo from neural crest cells and retains its connection with the central nervous system in the adult. The anatomical connection between these two components is generally much less definite in other vertebrates. In the cyclostomes, both tissues are present as isolated clusters of cells which are associated with the posterior cardinal veins and their branches. In other vertebrates there is often considerable mixing of the two, but in a haphazard fashion, and the straggling nature of the tissue often renders complete extirpation impossible. A functional reason for the close association between cortex and medulla in mammals has yet to be found.

Because of the straggling nature of 'medullary' tissue in most vertebrates, an alternative name is usually preferred. Medullary cells have a special affinity for chromium compounds, such as potassium dichromate and chromic acid, and they are therefore also referred to as *chromaffin cells*.

Chromaffin cells are essentially neurosecretory cells, being derived from neuroectoderm cells which have become aggregated into a special secretory tissue, and their secretion is therefore controlled by nerves of the sympathetic nervous system.

Two histologically different types of cell can be recognized in chromaffin tissue, which are known to secrete different, although closely related, hormones. They probably produce different quantities of their respective hormones according to the physiological state of the animal, and so they may be presumed to be independently innervated. The two hormones belong to a class of substances known as *catecholamines*, which are, like

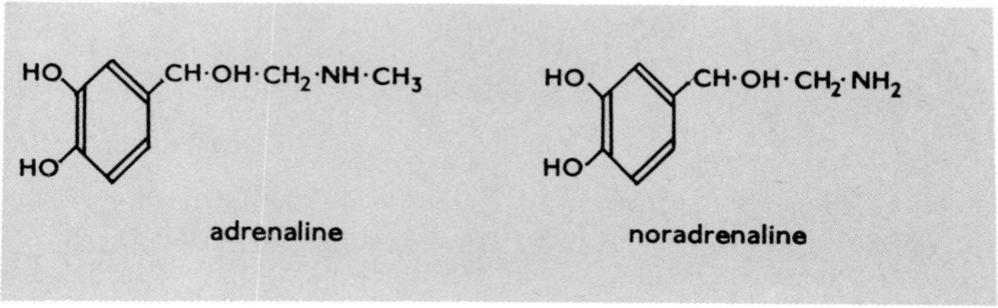

Fig. 13.5   Formulae of adrenaline and noradrenaline.

the thyroid hormones, derivatives of the amino acid tyrosine. *Noradrenaline* (norepinephrine) is very similar to *adrenaline* (epinephrine) (Fig. 13.5), but being without the methyl group of the former, it possesses one carbon atom less in its molecule. Noradrenaline is probably a precursor of adrenaline in the adrenaline-secreting cells, and it may be the primitive product of the chromaffin tissue, from which adrenaline was later developed, since it is the transmitter substance that is liberated by the adrenergic nerve endings of the autonomic nervous system. The substance *dopa* and *dopamine* are intermediates in the production of noradrenaline and adrenaline, and the latter may itself be an end-product in such tissues as the lungs, liver and intestines, where it may act as a 'local hormone'.

Both hormones cause constriction of the blood vessels of the skin and gut, and dilation of the vessels supplying the muscles. Adrenaline also dilates the blood vessels of the heart, liver and brain, and relaxes smooth muscle, thereby opening the bronchial passages to permit a greater volume of air to pass through them, and incidentally producing the 'sinking feeling' in the gut region well known to examinees. The net result of such changes is that an increased quantity of blood is pumped faster to heart, muscles, and brain, and it has been ensured that it is well oxygenated.

In this way the animal is enabled to meet emergency conditions—fright, flight or shock—in the proper physiological state.

This end is further achieved through the effect that the two hormones have on metabolism. They facilitate the breakdown of glycogen stored in liver and muscle to hexose units that are ready for oxidation, an action in which adrenaline is about four times as powerful as noradrenaline. Their mode of action is discussed later (§13.18).

13.8 The pancreas

The pancreas is a rather diffuse gland which is found in the region of the junction between the stomach and the duodenum. In nearly all vertebrates except the cyclostomes, the gland is a mixture of two histologically and functionally separate tissues. The bulk of the tissue is exocrine in function, and is drained by the pancreatic duct that opens into the duodenum. Scattered among the exocrine cells are patches of tissue named *Islets of Langerhans* after their discoverer, and these consist of endocrine cells that discharge their secretion directly into the blood. In the cyclostomes, the exocrine and endocrine cells are not joined in a common gland, although both form diffuse patches along the gut.

In mammals, two distinct types of cell are recognizable in the islet tissue, the A- or α-cells, which contain alcohol-soluble granules, and the B- or β-cells, whose granules are not soluble in alcohol. These two cell-types each secrete a different hormone, *glucagon* being secreted by the A-cells and *insulin* by the B-cells. The islets of all vertebrates contain B-cells, but the existence of A-cells is uncertain in cyclostomes and urodeles, and in other vertebrates there is considerable variation in the proportion of the two cell-types to one another. In reptiles and birds A-cells predominate, but in amphibians and mammals the B-cells are more plentiful. Teleost fishes often exhibit seasonal variation between the two. D-cells have also been described, which may secrete gastrin in addition to the pyloric cells of the stomach (§13.11).

Insulin and glucagon are both polypeptides of comparatively small molecular weight, 6000 in the case of insulin and 3485 in the case of glucagon. The two hormones are largely antagonistic in their action, which is primarily upon the metabolism of carbohydrate. Insulin promotes the formation of glycogen from glucose in the liver and muscles and probably also the diffusion of glucose across the cell membrane. In addition, it reduces the production of glucose from non-carbohydrate sources such as amino acids, and thereby favours an increase in the incorporation of amino acids into the peripheral tissues. Thus, the overall effect of insulin is to lower the glucose level of the blood; although in fishes, its effect on amino-acid metabolism may be its primary function.

Mammals typically exhibit a constant blood glucose level, but hyper-function of the islet β-cells results in a massive decrease in this level, to which the name *hypoglycaemia* is given. Conversely, loss of function by the β-cells leads to a considerable increase in blood glucose, or *hypergly-caemia*. The latter condition is known as *diabetes mellitus*, often popularly called 'sugar diabetes' because of the large amount of sugar in the blood. When the supply of insulin is inadequate fat metabolism is also affected, because fat must be used instead of carbohydrate for release of energy through cellular oxidation. The respiratory quotient (§4.6) drops from a value nearer unity to about 0.7 as a result. The large increase in the amount of acetyl co-enzyme-A which accumulates from the breakdown of the fat is too great to be dealt with by the cellular TCA cycle, and instead of being combined with oxaloacetic acid in the first stage of that cycle, it passes to the liver where it is condensed to form acetoacetic acid, and it may be excreted as such or as degradation products, which are together known as *ketone bodies*. The excess glucose swamps the kidney and is excreted in conjunction with these.

The effect of glucagon is to cause a rise in the blood glucose level by mobilizing the glycogen in the liver, which is broken down to glucose (see §13.18). The pancreatic hormones are secreted into the hepatic portal vein, so that the liver forms a barrier between the hormones and the body cells. The liver destroys glucagon before it can get into the body circulation, and its metabolic action is exerted solely upon the liver cells. It would seem that the ability of the liver to destroy the hormone has resulted in failure of the muscle cells to be sensitive to it. Because some insulin does get through the liver, the muscles are susceptible to its action. However, a good deal of the insulin which reaches the liver from the pancreas is destroyed, and there can be little doubt that the pancreatic hormones evolved pri-marily with the liver as their target organ.

13.9 The adrenal cortex

It is one of the many oddities of biological nomenclature that although the preferred term chromaffin tissue is available to describe medullary tissue, the other part of the mammalian adrenal gland retains its name throughout the vertebrates, and the scattered corresponding tissue of lower groups is also called cortical tissue (sometimes the interrenal tissue), while the hormones it secretes are universally known as *corticoids*.

Many active substances can be extracted from the cortical tissue, but all possess a basically similar chemical structure, which they share with the gonadal hormones. This consists of three rings containing 6 carbon atoms each (hexane rings) and a fourth ring with 5 carbon atoms (pentane ring). It should be noted that the hexane rings are not equivalent to benzene rings, but are completely saturated, there being no double bonds between any of the carbon atoms (Fig. 13.6). The cortical hormones are often called *corticosteroids* because they can undergo a type of stereo-isomerism, their configuration being variable without change in their empirical chemical formula. The parent substance of both the cortical steroids and the related sex hormones is probably *cholesterol* (Fig. 13.6).

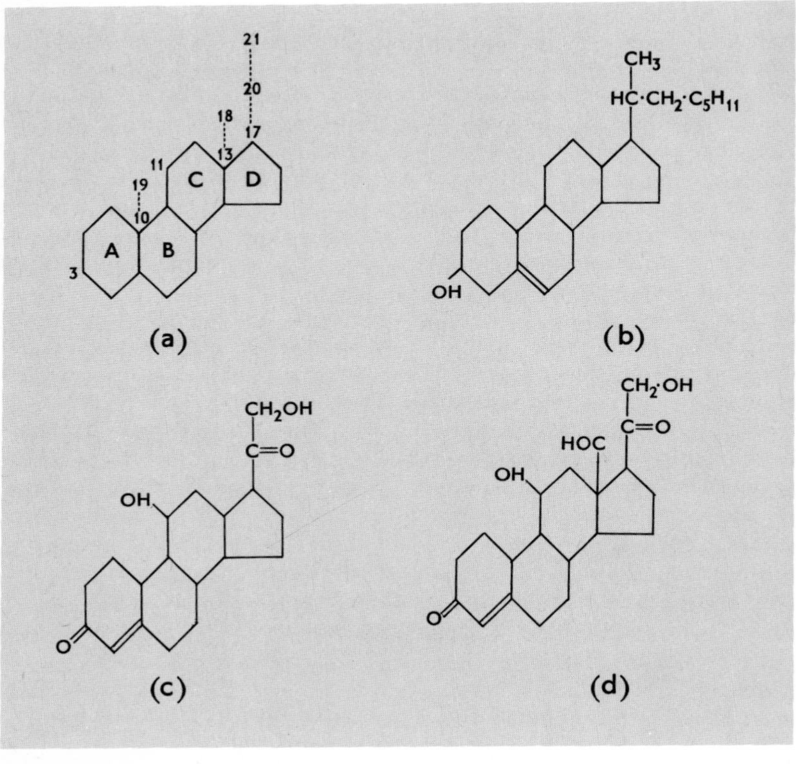

Fig. 13.6 Corticosteroids. (a) The hypothetical parent structure; the carbon atoms affected in the production of different steroids are numbered. (b) Cholesterol, the probable source of steroids. (c) Corticosterone, a gluco-corticoid. (d) Aldosterone, a mineralocorticoid with a fairly strong glucocorticoid activity, has an aldehyde group attached to the C-18 atom.

Many different steroids of varying activity have been extracted from the cortex, but it is probable that some of these are simply intermediate stages in the production of the definitive hormones. A further factor which can complicate the study of these steroids is that one steroid-secreting organ can modify the products of another, and very likely also its own products as they recirculate through it.

Three major groups of hormones are secreted by the cortex. The grouping is to some extent arbitrary, since a steroid which is most powerful in action in a particular direction will often exhibit weaker activity in other directions. The three groups are known as *mineralocorticoids, glucocorticoids* and *androgens* respectively, and their names give a clue to their functions. Glucocorticoids usually have a hydroxyl group or an oxygen atom attached to the carbon atom in position 11 (Fig. 13.6), whereas mineralocorticoids do not. This structural distinction is not complete, because *aldosterone*, with an —OH group in the C-11 position is more powerful as a mineralocorticoid than as a glucocorticoid. However, it has an unusual structure compared with other steroids (Fig. 13.6), which may account for the apparent anomaly. It is sometimes separated into a distinct class of its own. The androgens have oxygen or hydroxyl groups in the C-3 and C-17 positions. The amount secreted by the cortex is probably small enough for it not to be significant in relation to the output of the gonads, and they will be dealt with more fully with the other gonadal hormones.

The mammalian adrenal cortex is divided into three regions: the outer *zona glomerulosa*; the middle *zona fasciculata*; and the inner *zona reticularis*. These zones are believed to secrete the mineralocorticoids, glucocorticoids and androgens respectively. This is a mammalian specialization, since the cortical tissue of other vertebrates consists of cells similar in appearance to those of the mammalian zona fasciculata, although they secrete more than glucocorticoids.

The functions of the adrenal cortex are difficult to express in concise terms. The mineralocorticoids affect the salt and water balance of animals. In the mammals, a lack of these hormones results in a fall in blood sodium, chloride and bicarbonate levels due to their increased excretion by the kidney, and a rise in blood potassium due to its leakage from the cells. The water loss is such that it leads to an increase in the overall concentration of the blood, and the loss of bicarbonate results in the blood becoming more acid. The action of the mineralocorticoids must be seen in relation to the secretion of vasopressin by the neurohypophysis, which acts in an opposite manner to the mineralocorticoids; and also to the less powerful effect of glucocorticoids in inhibiting water resorption by the kidney. In addition, an antimineralocorticoid factor has been isolated from some mammalian adrenals, which is more like progesterone (see below) than the other corticoids, and its action is to inhibit sodium resorption by the kidney. The complex interrelationships involved in salt and water balance in mammals are summarized in Fig. 13.7.

Glucocorticoids promote the conversion of fats and proteins to glycogen.

Protein within the cells is broken down by deamination, and fats are mobilized, and not only are they converted into glycogen, but the peripheral utilization of glucose is reduced as well. Rather large amounts of glucocorticoids are required to drive these reactions to any extent, and whether they are significant for the normal functioning of the animal is

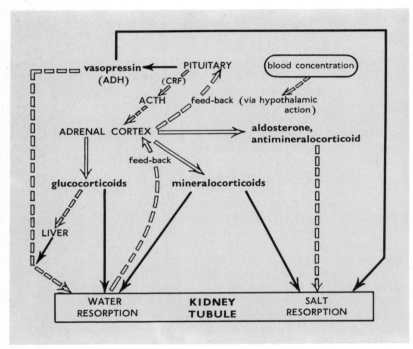

Fig. 13.7 Summary of hormonal control of water metabolism in mammals. The feedback to the adrenal cortex from the kidney comprises the volume of body fluid, its potassium concentration, and a glomerulotropin produced by the kidney. Broken white lines indicate an activating influence, solid black lines an inhibitory one.

not clear. The hormonal control of carbohydrate metabolism in mammals is shown in Fig. 13.8.

A number of other effects seem to be due to cortical hormones. It is well established that glucocorticoids favour the breakdown of connective tissue and inhibit its replacement, whereas mineralocorticoids have the opposite effect, which is enhanced by sodium ions and the growth hormone. The two groups of steroids are also antagonistic in relation to tissue inflammation. This is inhibited by glucocorticoids, and they also help to produce favourable conditions for the production of anti-bodies and have an anti-allergic effect.

The rather diffuse actions of the corticosteriods have led to the formulation of the concept of *stress*. The term is applied to any factor that subjects an animal to extreme conditions, such as shock, undue exercise, starvation, temperature extremes and infection. There is no doubt that adrenalectomized animals have a very poor toleration of stress, nor that stress

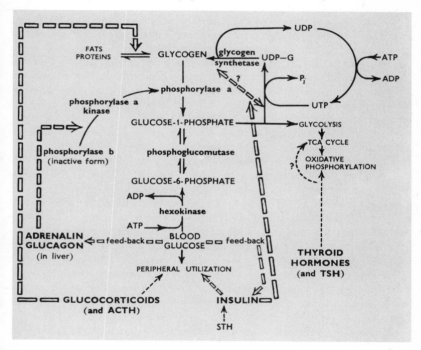

Fig. 13.8   Diagram to illustrate the hormonal control of carbohydrate metabolism in mammals. Broken lines, activation; dotted lines, inhibition. It is probable that the major control lies with the pancreatic hormones insulin and glucagon, which adjust for the effects produced by the other hormones.

activates the pituitary–adrenal axis with the result that there is an increased output of ACTH, corticosteroids and adrenaline. Some authorities have seen the cortical response to stress as central to that of the body as a whole, others as a merely contributory part of it. In the form in which it is often stated, the concept of stress is too simple, for it views the reaction of the endocrine system in isolation from that of the nervous system. Nor is there any evidence that, as has been claimed, prolonged exposure to stress may lead to a hormonal imbalance which may lead to the development of bodily malfunction in the form of ulcers, arteriosclerosis and the like.

Information about the nature and effects of cortical hormones in lower vertebrates is very sparse. The cortical tissue seems to be necessary for the maintenance of electrolyte balance in fishes and frogs, but not in reptiles. In the freshwater trout, mineralocorticoids stimulate the retention of sodium by the kidney and its outward excretion across the gills. It is interesting to note that the maturation of the salmon and the metamorphosis of the lamprey is accompanied by a considerable growth of the cortical tissue. The mammalian growth hormone has some effect on the growth of fish, and does raise their tolerance to salinity changes. In interpreting these facts we are always faced with the possibility of a change in hormonal function during evolution, or at least of selection of different aspects of hormonal function in different groups. It is this kind of presumed lability in the evolution of hormonal systems which makes deductions from experiments in which mammalian hormones are administered to non-mammalian animals so dubious and difficult to interpret.

The secretion of mineralocorticoids, including aldosterone, is not dependent on a trophic hormone from the pituitary. Glucocorticoid secretion, however, requires stimulation by the adrenocorticotrophic hormone (ACTH) from the pars distalis for its release. Growth hormone is thought to act synergistically with ACTH (§13.4). There is a direct feedback relationship between the production of ACTH and the secretion of glucocorticoids, but it is not known whether the glucocorticoids affect the pars distalis directly through their level in the blood, or indirectly through a secretion from the hypothalamus, or both. Work on stress leaves no doubt that the hypothalamus can influence the output of ACTH, and a specific hypothalamic substance, corticotrophin-releasing factor (CRF) has been identified.

Although the secretion of mineralocorticoids is independent of the pars distalis, the secretion of adosterone is inversely related to changes in the blood volume, and this relationship appears to be mediated through stretch receptors in the right auricle, suggesting a nervous reflex operating in some way unknown. Apart from blood volume, two other factors affect the secretion of aldosterone. If the blood potassium concentration rises, the secretion of aldosterone increases, potassium having the opposite effect, probably the result of a direct influence on the secretory cells.

The second of these factors involves a remarkable system by which the kidney helps to regulate blood pressure. The glomerulus of the kidney is the source of an enzyme *renin*, whose production varies with glomerular blood pressure. This is released into the blood, which also contains a protein formed by the liver, upon which it acts to produce the substance *angiotensin I*. Another enzyme, present in the blood, then transforms this into the active *angiotensin II*. Angiotensin II is an octapeptide like vaso-pressin and oxytocin, and it both produces arteriolar constriction, so raising blood pressure, and acts on the zona glomerulosa of the adrenal cortex to cause an increased output of aldosterone. The aldosterone causes the distal tubules of the kidney to absorb more water into the blood, raising the blood pressure further.

13.10   The sex hormones

In many environments, conditions favourable for reproduction are commonly restricted to one part of the year, and sexual activity in such cases often exhibits a rhythmicity in relation to the breeding period. Hormones are an important factor in breeding rhythmicity, in the con-trol of which they interact with environmental influences such as light and temperature. Sexual cycles of this sort are probably secondary specializations of the hormonal systems that govern the determination and maintenance of sexual activity and behaviour.

In mammals, the testis and the ovary both secrete hormones. The gonadal rudiments arise close to those that produce the adrenal cortex, and it is not surprising to find that the sex hormones are steroids with the same basic chemical structure as the corticosteroids (Fig. 13.6). As we have already seen, the adrenal cortex produces some male sex hormones or androgens, but the primary source of androgens is undoubtedly the testis. The testicular hormones, principal among which is *testosterone*, are secreted by groups of modified epithelial cells lying between the seminiferous tubules, the *interstitial* tissue. Interstitial tissue of the mam-malian type is lacking in a number of vertebrates including some teleosts, urodeles and turtles, but in these cases the seminiferous tubules are surrounded by a ring of cells which are thought to be homologous with interstitial tissue.

The principal ovarian hormone is oestradiol-17, and this and others which have been isolated are known collectively as *oestrogens*. The large cells of the theca interna are the most likely source of ovarian hormones, although some authorities think the granulosa and interstitial cells may do so. When the developing follicle is ripe ovulation takes place, which involves the discharge of the oocyte it contains into the fallopian tube. After ovulation, the discharged follicle cells become transformed into a characteristic body known as the *corpus luteum*, a temporary endocrine organ that secretes the hormone *progesterone*.

The oestrogens and androgens are responsible for the production and maintenance of the so-called *secondary sex characters*, in addition to being involved in the maintenance of sexual periodicity. The secondary sex characters are those concerned with recognition of the sexes and mating, and the birth and nutrition of the young, but not the maturation and movement of the germ cells. The term is arbitrary and tends to give the impression that these characters are somehow of secondary importance, and is better avoided.

Other sex hormones are produced by the placenta, which augments the supply of progesterone and reinforces the supply of one of the pituitary hormones with a similar substance; and by the pituitary gland itself. Extirpation of the pituitary gland results in a marked atrophy of the male interstitial tissue and the ovarian follicles of the female. The reason is that the pars distalis secretes two *gonadotrophins* which are directly concerned with the maturation of the germ cells. These are the follicle-stimulating hormone (FSH) and the interstitial cell-stimulating hormone (ICSH). ICSH maintains the secretory activity of the interstitial cells of the male testis, and is thereby responsible for the continuing output of testosterone, which must be present if the later stages of spermatogenesis are to take place. In the female, FSH promotes the development of the ovarian follicles, but these will not secrete oestrogens nor undergo ovulation and form a corpus luteum unless ICSH is also present. For this reason, ICSH was formerly known as luteinizing hormone, but because of its activity in the male, this name has now been abandoned.

As the follicles grow under the influence of FSH and ICSH, they produce an increasing quantity of oestrogens. These cause swelling and softening of the walls of the uterus in preparation for the implantation of one or more ova, an increase in protein synthesis, and the retention of salts and water. High concentrations of oestrogens tend to inhibit the secretion of FSH but to facilitate the output of ICSH, with the result that as the follicle approaches ovulation the development of further follicles is suppressed, but there is an increasing tendency for ovulation to take place, followed by luteinization due to the increase in ICSH. The drop in the level of FSH leads to a drop in the quantity of oestrogens produced, but their secretion does not cease altogether.

The progesterone secreted by the corpus luteum prepares the animal for pregnancy by inhibiting the secretion of ICSH, but not of FSH or oestrogens, because the presence of the latter is necessary for the functioning of progesterone. The inhibition of further ovulation is therefore achieved by the supression of ICSH and not of FSH. Progesterone sensitizes the epithelial lining of the uterus to the presence of the ovum, and encourages its proliferation and hence the formation of the placenta. At the same time, it reduces the sensitivity of the uterine wall to oxytocin, and so causes a reduction in its motility, which would otherwise endanger the implantation and retention of the ovum. Some workers think that parturition is more the result of the release of this block on motility by progesterone, than of the action of oxytocin. In the later stages of pregnancy, the placenta aids in its own maintenance by secreting progesterone and a hormone that promotes the continued activity of the corpus luteum. The hormone is probably the same as, or similar to, the third gonadotrophin from the pars distalis, *prolactin*, which not only helps to keep the corpus luteum in being, but also causes the mammary glands to enlarge and prepare for the production of milk. After birth, the same hormone facilitates milk production, and its ejaculation on sucking.

If the ovum is not fertilized, the corpus luteum breaks down following the failure of the ovum to implant, and the development of further follicles begins. Hence in the absence of fertilization regression to the situation with which the process originally began takes place, and the series of changes forms a complete cycle. This is known as an *oestrus cycle*, taking its name from the period of maximum breeding activity, *oestrus* that usually follows closely on ovulation. Oestrus is the period when the animal is said to be on *heat*, a time when it is at its most attractive to the male and most receptive to it.

The period covered by the oestrus cycle varies in different mammals, and so does the number of cycles which can take place in one breeding season. At one end of the scale is the silver fox, which has only one oestrus cycle per breeding season, and in which oestrus itself lasts only five days. The majority of mammals, of which the sheep and cow are examples, undergo several cycles during a restricted part of the year. Others, like rats and mice, domestic animals and the higher primates, breed without regard to season, presumably because seasonal factors are less important in their cases, and in these there is a succession of oestrus cycles throughout the year. Those mammals, which have only short oestrus cycles, such as rats and mice with a cycle of 4–6 days, are without a luteal phase, and presumably the placenta secretes all the progesterone required.

The higher primates have several peculiarities by comparison with other mammals. They do not really possess a period of heat, and their females are potentially receptive to the males at all times. Furthermore, if the ovum is not fertilized, the endometrium of the uterus (the epithelium and underlying connective tissue), sloughs off and is voided to the outside, a phenomenon that is observed outside the primates only in the elephant shrew. The shedding of the endometrium in this way is known as *menstruation*, and the oestrus cycle is termed the *menstrual cycle* in these cases.

Another variable feature in the mammals concerns ovulation. Many ovulate spontaneously as part of the oestrus cycle, but others require the stimulus of copulation before they will ovulate. Such *induced ovulators* include the ferret, the cat and the rabbit, and the process must be a neuro-endocrine one, initiated by a nervous reflex linked with the hypothalamus, which in turn induces the production of more ICSH, leading to ovulation.

Oestrus is clearly a process related especially to mammalian reproduction, being bound up with the need for ensuring optimum conditions for the implantation of the fertilized ovum in the uterus, and for suckling. But it would be surprising if some of the functions of the mammalian sex hormones were not similar in non-mammalian vertebrates, as for example in the determination of male and female characteristics. Much remains to be discovered about the situation in these animals, but what is already known suggests that there may be considerable differences of detail. In birds, the testes secrete androgens, but the ovaries secrete androgens and progesterone (or related compounds known as *progestins*) as well as oestrogens. Both the androgens and the oestrogens stimulate the development of the ovary, and the progestins stimulate the production of ICSH by the pituitary, the opposite effect to that in mammals. The progestins are probably secreted by the interstitial cells of the ovary, and the follicular remains after ovulation are probably not comparable with the corpora lutea of mammals. There is also considerable variation in the extent to which the secondary sexual characters are hormonally determined in birds. The plumage of the sparrow is determined genetically, the blackening of feathers and beak in the male finch *Steganura* is under gonadotropic control, and in the weaver finch *Euplectes* feather colour is controlled in this way but the colour of the beak results from the action of androgen. Internally, the sex hormones are much concerned in the serial production of the large yolky eggs that are characteristic of birds. Oestrogens facilitate the absorption of food from the intestine and the mobilization of food reserves for the formation of yolk, the deposition of which in the egg is controlled by gonadotrophins. Oestrogen and progesterone, and possibly androgen, promote the secretion of albumen by the oviduct.

The birds of temperate areas, like many of the mammals in the same regions, breed for only part of the year. Their gonads undergo regression after the breeding season has ended, when sexual behaviour can be seen to have ceased, and the birds form flocks rather than pairs. In the males, there is a more or less complete exhaustion of the hormone-producing capacity of the interstitial tissue, some of which breaks down, and the cells in the seminiferous tubules produce a mass of lipid tissue containing cholesterol. As the days begin to lengthen with the approach of spring, the interstitial tissue regenerates at the expense of the lipid material, which gradually disappears. The gonadal hormones begin to be secreted again, and there is a return of sexual behaviour such as territory and song. By springtime, the testes are mature and contain abundant spermatozoa. The female lags behind the male in the resumption of sexual activity. She requires special stimulation by the male, in the form of courtship and display, before ovulation and the final stages of development of the oocyte occur.

In Amphibia, gonadotrophins appear to be necessary for the later stages of gametogenesis and for ovulation, possibly acting through the regulation of hormonal secretion by the ovary. In *Rana temporaria* ovulation, spawning and fertilization occur immediately after the end of hibernation, in early March, thus ensuring that the tadpoles grow and metamorphose under favourable conditions. Gametogenesis continues from April until October, and the gonads are therefore fully mature when hibernation begins. They are maintained in a mature state throughout hibernation, which makes it possible for reproduction to begin immediately hibernation ends. This cycle appears to be innate in *R. temporaria*, but in some other cases, of which *R. esculenta* is an example, gametogenesis is continuous instead of being seasonal, which suggests that external factors related to the seasons may help to control the cycle in some members of the group.

In fishes also, a supply of gonadotrophins is necessary for the later stages of gametogenesis. As in birds, the interstitial tissue of the testes becomes exhausted after spawning, and cholesterol-rich material accumulates in the tubules. Since the same changes have been observed in reptile testes, the cycle of regression and regeneration of the interstitial cells may be a basic feature of vertebrate reproduction which has been lost in the mammals. The sex hormones secreted by the gonads of fishes have been little investigated, but testosterone and oestrogens have been found in the blood of different varieties of salmon. All that is known about cyclostomes is that hypophysectomy prevents the proper growth of the ovary in these animals, which suggests that gonadotrophins are secreted by the pituitary which play an essential role in the maturation of the ovary.

Apart from these internal, or *endogenous* factors, there are *exogenous* influences that affect the reproductive cycles of fishes to varying degrees. The minnow *Phoxinus phoxinus* breeds in Lake Windermere from May to July. Immediately breeding ceases gametogenesis occurs, and although it stops during the winter, the gonads are maintained in a relatively mature condition during this period, as in the case of the frog. In the spring, maturation is rapidly completed, its rapidity being subject to the influence of temperature and photoperiod (length of day). In *Gasterosteus aculeatus*, maturation of the gonads continues even through the winter, and in this case photoperiod seems not to affect the process, although temperature does.

13.11 Hormones and digestion

Historically, some of the hormones concerned with the digestion of food were among the first hormones to be discovered. They are 'local' hormones, being produced by a localized area of the gut and acting upon the gut itself or on one of its accessory glands.

When a mammal anticipates that food is going to be available, or when its taste buds are stimulated by the actual presence of food, the stomach begins to secrete a gastric juice that is rich in pepsin. This is a reflex action which depends on sight and smell as stimuli and has as its effectors the secretory peptic cells of the stomach lining. When the food actually reaches the stomach it stimulates the pyloric mucosa to secrete a hormone *gastrin*, which has a strong stimulating effect on the acid-secreting parietal cells of the stomach.

The duodenal wall secretes several hormones when food is passed into it from the stomach. *Secretin*, the first hormone to be discovered (in 1902), circulates in the blood and when it reaches the pancreas it causes the exocrine cells of the gland to secrete the fluid base of the pancreatic juice. A second duodenal hormone, *pancreomyzin*, follows the same path to evoke the production of the pancreatic enzymes. The gall bladder is caused to undergo rhythmic contractions by another duodenal hormone *cholecystokinin*, and these result in an increase in the rate of flow of bile into the duodenum. Secretin may also act on the gall bladder to some extent and increase the quantity of bile produced.

These hormones are all concerned with the mobilization of the various fluids needed for the proper digestion of the food. In addition, the small intestine appears to secrete a hormone with a rather different function, that of inhibiting gastric secretion. As emptying of the stomach tends to be delayed by fatty food, and it seems to be the presence of fat that is the stimulus for the production of this hormone, *enterogastrone*, its action is presumed to be intended to prevent the over-secretion of gastric juice in these circumstances.

There is no evidence for either nervous or hormonal control of gastric secretion in the lower vertebrates, but secretin or some similar substance seems to be widespread, having been extracted from members of all the vertebrate classes. As in the mammals, its effect in other vertebrates is to stimulate the secretion of pancreatic juice, but it is not clear whether pancreomyzin is also produced by these animals.

13.12 Other hormone-like substances

The pineal gland has already been referred to in §13.3 as a source of melatonin. Apart from its activity in pigment dispersal and concentration, it has long been suspected that this gland plays some part in reproductive events, and especially that it may mediate the influence of photoperiod on the gonads in mammals and influence the growth of the gonads in fishes. Melatonin may play some part in these activities, but other factors may also be involved. Much more investigation of these probabilities is required.

Another gland that puzzled physiologists for many years is the *thymus* gland, derived from pharyngeal tissue and found in the thorax of vertebrates. It is now clearly established that its presence is necessary for the proper development of the immunological reactions of lymphoid cells, and it appears to do so by producing a blood-borne factor, but nothing more is known about it.

Prostaglandins are substances that have attracted a good deal of attention in recent years. They derive their name from the fact that they were originally discovered in human seminal fluid. Since that time, they have been found to be very widely distributed through the tissues of the body. They differ from all other known hormones, which are either amines, amino-acids, peptides, and polypeptides or steroids, in being fatty acids, and all those so far examined contain twenty carbon atoms.

They appear to be released from various nerve endings when these are stimulated. They exert a wide variety of physiological effects, including blood pressure regulation, lipid and carbohydrate metabolism and tissue inflammation in response to wounds or disease. Many varieties of prostaglandin are found in the same species, and this has complicated the investigation of their action, which is proceeding actively.

Pheromones are substances released into the environment to aid in the integration of behaviour between individuals or sometimes between the members of whole colonies. They are thus not strictly hormones, yet the exocrine glands that produce them are often stimulated internally by hormonal mechanisms mediated by the nervous system. They may be broadly grouped into two types, those that act on the nervous system or neuroendocrine system to cause rapid responses, and those that activate a long series of neuroendocrine events that develop slowly and require prolonged activation for their achievement. The first type consists of signalling or releasing pheromones, such as sex attractants, and trail and alarm substances, and they may be difficult to distinguish from other substances which are not due to a behavioural response. The second type is exemplified by the secretion produced by the mandibular gland of the honey bee queen, whose function is to both inhibit the development of ovaries by the workers and the construction by them of royal cells. This pheromone can also act as one of the first type, being used as a sex attractant in the nuptial flight. *Primer* pheromones, as the second type are called, and sex attractants have been identified in a wide variety of terrestrial animals, including mammals and probably not excepting man. An example of a mammalian pheromone is the one produced by male mice that shortens the length of the oestrus cycle.

13.13   Hormones in invertebrates

The majority of invertebrate endocrine organs are neurosecretory, like the hypothalamic neurones that secrete the hormones of the vertebrate neurohypophysis, and because of this they are frequently linked with the nervous system to form neuro-endocrine reflex arcs. In some cases the evidence for neurosecretion rests on histochemical tests, never a very reliable basis. But in other cases, it has been shown that the cells concerned do produce substances that affect specific organs or systems. Thus, the cerebral ganglia of flatworms contain neurosecretory cells whose products seem necessary for the regenerative processes of which this group is especially capable. Cells in the oral ring of echinoderms secrete substances that regulate such events as colour change, water content, locomotion and the shedding of gametes. Annelids have long been recognized to possess many neurosecretory cells, which are involved in such activities as the maturation of the gonads, the regeneration of the body, and changes in body form related to maturation and reproduction. Neurohaemal organs appear to be present in some annelids.

Much remains to be discovered about these and other invertebrate systems, investigation of which is often rendered difficult by the small size of the animal and the scattered nature of the neurosecretory cells, which are often embedded deep in some part of the nervous system. Where more discrete organs are available, progress has been more rapid. Such an organ is the optic gland of cephalopod molluscs. This gland has been shown to produce a gonadotrophin that induces ovarian and testicular enlargement in the appropriate sex and encourages yolk deposition in the female. The gland is regulated by inhibitory nerves which seem to be governed in their action by photoperiod. Generally speaking, it has been found easier to investigate the hormonal mechanisms of insects and crustaceans than those of other groups, and we shall concentrate on those of them that have been reasonably well elucidated.

The growth, development and moulting of insects are all under hormonal control. Insects hatch from their eggs in an immature larval form, and the transition from the larva to the adult is a process which requires both growth and differentiation. In hemimetabolous insects there is a gradual transition through a series of larval *instars* separated by moults, and each successive instar is more like the adult or *imago* than the preceding one. However, the changes undergone for the final moult are much greater than the earlier ones. In holometabolous insects there is also a series of larval moults, but between the final larval moult and the imago there is interposed a metamorphic stage, the *pupa*. During the development of the pupa the tissues are subjected to an almost complete reorganization which is much more fundamental than anything that takes place in a hemimetabolous insect. The pupal stage often involves a period of quiescence, or *diapause*.

13.14 Growth, moulting and reproduction in insects

The essential features of the control of these processes appear to be similar throughout the insects. Development and moulting are governed

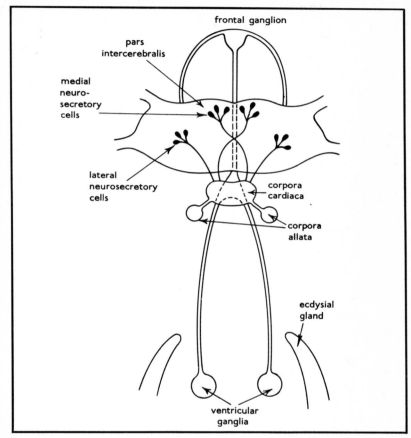

Fig. 13.9 Diagram of the main neurosecretory organs of insects. (After Gorbman and Bern, 1962, *A Textbook of Comparative Endocrinology*, Wiley, New York and London, modified from Hanstrom, Hormones in Invertebrates, Clarendon Press, Oxford)

by two interacting hormones, *ecdyson* and *neotenin*. Ecdyson is secreted by paired ecdysial or prothoracic glands whose position varies in different insects. These glands are under the control of an ecdysiotrophic or thoracotrophic hormone secreted by neurones of the *pars intercerebralis* in the protocerebrum (a diagram of insect endocrine organs is given in Fig. 13.9), and which is passed to paired bodies known as *corpora cardiaca*, that store the hormone until its release. Ecdyson is sometimes known as moulting hormone, and also as growth and differentiation hormone. It stimulates the production of adult structures and the onset of moulting. Since ecdyson is not needed in the adult, the ecdysial glands degenerate in most adult insects.

Neotenin, also known as the juvenile hormone, is secreted by the corpora allata (see Fig. 13.9). It is in some respects antagonistic in its action to ecdyson, since it suppresses any tendency to develop pupal or imaginal characters. It is therefore essential for the larval moults of both hemi- and holometabolous insects. In a sense, neotenin and ecdyson may be regarded as complementary in their action, since both are concerned in maturation processes, whether of individual juvenile stages or of metamorphosis into the adult. The development of larval stages therefore re- quires a high blood level of both hormones, although hemimetabolous insects, which begin to show imaginal characters several instars before the adult stage, have to produce progressively less neotenin in successive instars; and for pupation and the emergence of the imago the amount of neotenin must decrease, and the hormone may in some cases be absent for the emergence of the imago. In addition, there is a short-term reduc- tion in the level of neotenin preceding each moult, as might be expected.

Ecdyson exerts its effect on moulting by influencing the synthesis of protein by the epithelial DNA of the insect. Apart from this, it has been found to act on different genetic loci at different stages in development in some way not properly comprehended so as to make them sensitive to the subsequent action of neotenin in maintaining the larval pattern; and may also sensitize other, cytoplasmic processes, to the action of neotenin.

Diapause in larvae and pupae seems generally to be due to a failure to produce thoracotrophic hormone. Diapause eggs are due to the action of a hormone produced by the neurosecretory cells of the sub-oesophageal ganglia, that influences the developing eggs. Diapause is normally a seasonal phenomenon, intended to produce a phase which will overwinter without further development, and whether this be egg, larva or pupa it is normal for photoperiod to be the determining factor in producing dia- pause, acting through appropriate neuroendocrine reflexes. The termina- tion of diapause is normally determined by changes in the environmental temperature, also acting through such reflexes.

At one time there was thought to be no evidence that hormones play any part in the determination or maintenance of secondary sexual char- acters in insects. However, it is now known that the male characters, both primary and secondary, of the glow-worm *Lampyris* are induced by an 'androgenic' hormone produced by the testes, and this may well prove to be so in other insects. Apart from this, it is well established that the ripening of the eggs is in many cases under hormonal control. The median neuro-secretory cells of the cerebral ganglia secrete a hormone which is apparently metabolic in function. It may affect all body cells, but it certainly pro-motes the production of yolk proteins by the fat body. It probably acts synergistically with a hormone from the corpus allatum, which appears to be a gonadotrophin that facilitates the production of oocytes, perhaps by increasing their uptake of nutrient substances. This gonadotrophin also controls the development of the accessory reproductive glands that secrete protective substances or cement for attaching eggs to the substrate. In male insects, spermatogenesis appears to be independent of any hormonal influence, but the gonadotrophin from the corpus allatum controls the de-velopment of the accessory glands that produce seminal fluid and form spermatophores.

Cockroaches secrete an egg-case that contains a number of eggs. Pro-duction of the egg-case is relatively slow, and the structure is carried around for some time by the animal before it is ejected. While the egg-case is being secreted, the ripening of further eggs ceases, due to a halt in the release of egg-ripening hormone from the corpus allatum. This effect is accomplished by a hormone secreted by the follicular cells left behind by the ova being surrounded by the egg-case, the nearest approach to a corpus luteum known among the invertebrates. The luteal hormone acts through the brain via the pathway described in the previous paragraph.

The cuticle of insects on hatching from the egg and after each moult is white and soft. It becomes brown and hard through a process of *tanning*, which results from the action of a hormone *bursicon*. This hormone is secreted by the pars intercerebralis, and its production is triggered off through a nervous link with the prothoracic ganglion, presumably depen-dent on recognition of some tactile pattern by the newly emerged stage. Secretion of the hormone begins some 2–3 min after emergence, reaches its maximum activity after 30–60 min, and ceases completely about 10 hr later. It is thus closely geared in time to the pumping up of the intes-tine observed in many insects, by which they increase rapidly in size up to 20 min or so after emergence. The hormone is produced so as to reach maximum activity just after this process has been completed, thereby ensuring that the body reaches maximum size before the cuticle hardens. The egg-case of the cockroach is dropped ready tanned, but whether this is brought about by continued secretion of bursicon in this insect is not known.

13.15 Other insect hormones

Undoubtedly, many hormones of insects have yet to be discovered. In recent years, several have been found about which sufficient is known to be worth recording. A diuretic hormone has been demonstrated in blood-sucking insects and in the locust and cockroach, which is likely to be of general occurrence. It is produced by the corpus cardiacum, and acts on the malpighian tubules to cause an increase in their secretory activity, and on the rectum to inhibit reabsorption by it of the fluid so produced.

The corpus cardiacum is also the source of two metabolic hormones that cause the mobilization of glycogen in the fat body in the form of the disaccharide trehalose; and of a hormone that acts on the pericardial cells and gut cells to cause an increase in the rate of heart-beat and in the movements of the gut and malpighian tubules.

13.16 Growth, moulting and reproduction in crustaceans

Unlike insects, most crustaceans continue to moult throughout their adult life, and the hormones that control growth and moulting are produced continuously in the lifetime of the animal. Crustaceans resemble insects in that the hormone-producing cells are neurosecretory; these are found especially in the *medulla terminalis* of the brain, where some of them constitute the *medulla terminalis X-organ*. The hormones are transported from the medulla terminalis to an organ known as the *sinus gland*, which is situated at the base of the eye-stalk (Fig. 13.10). The X-organ produces a moult-inhibiting hormone, and probably also in some cases a moult-accelerating hormone. The other hormone concerned in moulting is secreted by the *Y-organ*, which is found in either the antennal or the maxillary segment, and is analogous to the ecdysial gland of insects. The hormone it secretes initiates moulting, and may even be identical with ecdyson.

In the intermoult period the inhibitory hormone from the X-organ prevents the secretion of moulting hormone by the Y-organ. As the time for moulting approaches, there is a gradual resorption of calcium and phosphate from the cuticle, which are deposited in the hepatopancreas, and in freshwater forms in the stomach as crystals called *gastroliths*. At the same time there is a general increase in metabolism, and any necessary regeneration is speeded up. These changes, which are often referred to as *proecdysis*, are believed to result from an increasing secretion of moulting hormone. The moult itself is accompanied by a sudden increase in size due to water absorption, the uptake of the water being controlled by another hormone from the sinus gland. The swelling caused by the uptake of water actually initiates the moult by splitting the cuticle.

The post-moult period *metecydysis* is characterized in general by activities that are the opposite of those occurring in premoult, and so there is a rapid deposition of chitin, which becomes impregnated with calcium and phosphate, and growth of new tissue. During this period, too, there is a rapid intake of calcium from the external medium and from food, because the quantity stored in the hepatopancreas is insufficient. The existence of a specific growth and differentiation hormone distinct from the hormones already referred to is still problematical, but there is no doubt that eye-stalk extracts affect the metabolism of calcium and phosphate and also

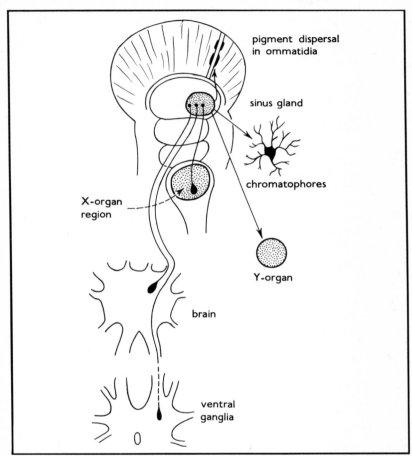

pigment dispersal in ommatidia

sinus gland

chromatophores

X-organ region

Y-organ

brain

ventral ganglia

Fig. 13.10 Diagram of neurosecretory system of decapod crustaceans. (After Welsh, 1961, in Waterman (ed.), *The Physiology of Crustacea*, Vol. 1, Academic Press, New York and London)

stimulate oxygen consumption in the premoult period, and it would be surprising if hormonal factors were not involved in post-moult.

Much remains to be discovered about the role of hormones in the reproduction of crustaceans. In *Carcinus* the Y-organ is essential for the development of the gonads but not for their maintenance and functioning, and this is thought to be the situation in other members of the group. The X-organ is the source of a gonad-inhibiting hormone that acts only in females, at least in crabs. It is secreted at the end of the breeding season and throughout the non-breeding season—autumn, winter and early spring.

Neither of these reproductive hormones affects the development of the secondary sexual characters. In a genetic male, the *androgenic glands* close to the testes secrete a hormone that causes the testes, the vasa deferentia and the male sexual characters to develop. If these glands are removed, the primary germ cells develop into oocytes; conversely, if androgenic glands are transplanted into genetic females, the ovaries are transformed into testes. Thus, the primary female characters develop in the absence of androgenic hormone, and the primary and secondary male characters develop when it is present. The secondary sexual characters of the female are determined by an ovarian hormone or hormones.

Reproduction and growth are antagonistic in decapod crustaceans (as in polychaets), i.e. moulting cannot be concurrent with egg-laying. During the moulting process the sinus gland secretes a hormone that inhibits yolk formation and oocyte production in the female; and prevents spermatogenesis in the male by inhibiting the activity of the androgenic glands.

13.17 Colour change in invertebrates

Change in overall body colour is often under the control of hormones in arthropods and other invertebrates. Except for the cephalopods, the chromatophores of invertebrates resemble those of vertebrates in possessing a permanent shape and containing a pigment that may be either concentrated or dispersed. The chromatophores of cephalopods are unique multicellular structures in which a central pigment cell is surrounded by radially arranged smooth muscle cells. When the muscle cells contract, the central cell is pulled out into an enlarged shape; and when they relax the elastic wall of the central cell contracts, and it becomes smaller (Fig. 13.11).

Cephalopod chromatophores are under nervous control, each muscle cell being innervated by a nerve fibre, and the colour changes in these animals are a matter of seconds only. In other invertebrates, the control of the chromatophores is often hormonal, and their colour changes are relatively slow. A slow change cannot always be taken to indicate that a hormonal mechanism is implicated, for colour changes in the echinoderm *Centrostephanus* and the leech *Placobdella* can take an hour or more for their completion, although they are undoubtedly accomplished by nervous mechanisms.

The pigmentation of insects is often permanently fixed. But in some cases, as in mantids, locusts and the stick insect *Carausius*, the pigmentation may alter under the influence of external factors such as temperature, humidity and light. In these insects the colour pattern is genetically determined and it is the intensity of the shade that alters. Nevertheless, it seems that at least some of the pigments are laid down in their genetically determined pattern through the agency of hormones, probably ecdyson and neotenin.

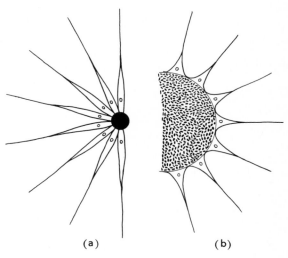

(a)   (b)

Fig. 13.11   Single chromatophore of *Sepia*; (a) contracted, (b) expanded. The muscle fibres are actually about four times longer than they are drawn. (After Tompsett, 1939, *Sepia*, L.M.B.C. Memoir, Liverpool University Press)

The brown and orange/red pigments of *Carausius* can be dispersed and concentrated within their chromatophores, but not the green and yellow ones. Dispersion is stimulated by a hormone from the tritocerebrum of the brain. This hormone is involved in the diurnal colour rhythm of the insect, in which the pigment is dispersed at night and concentrated during the day. The rhythm can be altered over a relatively long period by alteration of the external light pattern, but is changed rather quicker by alterations in humidity, increased humidity leading to darkening.

The chromatophores of prawns and shrimps are often polychromatic, with several pigments in a single cell, each of which can be independently dispersed and concentrated by its own pair of antagonistic hormones. Other crustaceans generally possess monochromatic or dichromatic chromatophores, although this does not of course mean that they necessarily have fewer kinds of pigments. The hormones that activate these chromatophores, known as *chromatophorotrophins*, are produced in a number of different places in the body apart from the eye-stalk—including the tritocerebral commissures, the circum-oesophageal commissures and the abdominal nerve cord. This fact, coupled with their mode of activation by a multiple hormone system, has made direct experimentation extremely difficult, and as a result the control mechanism for colour change has not been completely worked out in any crustacean.

It has also been shown that the migration of the pigment in the compound eye of crustaceans is under hormonal control. Each ommatidium of the compound eye contains three groups of pigments. Two are black, usually melanin, and contained within the retinal cells, one group being distal and the other proximal. The third group consists of a white reflecting pigment, probably guanine or related compounds. In good light, the proximal pigment moves down to meet the distal pigment, and the reflecting pigment takes up a position beneath the basement membrane of the retinula cells (Fig. 9.24). The lower half of the ommatidium is thus surrounded by a light-absorbing ring of pigment in bright light. In dim light the reflecting pigment moves to a position just above the basement membrane, and the distal pigment moves to the region of the crystalline cone.

Experiments have shown that the control of the movements of the pigments in crustacean eyes is a complex phenomenon, and may depend not only on both nervous and hormonal mechanisms, but also on a direct reaction to light by the retinal pigments themselves. The hormonal control concerned seems to be exercised by two hormones. The retinal pigment hormone (RPH) is responsible for the light-adapted state, and is released from the sinus gland, except in a few animals in which it is secreted by the sub-oesophageal gland. In the crayfish, and no doubt also in other crustaceans, the retinal pigments respond differentially to this hormone according to its concentration. Low amounts result in the movement of only the distal pigment, whereas higher concentrations cause the movements of all pigments.

The other hormone is derived from neurosecretory cells in the central nervous system, and is responsible for the dark-adapted state of the distal retinal pigment. The two hormones seem to be secreted in different amounts in a diurnal rhythm, no doubt in response to changes in the external light density.

Several metabolic hormones are known to occur in crustaceans, including a hyperglycaemic hormone whose effect is to increase the blood sugar level, but little more is known about them than that they do exist. Apart from these, the pericardial organs near the hearts of crustaceans are known to produce peptide secretions that both increase any activity in the accelerator nerves to the heart and directly stimulate the heart muscle, the latter probably being their primary role.

13.18. The mode of action of hormones

As we have seen, hormones are invariably either steroids on the one hand, or peptides, proteins, amines or amino-acids on the other hand. The discovery that a steroid hormone, ecdyson, could act directly on individual genes raised the question whether other steroids might also function in the same way. In the case of vertebrate steroid hormones, the effects they produce are usually multiple and it is difficult to sort out precise details. Nevertheless, it is clear that such hormones do indeed produce an increase in the formation of RNA by nuclear DNA, and hence in the synthesis of protein, and we may expect detailed examples to be worked out in the future.

Non-steroid hormones are believed to exert their effects in two other ways, quite apart from any action on DNA, which are not necessarily mutually exclusive. Both depend on the recognition of the hormone by specific receptors on the plasma membranes of the target cells, much as transmitter substances are recognized by receptors on the post-synaptic membrane. There is considerable evidence for such receptors in the case of adrenaline, and two types of adrenergic receptors have been distinguished in smooth muscle, the α-receptor causing contraction when stimulated by adrenaline, the β-receptor causing relaxation.

One effect that recognition by target cell receptors might produce would be a change in membrane permeability. Nor-adrenaline (§11.15) and neuro-muscular transmitters (§9.11) alter the permeability of the plasma membrane to certain ions, and there is no reason why the concept should not be extended to include changes in permeability to other substances. It is known that insulin may increase the permeability of cells to glucose and alter the transport of amino-acids and electrolytes across the plasma membrane, perhaps reflecting a general change in membrane permeability. Somatotrophin appears to facilitate the entrance of amino-acids into cells. It might be argued that these are secondary effects of the recognition of the hormones by the surface receptors, but it is surely significant that the result is one that facilitates the action of the hormone.

In recent years, much attention has been directed to a particular enzyme system within the cell known as the *adenyl cyclase* system, and evidence has built up that many hormones achieve their effect by acting upon this system, including the parathyroid hormone, ACTH, thyrotrophic hormone, noradrenaline, insulin, glucagon, melanophore-stimulating hormone, and ICSH. The adenyl cyclase system is shown in Fig. 13.12. It is widely distributed throughout the animal kingdom, and is probably located on the inner surface of the plasma membrane and possibly on other membranous components in some cells.

The operative feature of the adenyl cyclase system is the intracellular level of cyclic AMP, which acts as an agent that increases or decreases the

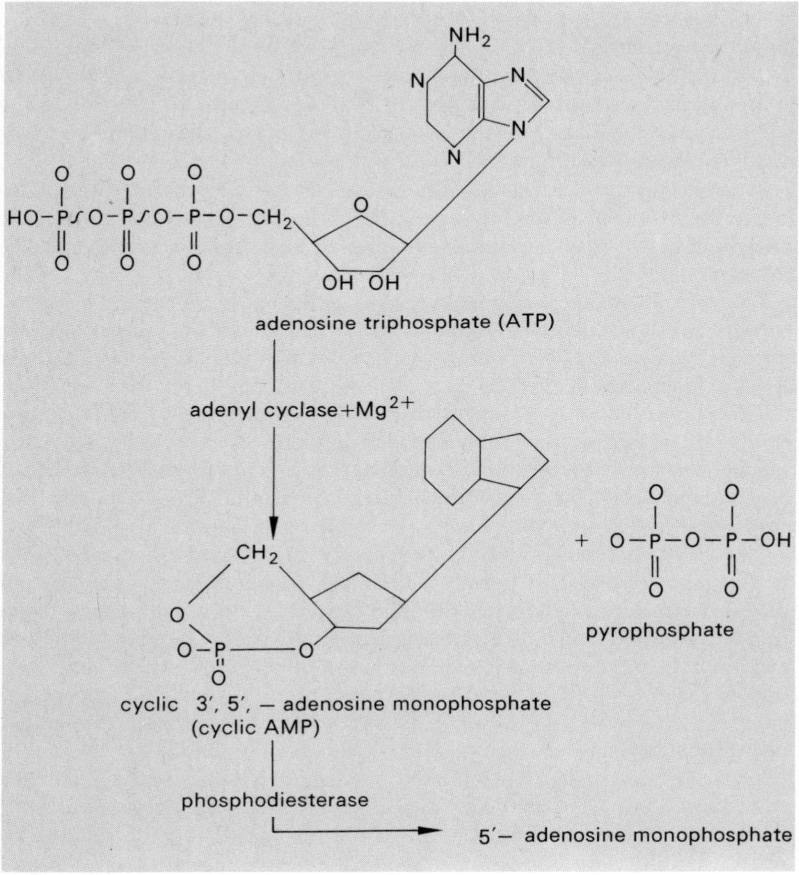

Fig. 13.12   Diagram to illustrate the adenyl cyclase system.

rates of reactions. As an example of the way it affects the action of hormones, we may take the action on the cells of the mammalian liver by the hormones insulin, glucagon and adrenaline. The level of glucose in the blood relative to the level of glycogen in the liver at any given time depends primarily on the action of these three hormones. Glycogen is broken down initially by the enzyme *phosphorylase a* (§3.15), which is itself converted from an inactive form, *phosphorylase b*, by the enzyme *phosphorylase b kinase*. The latter actually exists in two forms, and the activity of one of these forms is increased by a rise in the intracellular level of cyclic AMP, and it is this rise that is brought about by glucagon and adrenaline. Other enzymes are activated by the increased cyclic AMP at the same time, notably those that facilitate *gluconeogenesis*, the production of glucose from lactate, pyruvate and certain amino-acids.

The formation of glycogen from glucose is carried out by the enzyme *glycogen synthetase*, and this is also activated by a kinase whose rate of action is influenced by the intracellular level of cyclic AMP, but here the kinase is activated by a *decrease* in its level. Some authorities think that cyclic AMP may always work through such kinase systems.

It should be appreciated that cyclic AMP may not be the only factor involved in the functioning of those enzymes in whose working it may be implicated. For example, glycogen synthetase is also influenced in its activity by the blood glucose level and by glucocorticoids. Insulin has a direct effect on nuclear DNA, thereby influencing the output of at least three rate-limiting enzymes that take part in glycolysis, as well as its effect on permeability referred to above. When its level in the blood drops, the nucleus produces other enzymes which tend to reverse the breakdown of glucose.

The adenyl cyclase system is clearly a generalized one. The specificity of action of a given hormone must depend upon the type of cell surface receptor with which it reacts, and on the biochemical nature of the cell. It has also been postulated that cyclic AMP may act through specific subunits to which it may attach in effecting particular reactions (the analogy with co-enzymes springs to mind), but so far there is no definite evidence that such subunits exist.

14

Retrospect: Homeostasis

The uniqueness of living organisms lies in their ability to operate as an open thermodynamic system, through which energy is obtained from the environment and used to maintain an orderly set of characteristics we call 'life'. The uniqueness of a species or of an individual animal depends on its ability to maintain certain features that distinguish it from all other species or animals. Thus, the evolution of life and of specific plants and animals both required an ability to maintain a constancy in organization and functioning. It was the realization of this fact that led the nineteenth-century physiologist Claude Bernard to develop the idea that the fluid in which all the cells and tissues of the body are bathed constitutes an internal environment (*milieu intérieur*), which he contrasted with the external environment (*milieu extérieur*) of the whole animal. He saw the constancy of the internal environment as the factor that permitted independence of the external environment.

Claude Bernard recognized that the constancy of the internal environment must be the result of a balance between many different factors, actively maintained. The tendency for an animal to maintain itself in a given condition, or *steady-state*, was later given the name *homeostasis*. Homeostasis embraces Claude Bernard's concept, but it also possesses wider connotations. In previous chapters we have examined a number of homeostatic mechanisms designed to keep the internal environment constant, among them energy release and utilization, osmotic and ionic regulation, temperature regulation, and excretion. But it is obvious on the one hand that intracellular systems must also be maintained in a steady state; and on the other hand that instinctive behaviour is a kind of regulatory mechanism which tends to maintain the status quo. These and other systems, even including the maintenance of a constant population, are now also considered to be controlled by homeostatic mechanisms.

Although the detailed operation of these homeostatic systems is widely different, attempts have been made to discover whether there may not be fundamental principles that are common to them all. These attempts have generally begun by emphasizing that biological homeostatic systems have much in common with the control systems used by engineers, the function of which is to keep machines or circuits working at a steady predetermined level.

14.1 The analogy with control systems

The essential components of any control system are a *misalignment detector* that will signal the extent of any deviation from the required norm; and a *controller* that will translate the information from the misalignment detector into an output signal which will bring the system back to the required condition. The correction of the misalignment will require the expenditure of energy—in an engineering system a motor will often be used to drive the system—and in general the output from the controller will activate some *effector* system that is capable of taking effective action to drive the system back to its proper level. Clearly, in a properly functioning system the output from the controller ought to be proportional to the input to the misalignment detector, and this is most simply achieved by feeding back some of the output to the input. The misalignment detector will then measure the deviation from the fixed proportionality between the two. Such *output feed-back* is an essential feature of fully automatic systems such as biological homeostatic systems.

A control system of this type is shown in the form of a block diagram in Fig. 14.1. Its level of operation must be fixed by defining the range of

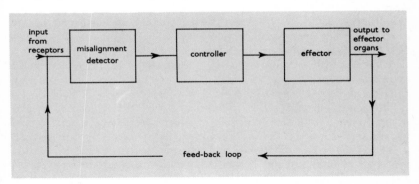

Fig. 14.1   Diagram to illustrate the main components of an automatic control system.

disturbance over which the misalignment detector will work, as in the setting of the contacts of a thermostat. It is a necessary feature of control systems that the force exerted to oppose the disturbance of the system is greater than the disturbance itself. This is because it must not only produce a force large enough to restore the system to normal, but also must have some reserve in hand to counter any forces that may oppose its action, whether predictable forces like the inertia of the system or forces that may be unpredictable and uncertain. These requirements are normally expressed by saying that

$$\text{force applied} \propto \text{misalignment} \times k \text{ (a constant)}$$

The constant k is referred to as the *stiffness* of the system. The greater the stiffness, the bigger is the opposing force in relation to the misalignment.

Stiffness will result in a tendency for the controller to over-compensate i.e. to exert more force at a given time than is necessary to correct the system, and so the output will tend to oscillate about a mean. The tendency to oscillate will be enhanced by the presence of feed-back. The passage of signals through the system must involve some element of physical delay, and this means that the feed-back will be out of phase with the input, an effect that will be increased by any inertia in the system. This again will produce a tendency to oscillate, which may theoretically be compensated for by a suitable phase advance somewhere in the system, although in practice this may be by no means easy.

The correspondence between a generalized control system of this kind and a reflex arc will already be evident to the reader. The input from the receptors is fed to the misalignment detector, which is the first stage of the central nervous system acting as controller. The output is fed, say, to a muscle, which contracts to a corresponding extent. The output to the effectors could be so arranged, by suitable connections within the central nervous system, that it was fed back to the input; although in this case the degree of contraction is also signalled back to the central nervous system from proprioceptors in the muscles and tendons. Such a system is essential if reflex arcs are to subserve automatic purposes such as the maintenance of posture.

The application of control systems analysis to other body mechanisms is not always so obvious. We shall consider two specific examples. Let us first consider a genetic locus that produces a particular mRNA that, in conjunction with ribosomes and the appropriate tRNA molecules, synthesizes a given protein. This protein will in due course be metabolized and the resulting metabolites will feed back to the genetic locus, or possibly to the mRNA, to control either the output or the activity of the latter (Fig. 14.2). Thus, if we take the case in which the metabolite affects the genetic locus as controller, the latter acts also as misalignment detector and adjusts the output (of mRNA) accordingly.

The other example concerns the regulation of ventilation in mammals,

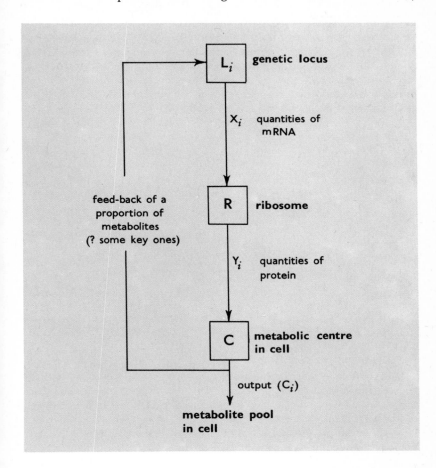

Fig. 14.2 Automatic control loop that might operate in the cellular control of protein synthesis. (Based on Goodwin, 1963, *Temporal Organization in Cells*, Academic Press, New York and London)

the aim of which is to keep the partial pressures of O_2 and CO_2 constant in the blood. It is known that an alteration in the content of either of these gases in the blood, or in the pH of the blood (related to the CO_2

through the bicarbonate transport system), results in an alteration in the pulmonary ventilation rate. The input to the misalignment detector is therefore formed by these parameters. The misalignment detector is once more part of the controller, being in this case the primary respiratory centre in the medulla oblongata, which establishes a fundamental ventilation rhythm by means of nervous impulses that are transmitted to the muscles of the rib-cage and the diaphragm. These nervous impulses may therefore be regarded as equivalent to the ventilation rate. Alteration of the ventilation rate leads to a change in the concentration of the blood gases, and the latter acts as a feed-back to the misalignment detector. The system is summarized in block form in Fig. 14.3.

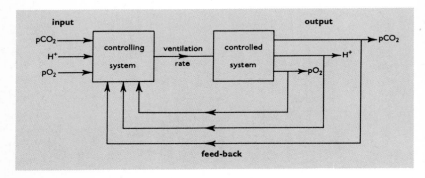

Fig. 14.3 The respiratory homeostat. (After Grodins, 1963, *Control Theory and Biological Systems*, Columbia University Press, New York)

14.2 The transfer function

It is possible to take the comparison between biological homeostats and engineering control systems a stage further and express it in mathematical terms. It will be seen that Fig. 14.3 tells us nothing about the nature of the misalignment detector or the controller. In fact, the controller contains a number of sub-systems in the form of interacting nervous centres, each with its own input and output. Nor does it specify the details of the controlled system, which includes gaseous exchange and transport and buffer systems. In effect, the analysis of homeostatic mechanisms in terms of control systems can be rather like the thermodynamic approach to a reaction. It is concerned with the overall dynamics of the situation rather than with the individual details. If we take this approach to its extreme, a homeostatic or control system can be simplified to an input, an output, and a *black box* that connects them. The black box may represent a host of complex reactions and interactions, but its nature is irrelevant for the purposes of this type of analysis since it is reflected in the relationship between the input and the output.

In its passage through a physical system, a signal may be altered in size and in timing (phase). The alteration in these parameters between input and output can be expressed mathematically by a differential equation known as the *transfer function*. If we denote the input as Q_i and the output as Q_o, then the transfer function describes the relationship Q_o/Q_i. In a control system composed of a series of sub-systems we can represent each of the latter by its own transfer function (Fig. 14.4). The *feed-forward* line A in Fig. 14.4 can be represented by an overall transfer function ϕ_A

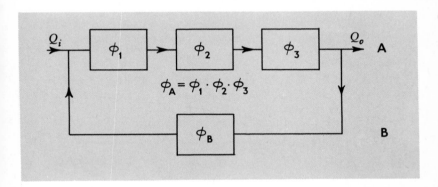

Fig. 14.4 Representation of control system expressed in terms of black boxes and transfer functions. For explanation see text. (Based on Machin, 1964, in Hughes (ed.), *Homeostasis and Feedback Mechanisms, Symp. Soc. exp. Biol.*, **18**, Cambridge University Press)

which will equal the product of all the individual transfer functions on that line, i.e.

$$\phi_A = \phi_1 \phi_2 \phi_3$$

and similarly, the feed-back loop B can be represented by a transfer function ϕ_B. These transfer functions can be combined into a single transfer function for the whole system, and if we denote the latter as ϕ_C it can be shown that

$$\phi_C = \frac{\phi_A}{1 + \phi_A \phi_B}$$

The simplest way to translate the transfer function into quantitative terms is to break open the loop and see what happens to the output when a given input is applied to it. It can be shown that for an open loop (ϕ_O) of this kind $\phi_O = \phi_A \phi_B$, and if, as is usual, $\phi_B = 1$

then
$$\phi_C = \frac{\phi_O}{1 + \phi_O}$$

which is a relatively simple expression to evaluate. Once this value is known, it is possible to calculate that for the feed-back loop. The input applied in determining the transfer function will be either a step-wise change or a sinusoidal one, the latter being rather simpler to employ in practice.

This kind of analysis is only suitable for a linear system, one in which the output always bears the same proportional relationship to the input. In a non-linear system, the transfer function varies with the input. Thus, if we are investigating the properties of a reflex system, we may find that the transfer function varies with the frequency of the input stimulus. It is still possible to analyse such a system in terms of its transfer function by restricting the analysis to a given frequency, and by doing this at a number of individual frequencies a picture of the system may be obtained. Otherwise, the analysis of non-linear systems is a complex affair, although not nearly so tedious as it was before the widespread availability of computers.

The transfer function is still largely a theoretical concept, for the studies of most biological control systems have scarcely progressed beyond the recognition of the fact that feed-back is present. However, a study of the respiratory homeostat represented by Fig. 14.3 has already thrown some insight into problems of ventilation in abnormal circumstances such as high altitudes and in strenuous exercise which are not obviously soluble by other means. Furthermore, a quantitative elucidation of behaviour can only be made in terms of mathematical relationships, among which the transfer function holds some promise.

14.3 Limitations of control systems analysis

The complexity of control theory when applied to non-linear systems raises the question whether there may not be some other mathematical way of expressing the organization of living organisms. The need to do so has been urged on other grounds. The nervous system is an apparently fruitful field for analogies with control systems, but it has been argued that the complex interconnections within the nervous system, and the element of uncertainty they introduce, make control theory inadequate and ambiguous as a description of their activity. In particular, nervous systems are seen to adjust themselves to changes in the environment, not merely during learned behaviour, but also in the reinforcement of innate patterns through experience, as in the stabilization of some locomotory patterns. It may therefore be said that, despite the apparently high degree of organization of most nervous systems, the connections between their individual elements are incompletely specified, grow in complexity with the development of internal organization, and hence function with adjustable probability. In these respects the nervous system may be compared with the embryonic differentiation of cells based on the outworking of pre-determined genetic patterns in a particular environment; and with the control of cellular processes by that same genetic pattern. For these reasons many workers consider that biological control theory must be developed to include a probability element that can only be expressed statistically.

Bibliography

Wherever possible, review articles and monographs are recommended for further reading, which will contain the necessary references to earlier work. In addition to the references specific to each chapter there are a number of books whose contents cover the subject matter of more than one chapter, and these are listed below.

GENERAL

DAVSON, H. (1970). *A Textbook of General Physiology*, 4th edn., 2 vols. Churchill, London.

GIESE, A. C. (1968). *Cell Physiology*, 3rd edn. W. B. Saunders, Philadelphia and London.

LEHNINGER, A. L. (1970). *Biochemistry. The Molecular Basis of Cell Structure and Function*. Worth Publishers Inc., New York.

PROSSER, C. L. and BROWN, F. A. (1961). *Comparative Animal Physiology*, 2nd edn. W. B. Saunders, Philadelphia and London.

The monthly periodical *Scientific American* often contains review articles summarizing the state of current knowledge in specific biological fields.

ENERGY AND FOOD

BOURNE, G. H. and KIDDER, G. W., eds. (1953). *The Biochemistry and Physiology of Nutrition*. Academic Press, New York and London.

DICKERSON, R. E. and GEIS, I. (1970). *The Structure and Action of Proteins*. Harper & Row, London and New York.

JENNINGS, J. B. (1972). *Feeding, Digestion and Assimilation in Animals*. 2nd edn. MacMillan, London.

KLOTZ, I. M. (1967). *Energy Changes in Biochemical Reactions*. Academic Press, New York and London.

SPEAKMAN, J. C. (1966). *Molecules*. McGraw-Hill, New York and Maidenhead.

CELLULAR ORGANIZATION AND FUNCTION

BALDWIN, E. (1967). *Dynamic Aspects of Biochemistry*. 5th edn. University Press, Cambridge.

FINEAN, J. B. (1973). *Membranes and their Cellular Function*. Blackwell, Oxford.

INGRAM, V. M. (1972). *The Biosynthesis of Macromolecules*. 2nd edn. Benjamin, New York and Amsterdam.

LEHNINGER, A. L. (1972). *Bioenergetics*. 2nd edn. Benjamin, New York and Amsterdam.

RESPIRATION AND METABOLISM

BENEDICT, F. G. (1938). *Vital Energetics*. Carnegie Inst., Washington, Publ. 503.

JONES, J. D. (1972). *The Comparative Physiology of Respiration*. Edward Arnold, London.

STEEN, J. B. (1971). *The Comparative Physiology of Respiratory Mechanisms.* Academic Press, London and New York.

ZEUTHEN, E. (1953). Oxygen Uptake as Related to Body Size. *Quart. Rev. Biol.* **28**, 1.

ZEUTHEN, E. (1955). Comparative Physiology of Respiration. *Ann. Rev. Physiol.*, **17**, 459.

EXCRETION

CAMPBELL, J. W. and GOLDSTEIN. L. eds. (1972). *Nitrogen Metabolism and the Environment.* Academic Press, London and New York.

KEILIN, J. (1959). Uric Acid and Guanine Excretion. *Biol. Rev.*, **34**, 265.

MARTIN, A. W. (1958). Comparative Physiology of Excretion. *Ann. Rev. Physiol.*, **20**, 225.

See also the reference to RIEGEL, J. A. in the section below.

OSMOTIC AND IONIC REGULATION

BEADLE, L. C. (1957). Comparative Physiology of Osmotic and Ionic Regulation in Aquatic Animals. *Ann. Rev. Physiol.*, **19**, 329.

EDNEY, E. B. (1957). *Water Relations of Terrestrial Arthropods.* University Press, Cambridge.

KROGH, A. (1939). *Osmotic Regulation in Aquatic Animals.* University Press, Cambridge.

POTTS, W. T. W. and PARRY, G. (1964). *Osmotic and Ionic Regulation in Animals.* Pergamon, Oxford and New York.

REMANE, A. and SCHLIEPER, C. (1972). *Biology of Brackish Water,* 2nd rev. edn. (vol. 25 of series *Die Binnengewässer,* ed. Elster and Ohle). E. Schweizerbart'she Verlagsbuchshandlung, Stuttgart.

RIEGEL, J. A. (1972). *Comparative Physiology of Renal Excretion.* Oliver & Boyd, Edinburgh.

ROBERTSON, J. D. (1957). The Habitat of the Early Vertebrates. *Biol. Rev.*, **32**, 156.

SHAW, J. and STOBBART, R. H. (1963). Osmotic and Ionic Regulation in Insects. *Adv. Ins. Physiol.*, **1**, 315.

HYDROGEN ION CONCENTRATION (pH)

THORPE, W. V. (1970). Acidity and Alkalinity. In *Biochemistry for Medical Students.* 9th edn. Churchill, London.

TEMPERATURE RELATIONS

BLIGH, J. (1973). *Temperature Regulation in Mammals and other Vertebrates.* North-Holland, Amsterdam.

JOHANSEN, K. (1962). Evolution of Temperature Regulation in Mammals. In *Comparative Physiology of Temperature Regulation.* ed. HANNON, J. P. and VIERECK, E. Arctic Aeromed. Lab., Fort Wainwright, Alaska.

KAYSER, CH. (1961). *The Physiology of Natural Hibernation.* Pergamon, Oxford and New York.

MERYMAN, H. T., ed. (1966). *Cryobiology.* Academic Press, New York and London.

ROSE, A. H., ed. (1967). *Thermobiology.* Academic Press, New York and London.

SCHMIDT-NIELSEN, K. (1964). *Desert Animals.* Oxford University Press, London.
SMITH, A. U. (1961). *Biological Effects of Freezing and Supercooling.* Edward Arnold, London.

NERVOUS COMMUNICATION

BURTT, E. T. and CATTON, W. B. (1966). Image Formation and Sensory Transmission in the Compound Eye. *Adv. Ins. Physiol.* **3**, 1.
BUSNEL, R. G., ed. (1963). *Acoustic Behaviour of Animals.* Elsevier, London and New York.
DAVIS, H. (1961). Some principles of Receptor Action. *Physiol. Rev.*, **41**, 391.
ECCLES, J. C. (1964). *The Physiology of Synapses.* Verlag-Springer, Berlin.
HODGKIN, A. L. (1964). *The Conduction of the Nervous Impulse.* Liverpool University Press.
HUBBARD, J. I., LLINAS, R., and QUASTEL, D. M. J. (1969). *Electrophysiological Analysis of Synaptic Transmission* (Physiological Society Monograph no 19). Edward Arnold, London.
KATZ, B. (1966). *Nerve, Muscle and Synapse.* McGraw-Hill, New York and Maidenhead.
LANGER, H., ed. (1973). *Biochemistry and Physiology of Visual Pigments.* Springer-Verlag, Berlin.
LOEWENSTEIN, W. R., ed. (1971). *Principles of Receptor Physiology.* (vol. 1 of Handbook of Sensory Physiology). Springer-Verlag, Berlin.
(Subsequent volumes in this series may also be consulted for specialist articles on specific sensory modalities.)
OCHS, W. (1965). *Elements of Neurophysiology.* Wiley, New York and London.
OHLOFF, G. and THOMAS, A. F. eds. (1971). *Gustation and Olfaction.* Academic Press, London and New York.
TANSLEY, K. (1965). *Vision in Vertebrates.* Chapman & Hall, London and New York.
WOLKEN, J. J. (1971). *Invertebrate Photoreceptors.* Academic Press, London and New York.

NERVOUS INTEGRATION

BULLOCK, T. H. and HORRIDGE, G. A. (1965). *Structure and Function in the Nervous System of Invertebrates,* 2 vols. W. H. Freeman, San Francisco and London.
KONORSKI, J. (1948). *Conditioned Reflexes and Neuron Organization.* Oxford University Press, London.
LEIBOVIC, K. N. (1972). *Nervous System Theory: An Introductory Study.* Academic Press, London and New York.
ROMER, A. S. (1962). The Nervous System. In *The Vertebrate Body,* 3rd edn. W. B. Saunders, Philadelphia and London.
WELLS, M. J. (1962). *Brain and Behaviour in Cephalopods,* Heinemann, London.
YOUNG, J. Z. (1965). *A Model of the Brain.* Oxford University Press, London.

MUSCLES

BENDALL, J. R. (1969). *Muscles, Molecules and Movement.* Heinemann, London.

BOURNE, G. H., ed. (1972). *Structure and Function of Muscle*, 2nd edn. 3 vols. Academic Press, New York and London.

BULBRING, E. and NEEDHAM, D. (1973). Organizers of *A Discussion on recent developments in vertebrate smooth muscle physiology*. *Phil. Trans. R. Soc.* B 265, 1.

CHAPMAN, G. (1958). The Hydrostatic Skeleton in the Invertebrates. *Biol. Rev.* 33, 338.

DAVIES, R. E. (1965). Muscle Contraction. In *Essays in Biochemistry*, ed. CAMPBELL, P. N. and GREVILLE, G. D. Vol. 1, p. 29. Academic Press, New York and London.

HOYLE, G. (1962). Neuromuscular Physiology. *Adv. Comp. Physiol. Biochem.*, 1, 117.

HUXLEY, H. E. (1964). Leader of Symposium on Sliding Filament Theory. *Proc. Roy. Soc.*, B 160, 434.

WILKIE, D. R. (1968). *Muscle*. (Studies in Biology no. 11). Edward Arnold, London.

AMOEBOID AND CILIARY MOVEMENT

HAYASHI, T. (1961). How Cells Move. *Sci. Amer.*, 205 (9).

JEON, K. W., ed. (1973). *The Biology of Amoeba*. Academic Press, London and New York.

SLEIGH, M. A. (1971). Cilia. In *Endeavour*, 30, 11.

HORMONES

BARRINGTON, E. J. W. (1964). *Hormones and Evolution*. English Universities' Press, London.

DAVIDSON, E. H. (1965). Hormones and Genes. *Sci. Amer.*, 212 (2), 36.

GABE, M. (1966). *Neurosecretion*. Pergamon, Oxford and New York.

GORBMAN, A. and BERN, H. A. (1962). *A Text Book of Comparative Endocrinology*. Wiley, New York and London.

HIGHNAM, K. C. and HILL, L. (1969). *The Comparative Endocrinology of the Invertebrates*. Edward Arnold, London.

ROBISON, G. A., BUTCHER, R. W. and SUTHERLAND, E. W. (1971). *Cyclic AMP*. Academic Press, London and New York.

TURNER, C. D. and BAGNARA, J. T. (1971). *General Endocrinology*, 5th edn. W. B. Saunders Co., Philadelphia and London.

RESTROSPECT: HOMEOSTASIS

BAYLISS, L. E. (1966). *Living Control Systems*. English Universities Press, London.

GOODWIN, B. C. (1963). *Temporal Organization in Cells*. Academic Press, New York and London.

GRODINS, F. S. (1963). *Control Theory and Biological Systems*. Columbia University Press, New York.

HUGHES, G. M., ed. (1964). *Homeostasis and Feed-Back Mechanisms. Symp. Soc. exp. Biol.*, 18, University Press, Cambridge.

MILSUM, J. H. (1966). *Biological Control Systems Analysis*. McGraw-Hill, New York and Maidenhead.

Index